Relaxed Barrier Function Based Model Predictive Control

Theory and Algorithms

Von der Fakultät Konstruktions-, Produktions- und Fahrzeugtechnik
und dem Stuttgart Research Centre for Simulation Technology
der Universität Stuttgart zur Erlangung der Würde eines
Doktor-Ingenieurs (Dr.-Ing.) genehmigte Abhandlung

Vorgelegt von

Christian Feller

aus Flensburg

Hauptberichter: Prof. Dr.-Ing. Christian Ebenbauer
Mitberichter: Prof. Dr. rer. nat. Moritz Diehl
Prof. Gabriele Pannocchia, Ph.D.

Tag der mündlichen Prüfung: 27. Juni 2017

Institut für Systemtheorie und Regelungstechnik
der Universität Stuttgart

2017

Bibliografische Information der Deutschen Nationalbibliothek

Die Deutsche Nationalbibliothek verzeichnet diese Publikation in der Deutschen Nationalbibliografie; detaillierte bibliografische Daten sind im Internet über http://dnb.d-nb.de abrufbar.

D93

©Copyright Logos Verlag Berlin GmbH 2017
Alle Rechte vorbehalten.

ISBN 978-3-8325-4544-4

Logos Verlag Berlin GmbH
Comeniushof, Gubener Str. 47,
10243 Berlin
Tel.: +49 (0)30 42 85 10 90
Fax: +49 (0)30 42 85 10 92
INTERNET: http://www.logos-verlag.de

Für Luisa

Acknowledgements

This thesis is the result of my time as a research and teaching assistant at the Institute for Systems Theory and Automatic Control (IST) at the University of Stuttgart. During this time, I was fortunate to meet, live, and collaborate with many interesting and unique people who supported me and my research in various ways and to whom I want to express my sincere gratitude.

First and foremost, I want to thank my advisor Prof. Dr.-Ing. Christian Ebenbauer for giving me the chance to be part of his research group as well as for his motivating, skillful, and inspiring advice during my time as a PhD student. I am very grateful for the freedom to develop and pursue own research ideas, the opportunity to present and discuss our results at various international conferences and, above all, for the many hours of brainstorming, thinking, discussing, and, well, just doing research, in which he became a both challenging and understanding mentor as well as a valued friend.

I also thank Prof. Dr.-Ing. Frank Allgöwer for sparking my interest in systems and control theory and for establishing and continuously developing the unique and fruitful academic environment that I had the chance to enjoy during my time at the IST.

I thank Prof. Dr. rer. nat. Moritz Diehl and Prof. Gabriele Pannocchia for the interest they showed in my research as well as for serving as members of my doctoral examination committee. Their valuable and helpful comments were greatly appreciated and definitely led to an improved final version of the thesis.

I am also very grateful to Prof. Francesco Borrelli and Dr.-Ing. Georg Schildbach who invited me to Berkeley and gave me the opportunity to implement my control algorithms on a challenging real-time application. In this context, I also want to thank Ashwin Carvalho, Matthias Soppert, Arnav Sharma, Ziya Ercan and Andreas Hansen for their help with the car as well as for the great time we had together.

Special thanks go to all my colleagues and fellow PhD students at the IST. Although a list of all the talks, discussions, conference trips, parties, and fun activities we had together is far too long to repeat it here, I will remember and cherish them for the rest of my life. Particular thanks go to my long-term office mate Florian Bayer as well as to Florian Brunner, Hans-Bernd Dürr, Gregor Goebel, Simon Michalowsky, Matthias Müller, Jan Maximilian Montenbruck, and all past and present members of the MPC group.

I will be eternally grateful for the unconditional love and support of my parents, my three sisters, and my entire family. Throughout my whole life you have been a source of motivation and encouragement and none of my achievements would have been possible without you. Finally, I thank my wonderful partner Luisa for her love, patience, and inspiration and for boarding the train to Stuttgart so many times.

Flensburg, July 2017
Christian Feller

Table of Contents

Abstract

In this thesis, we present a framework for the design, analysis, and implementation of model predictive control (MPC) approaches and algorithms based on relaxed logarithmic barrier functions. In particular, we introduce the novel concept of *relaxed barrier function based model predictive control* and present a comprehensive theoretical framework for the design and analysis of relaxed barrier function based MPC schemes for discrete-time linear systems – covering thereby important systems theoretic aspects such as asymptotic stability, constraint satisfaction, closed-loop robustness, and output feedback. Moreover, we propose, analyze, and test different classes of numerically efficient MPC iteration schemes and algorithms that, based on the proposed barrier function based MPC formulation, allow us to guarantee important systems theoretic properties of the resulting overall closed-loop system. Instead of treating the underlying optimization as an idealized static map, a key motive of the MPC results and algorithms presented in this thesis is to design, analyze, and implement the interconnected dynamics of controlled plant and optimization algorithm in an integrated barrier function based framework, and to study the properties of the resulting overall closed-loop system both from a systems theoretic and algorithmic perspective. One of the overall main results is a novel class of barrier function based *anytime MPC algorithms* that provide important stability, robustness, and constraint satisfaction guarantees *independently* of the number of optimization algorithm iterations that are performed at each sampling step while allowing at the same time for a fast and efficient numerical implementation with desirable algorithmic properties. The obtained theoretical results are illustrated by various numerical examples and benchmark tests as well as by an experimental case study in which the proposed class of barrier function based MPC algorithms is applied to the predictive control of a self-driving car.

Deutsche Kurzfassung

In dieser Arbeit wird ein systematischer Ansatz für den Entwurf, die Analyse und die algorithmische Implementierung modellbasierter prädiktiver Regelungsverfahren (engl. Model Predictive Control, MPC) basierend auf relaxierten logarithmischen Barrierefunktionen vorgestellt. Insbesondere führen wir das neuartige Konzept der prädiktiven Regelung basierend auf relaxierten Barrierefunktionen ein, stellen ein umfassendes methodisches Grundgerüst für den Entwurf und die systemtheoretische Analyse barrierefunktionenbasierter prädiktiver Regelungsverfahren für zeitdiskrete lineare Systeme vor, und präsentieren und testen darüber hinaus verschiedene Klassen numerisch effizienter MPC-Iterationsverfahren und Algorithmen, welche inhärent wichtige Eigenschaften des geschlossenen Regelkreises garantieren. Im Gegensatz zu konventionellen MPC-Verfahren, welche die unterlagerte Optimierung üblicherweise vernachlässigen und als statische Abbildung abstrahieren, besteht einer der Grundgedanken der in dieser Arbeit vorgestellten MPC-Algorithmen darin, Entwurf, Analyse und Implementierung der verkoppelten Dynamiken von geregelter Strecke und zugehörigem Optimierungsalgorithmus im Rahmen eines integrativen, barrierefunktionenbasierten Gesamtansatzes zu vereinen und dabei sowohl systemtheoretische als auch algorithmische Eigenschaften des resultierenden geschlossenen Regelkreises zu untersuchen. Ein übergeordnetes Hauptresultat der Arbeit ist hierbei eine neue Klasse barrierefunktionenebasierter *Anytime-MPC-Algorithmen*, welche wichtige Eigenschaften des geschlossenen Regelkreises in Bezug auf Stabilität, Robustheit oder die Einhaltung von Eingangs- und Zustandsbeschränkungen garantieren – und zwar unabhängig von der Anzahl der pro Abtastschritt ausgeführten Iterationen des unterlagerten Optimierungsalgorithmus. Die vorgestellten theoretischen Konzepte und Ergebnisse werden durch verschiedene numerische Simulationsbeispiele und Benchmark-Tests sowie anhand eines realen Echtzeit-Experiments veranschaulicht, in welchem die vorgestellte Klasse barrierefunktionenbasierter MPC-Algorithmen für die prädiktive Trajektorienfolgeregelung eines autonomen Fahrzeugs eingesetzt wird.

Chapter 1
Introduction

Motivation and problem statement

Controlling the behavior of a complex dynamical system in a stable or even optimal way while satisfying at the same time certain constraints on the underlying system state and control input variables is a fundamental and important decision-making problem that all of us are constantly encountering in our everyday lives. Imagining for example a car on a curvy road, the driver needs to decide every few milliseconds not only on the correct positions of throttle control and steering wheel but also whether the current driving situation requires some form of emergency maneuver. In addition, the driver might want to optimize the operation of the vehicle with respect to some individual performance criterion, related for example to the expected arrival time, the driving comfort, or the incurred fuel consumption. From a control theory perspective, the driver therefore needs to repeatedly solve a constrained optimal control problem in real-time. While the underlying decision-making problem is in this example solved unconsciously and mainly based on the driver's longtime experience, there are in fact many technical applications in which similar constrained control and decision-making problems need to be solved in a constructive and reliable way by means of suitably defined control algorithms. For example, this may include the problem of stabilizing a chemical plant at a given (optimal) set point, the optimal distributed operation of traffic, transport, and electrical networks, or simply dealing with the above car example in an autonomous driving context.

One approach to handle such constrained optimal control problems is provided by the concept of *model predictive control (MPC)*, often also referred to as *receding horizon control*. In this optimization-based control strategy, a suitable control input is obtained at each sampling instant by solving in a receding horizon fashion a finite-horizon open-loop optimal control problem that, based on the current system state and a prediction of the future system behavior, aims to minimize a given performance criterion (or *cost function*) while taking potential constraints on the state and input variables explicitly into account. Only the first part of the computed optimal input sequence is applied to the plant, and the procedure is repeated at the next sampling instant over a shifted prediction horizon. Due to its ability to handle complex and possibly nonlinear multivariable systems in a constructive and unified way while at the same time allowing to incorporate both a user-defined performance criterion and state and input constraints directly into the control design, MPC has received a lot of attention in both academia and industrial applications during the last decades, see for example (Mayne et al., 2000; Qin and Badgwell, 2003; Mayne, 2014). A particularly interesting development is that, while having originally been used predominantly in the process industries, MPC is by now also more and more applied to challenging control problems from other areas such as power electronics (Cortés et al., 2008), building climate control (Oldewurtel et al., 2012; Ma et al., 2012), autonomous driving and vehicle stabilization (Falcone, 2007; Di Cairano et al., 2013; Carvalho et al., 2015), spacecraft attitude control (Manikonda et al., 1999; Saponara et al., 2011; Guiggiani et al., 2015), airborne power generation (Gros et al., 2013; Ahrens et al., 2013), ship collision avoidance (Johansen et al., 2016), dynamic option hedging (Primbs, 2009; Bemporad et al., 2014), or optimal drug therapy (Hovorka et al., 2004; Pannocchia et al., 2010).

However, while there is an ever-growing number of both MPC applications and suitable tailored optimization algorithms and the theoretical foundation of linear and nonlinear MPC can by now be considered to be quite mature, there is often still a significant gap between MPC theory and MPC practice – or, to use the distinction made in this thesis, between *MPC schemes* and *MPC algorithms*. In particular, when analyzing systems theoretic properties of the closed-loop system, conventional MPC schemes usually assume that the exact optimal control input, i.e., the exact solution to the associated constrained open-loop optimal control problem, is available immediately upon measurement of the current system state. While this assumption is typically admissible when considering processes with slow system dynamics (implying some form of time-scale separation), it might not be valid when considering practical applications in which the underlying optimization algorithm needs to operate on the same time-scale as the controlled plant – for example due to fast system dynamic and/or the use of low-cost hardware. In this case, the fact that underlying optimization-based controller is *itself* a dynamical system needs to be taken into account in the design and analysis of the overall closed loop. Thus, particularly in the context of safety-critical applications, there is a well-founded need for the development of numerically efficient optimization algorithms and tailored MPC approaches that allow guaranteeing important systems theoretic and algorithmic properties of the closed-loop system even when applying possibly only suboptimal control inputs to the plant. This need has also been emphasized in the recent survey paper by Mayne (2014) as well as by the growing number of publications in this area of research, see for example (Diehl et al., 2005, 2007; Graichen, 2012; Zeilinger et al., 2014b; Rubagotti et al., 2014; Bemporad et al., 2015) as well as the discussion in Chapters 2 and 4 of this thesis. Furthermore, in order to ensure recursive feasibility of the underlying optimization problem as well as asymptotic stability of the resulting closed-loop system, MPC schemes from the literature usually require (even in the linear case) the inclusion of additional terminal set constraints. Such additional stabilizing constraints are, however, often not desired and typically also not used in practice (Grüne, 2012). As a consequence, there is a need for conceptually simple and terminal set free MPC schemes that nevertheless allow providing guarantees on closed-loop stability. Finally, an additional and often-cited obstacle in transferring theoretical MPC concepts from the literature to actual real-world applications is given by the fact that large parts of the MPC literature rely on the assumption that no (or at least only partially known) disturbances are acting on the system and that the exact full system state is accessible at each sampling instant. As pointed out by Mayne (2014), the main reason for this simplistic assumption is that when allowing for disturbances or inexact state estimates, the underlying MPC scheme needs to be designed in such a way that it is robust (in particular in the sense of recursive feasibility) against potentially unknown internal and external disturbances. Even in the case of linear systems this is often only possible under rather strong assumptions. While MPC is in practice often successfully employed by designing the MPC scheme for the nominal system and then simply replacing the exact system state in a certainty equivalence fashion with the corresponding state estimate, this intuitive approach will in general destroy all the theoretical properties of the underlying nominal MPC scheme. Consequently, there is a strong need for the development of MPC schemes and algorithms that allow for rigorous systems theoretic guarantees also in the presence of unknown external disturbances and/or state estimation errors.

Summarizing, the above discussion illustrates that there is a strong need for the development of novel predictive control approaches that allow us to bridge the gap between MPC theory and MPC practice, taking thereby the step from *MPC schemes* to *MPC algorithms*. More precisely, a both challenging and important research goal is to develop a framework for the integrated design, analysis, and implementation of model predictive control algorithms that rely on a conceptually simple and intuitive problem formulation, allow for a numerically efficient implementation with hard systems theoretic and algorithmic guarantees, and are able to cope with unknown external disturbances and estimation errors. In this thesis, such an integrated systems theoretic and algorithmic framework is presented based on the novel concept of *relaxed barrier function based model predictive control*.

The results presented in this thesis are based on two main conceptual key ideas. The first one is to handle potential constraints on the state and input variables by means of so-called *relaxed barrier functions*. Using a concept from barrier function based optimization, this term refers to a conventional (e.g. logarithmic) barrier function that is smoothly extended by a globally defined penalizing term as soon as the underlying variable or constraint gets closer than some distance δ to the border of the respective feasible set (Ben Tal et al., 1992; Hauser and Saccon, 2006). The scalar parameter δ is called the *relaxation parameter* of the relaxed barrier and can in principle be used to approximate the original barrier function arbitrarily close. A graphical illustration of this concept is given in Figure 1.1 on the next page. Moreover, the second key idea is to design tailored optimization algorithms for the resulting relaxed barrier function based MPC problem formulation and to analyze the overall closed-loop system consisting of both plant and optimization algorithm dynamics in an integrated fashion, acknowledging thereby the fact that the underlying optimization-based controller is itself a dynamical system. See again Figure 1.1 for a schematic illustration of the principle idea.

While the use of relaxed barrier functions inherently allows for an unconstrained, globally defined, and strongly convex formulation of the associated optimization problem, it is neither immediately clear how important systems theoretic properties like stability, robustness, or constraint satisfaction can be ensured in the resulting MPC schemes, nor

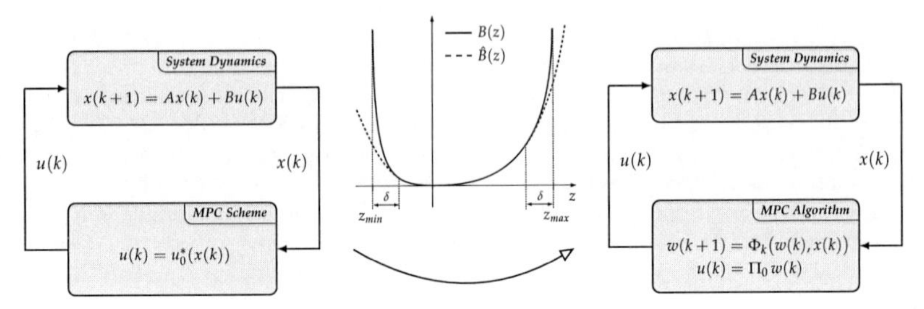

Figure 1.1. *Middle*: Basic idea of a relaxed barrier function for a scalar constraint of the form $z_{min} \leq z \leq z_{max}$. For small values of the relaxation parameter $\delta \in \mathbb{R}_{++}$, the relaxed barrier function $\hat{B}(z)$ approximates the original barrier function $B(z)$ arbitrarily close. One of the main goals of the thesis is to take the step from idealizing MPC schemes (*left*) to MPC algorithms (*right*) which allow one to analyze and implement the overall closed-loop system dynamics as an interconnection of the controlled plant and the dynamics of the underlying MPC algorithm. The operator Φ_k captures the dynamics of the employed iterative optimization procedure, while multiplication with the projection matrix Π_0 is used to select the first block element from the current optimizer state $w(k)$.

whether and to which extent these properties may actually carry over to the corresponding algorithmic implementation. Concerning this problem setup, the main goal of this thesis is to demonstrate that the proposed novel relaxed barrier function based approach offers a powerful MPC framework that allows to design, analyze, and implement stabilizing MPC schemes and algorithms in a holistic and integrated fashion – taking algorithmic aspects into account in the MPC design and, vice versa, allowing for a rigorous systems theoretic analysis of the actually implemented MPC algorithms. The overall main result is a novel class of conceptually simple barrier function based anytime MPC algorithms with important stability, robustness, and real-time properties. Such algorithms may in particular be useful in the context of safety critical applications that often require guaranteeing a well-defined and stable behavior of the closed-loop system both from a systems theoretic and algorithmic perspective.

Contributions and structure of the thesis

In the following, we briefly highlight the main contributions of the thesis before discussing in more detail an outline of the closely related overall thesis structure.

Main contributions

From a broader conceptual point of view, the thesis comprises three main contributions – each of them related to one of the three central chapters of the thesis.

- In the first main part of the thesis, Chapter 3, we introduce the novel concept of relaxed barrier function based model predictive control and present a self-contained systems theoretic framework for the design and analysis of relaxed barrier function based MPC schemes for discrete-time linear systems subject to polytopic input and state constraints. Both terminal set based and terminal set free approaches are presented, leading for example to conceptually simple linear MPC schemes with a purely quadratic terminal cost or no terminal set and terminal cost at all. In each case, constructive design approaches for choosing the respective problem parameters are provided, and asymptotic stability of the resulting closed-loop system is proven based on Lyapunov stability theory. Furthermore, we analyze the constraint satisfaction properties of the relaxed MPC formulation and show that exact or approximate constraint satisfaction can in many cases be achieved by choosing the underlying relaxation parameter sufficiently small. Finally, we investigate the robustness properties of the proposed relaxed barrier function based MPC schemes and show that the resulting closed-loop system is input-to-state stable with respect to arbitrary additive disturbances. Based on this result, we then present a novel separation principle that allows us to combine the considered class of relaxed barrier function based MPC schemes in a certainty equivalence output feedback fashion with suitable state estimation procedures such as the Kalman filter.

- In the second main part of the thesis, Chapter 4, we complement and significantly extend the systems theoretic framework from Chapter 3 by providing tailored optimization algorithms and MPC iteration schemes that allow for an efficient and stability preserving implementation of the respective relaxed barrier function based MPC approaches. In particular, we show that the relaxed barrier function based formulation reduces the MPC problem to the minimization of a globally defined, twice continuously differentiable, and strongly convex cost function which satisfies in addition important Lipschitz conditions on its associated gradient and Hessian. This allows to apply many efficient and well-known optimization procedures like (accelerated) gradient methods or the Newton method, including in particular also the convergence results and complexity estimates that are available for these methods. Furthermore, by exploiting properties of the relaxed MPC problem formulation and by analyzing the interconnected dynamics of plant and optimization algorithm in an integrated fashion, we show that it is possible to derive barrier function based *anytime MPC algorithms* that guarantee many desirable systems theoretic properties *independently* of the number of performed optimization algorithm iterations. Again, asymptotic stability and robustness properties are analyzed in detail based on Lyapunov techniques and input-to-state stability concepts. Moreover, we show that when keeping the predicted system states as optimization variables, the numerical complexity of the underlying optimizer iteration can be reduced significantly by exploiting the sparsity structure of the resulting equality constrained optimization problem. Based on numerical simulations and benchmark tests, the proposed class of MPC algorithms is shown to be competitive with existing fast MPC solvers.

- In addition, all the presented theoretical and algorithmic results are illustrated and discussed by means of extensive numerical simulations as well as by a practical real-time application. In particular, we present in Chapter 5 an experimental case study that illustrates and confirms the applicability and versatility of the proposed relaxed barrier function based MPC approach by applying it in an autonomous-driving test scenario to the predictive path-following control of a self-driving car.

Beside these main contributions, we tried to render the thesis self-contained by including and discussing in addition several interesting auxiliary results, e.g., an explicit solution to the finite-horizon LQR problem with zero terminal state, and an input-to-state stability proof for the discrete-time Kalman filter, and a generalized framework for the constructive design of both ellipsoidal and polytopic terminal sets that is actually applicable to both nonrelaxed and relaxed barrier function based MPC.

Thesis structure

Chapter 2: Backgound. In this chapter, we introduce the necessary background for the concepts and results presented in this thesis. In particular, we briefly summarize the principle idea of model predictive control, discuss some background on barrier function methods for convex optimization, and revisit in some more detail the existing concept of (nonrelaxed) barrier function based model predictive control as proposed by Wills (2003) and Wills and Heath (2004).

Chapter 3: Relaxed Barrier Function Based MPC Theory. In this chapter, we introduce and discuss in detail the abovementioned systems theoretic framework for linear model predictive control based on relaxed barrier functions. After defining the concept of relaxed logarithmic barrier functions in a more rigorous way, we formulate the relaxed barrier function based MPC problem and present different constructive design approaches for choosing the underlying terminal cost function term. Both asymptotic stability of the origin, satisfaction of input and state constraints, and robustness properties of the closed-loop system are analyzed in detail. In addition, it is shown how the obtained robust stability result can be employed in the context of estimation-based certainty equivalence output feedback, including the cases of Luenberger observers and Kalman filtering.
The results presented in this chapter are partially based on the preliminary works (Feller and Ebenbauer, 2013, 2014a,b, 2015a,b, 2017a) and (Feller, Ouerghi, and Ebenbauer, 2016). All the technical proofs are provided in Appendix C.

Chapter 4: Relaxed Barrier Function Based MPC Algorithms. In this chapter, we complement the systems theoretic framework from Chapter 3 with an algorithmic counterpart. In particular, after analyzing structural convexity and regularity properties of the resulting relaxed barrier function based optimization problem, we introduce a novel class of MPC iteration schemes that combine a suitable warm-starting procedure with descent direction optimization methods and allow us to analyze the overall closed-loop system (consisting

of both the plant and optimization algorithm dynamics) in an integrated fashion. The same systems theoretic properties as in Chapter 3 are discussed, that is, asymptotic stability of the origin, satisfaction of input and state constraints, and robustness with respect to additive disturbances and observer errors. Furthermore, a similar iteration scheme is proposed for the associated non-condensed, equality constrained formulation and it is shown how the resulting sparsity structure can be exploited in the underlying search direction computation. The proposed anytime MPC algorithms for both the condensed and the sparse formulation are compared in a benchmark study to two existing fast MPC solvers, namely qpOASES and FORCES Pro. In addition, it is briefly discussed how the relaxed barrier function formulation may be used for the design of so-called continuous-time MPC algorithms that asymptotically track the associated parametrized optimal solution by integrating a suitably defined ordinary differential equation in continuous time.
The results presented in this chapter are partially based on (Feller and Ebenbauer, 2013, 2014a, 2016, 2017b). All the technical proofs are again provided in Appendix C.

Chapter 5: Experimental Case Study. In this chapter, we present the abovementioned experimental case study, in which the relaxed barrier function based MPC approach is applied to the predictive path-following control of a self-driving car. After introducing the employed kinematic vehicle model and extending the proposed class of barrier function based MPC algorithms to a suitable linear time-varying tracking formulation, the performance of the resulting predictive controller is discussed both by means of numerical simulations and an experimental test scenario.

Chapter 6: Conclusions. In this chapter, we summarize the main results of the thesis and give an outlook on some possible directions for future research.

Appendices. In Appendix A, we briefly recall some fundamental stability concepts that are used throughout the thesis. In particular, we focus on the concepts of *(global) asymptotic stability* and *input-to-state stability*, as well as on their characterizations in terms of comparison and Lyapunov functions. In Appendix B, we summarize and discuss some auxiliary results related to the results presented in the main chapters. Finally, Appendix C contains all the technical proofs.

Notation and definitions

Throughout the thesis, we mainly use standard notation and definitions from the related control, convex optimization, and MPC literature. A comprehensive list of the employed notation and some frequently used acronyms is given on the pages 247 and 248. All other definitions are introduced where appropriate throughout the thesis, particularly those that are of key importance to the presented material or which go beyond the standard definitions available in the literature. As outlined above, a detailed summary of the required stability concepts and definitions is given in Appendix A.

Chapter 2

Background

In this chapter, we give a brief and introductory overview of the most important concepts and results from the literature that are related to the work presented in this thesis. In particular, we briefly summarize the general principle of model predictive control, discuss some background material on barrier function methods for convex optimization, and, finally, elaborate on the related concept of barrier function based model predictive control.

2.1 Model predictive control

The concept of model predictive control (MPC) has become one of the most attractive control principles when it comes to the control of multivariable processes that are subject to input, output, or state constraints. In contrast to many conventional control design approaches, which often aim for an *explicit* control law in form of an explicit map from the state or output to the input space, the applied control is in the MPC framework defined *implicitly* via the solution of a finite-horizon open-loop optimal control problem that is parametrized by the current system state. Based on a model of the system to be controlled, this approach allows to take a user-defined performance criterion as well as constraints on the underlying state and input variables explicitly into account when choosing the most appropriate, i.e. optimal, control input. In the following, we briefly introduce and summarize some of the main ideas and results that characterize and underpin the framework of model predictive control. In accordance with the overall focus of this thesis, we limit our discussion to the set-point stabilization problem and highlight in particular some aspects of algorithmic MPC implementations for discrete-time linear systems. More detailed expositions may be found in the cited references as well as in the textbooks by Maciejowski (2002), Rawlings and Mayne (2009), and Grüne and Pannek (2011).

Let us start with the problem formulation for the general nonlinear case. Accounting for the fact that the applied control input is typically piecewise constant, the system to be controlled may be modeled by a difference equation of the form

$$x(k+1) = f(x(k), u(k)), \tag{2.1}$$

where $x(k) \in \mathbb{R}^n$ and $u(k) \in \mathbb{R}^m$ denote the system state and system input at time $k \in \mathbb{N}$, respectively, and $f : \mathbb{R}^n \times \mathbb{R}^m \to \mathbb{R}^n$ describes the transition to the associated successor state, starting from the initial condition $x(0) \in \mathbb{R}^n$. We assume that $f(0, 0) = 0$, i.e., that the origin is an equilibrium point of the dynamical system (2.1). The goal is to control the system state to the equilibrium point $(x_s, u_s) = (0, 0)$ while satisfying at each sampling instant $k \in \mathbb{N}$ pointwise input and state constraints of the form

$$(x(k), u(k)) \in \mathcal{X} \times \mathcal{U}, \tag{2.2}$$

where $\mathcal{X} \subseteq \mathbb{R}^n$ and $\mathcal{U} \subseteq \mathbb{R}^m$ are given constraint sets, typically convex and each containing the respective origin in its interior. Note that the problem of stabilizing an arbitrary equilibrium point $(x_s, u_s) \neq (0, 0)$ can always be reduced to the above problem formulation by a suitable change of coordinates.

The basic idea of model predictive control is to solve at each sampling instant a suitably chosen open-loop optimal control problem which, based on (2.1), aims to minimize a user-defined performance criterion over a finite prediction horizon, while at the same time ensuring that the constraints (2.2) are satisfied. In particular, most existing MPC schemes are based on a finite-horizon open-loop optimal control problem of the form

$$J_N^*(x) = \min_{u} \sum_{k=0}^{N-1} \ell(x_k, u_k) + F(x_N) \tag{2.3a}$$

$$\text{s.t. } x_{k+1} = f(x_k, u_k),\ x_0 = x, \tag{2.3b}$$

$$x_k \in \mathcal{X},\ k = 0, \ldots, N-1,\ x_N \in \mathcal{X}_f, \tag{2.3c}$$

$$u_k \in \mathcal{U},\ k = 0, \ldots, N-1, \tag{2.3d}$$

where $x = x(k)$ denotes the current system state, $N \in \mathbb{N}_+$ is the given, finite prediction horizon, and $\ell : \mathbb{R}^n \times \mathbb{R}^m \to \mathbb{R}_+$ and $F : \mathbb{R}^n \to \mathbb{R}_+$ refer to suitably chosen stage and terminal cost functions, respectively. Furthermore, $\boldsymbol{u} = \{u_0, \ldots, u_{N-1}\}$ denotes the sequence of open-loop control inputs, while $\mathcal{X}_f$ refers to a closed and typically convex *terminal set* that, as discussed in the following, may be used to guarantee stability properties of the closed-loop system. Note that we make use of subindices to distinguish open-loop predictions x_k, u_k from actual state and input realizations $x(k), u(k)$.

The control law is obtained by solving (2.3) at each sampling instant $k \in \mathbb{N}$ and applying $u(k) = u_0^*(x(k))$ in a receding horizon fashion. In particular, only the first element of the computed optimal open-loop input sequence $\boldsymbol{u}^*(x(k))$ is applied to the plant, and the optimization is repeated at the next sampling instant for the updated system state and over a shifted horizon. A graphical illustration of this receding horizon policy is given in Figure 2.1. Consequently, the resulting closed-loop system dynamics are given by

$$x(k+1) = f(x(k), u_0^*(x(k))). \tag{2.4}$$

Note that, up to now, we tacitly assumed that there exists at each sampling instant a solution to problem (2.3), i.e., that the MPC problem is *recursively feasible*, and that the corresponding minimum can always be attained. However, even in the case of feasible initialization, this is not automatically the case. In fact, at least starting from the 1980s, a vast amount of the available MPC literature has been concerned with investigating and

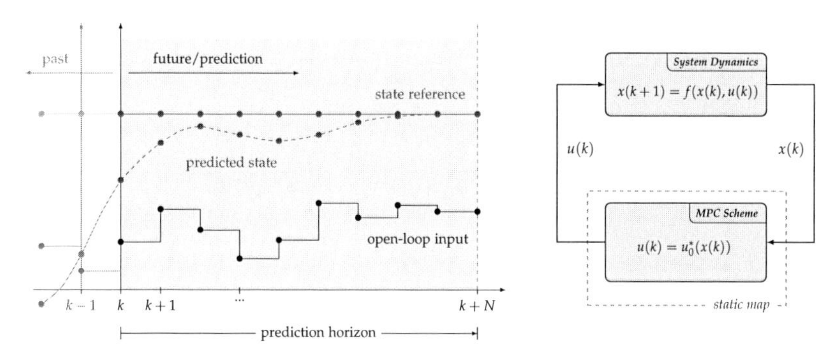

Figure 2.1. *Left:* MPC as repeated open-loop optimal control over a receding horizon. *Right:* The resulting idealized closed-loop system with instantaneous MPC feedback.

formulating constructive design approaches that allow to ensure, at least for a certain region of attraction, that problem (2.3) recursively admits a feasible solution, that the associated optimal solution $\boldsymbol{u}^*(x(k))$ is well-defined, and that the resulting feedback law indeed leads to asymptotic stabilization of the origin of system (2.4). In particular, the following set of sufficient conditions have been distilled in the seminal paper by Mayne et al. (2000) as part of a unifying MPC framework.

Assumption 2.1. *The function $f : \mathbb{R}^n \times \mathbb{R}^m \to \mathbb{R}^n$ is continuous. Furthermore, the set $\mathcal{X}$ is closed, the set $\mathcal{U}$ is compact, and both $\mathcal{X}$ and $\mathcal{U}$ contain the respective origin in their interior.*

Assumption 2.2. *The stage cost $\ell : \mathbb{R}^n \times \mathbb{R}^m \to \mathbb{R}_+$ is continuous, satisfies $\ell(0,0) = 0$, and there exists $c \in \mathbb{R}_{++}$ such that $\ell(x,u) \geq c\|(x,u)\|^2$ for all $(x,u) \in \mathbb{R}^n \times \mathbb{R}^m$. Furthermore, the terminal cost $F : \mathbb{R}^n \to \mathbb{R}_+$ is continuous, $F(0) = 0$, and $F(x) \geq 0$ for all $x \in \mathbb{R}^n$.*

Assumption 2.3. *The terminal set $\mathcal{X}_f$ is closed, satisfies $\mathcal{X}_f \subseteq \mathcal{X}$, and $0 \in \mathcal{X}_f^\circ$. Furthermore, there exists an auxiliary control law $k_f : \mathcal{X}_f \to \mathbb{R}^m$ such that for any $x \in \mathcal{X}_f$ the following holds: i) $k_f(x) \in \mathcal{U}$, ii) $f(x, k_f(x)) \in \mathcal{X}_f$, and iii) $F(f(x, k_f(x))) - F(x) \leq -\ell(x, k_f(x))$.*

Note that Assumptions 2.1 and 2.2 ensure that the optimal solution $\boldsymbol{u}^*(x)$ is attained for any x for which a feasible solution to problem (2.3) exists. Furthermore, Assumption 2.3 essentially implies that there exists an auxiliary controller which is feasible inside the terminal set, which renders the terminal set positive invariant, and for which the terminal cost serves as a local control Lyapunov function (CLF) within the set $\mathcal{X}_f$. Constructive approaches on how to obtain suitable choices for $F(\cdot)$, $k_f(\cdot)$, and $\mathcal{X}_f$ are summarized in both Mayne et al. (2000) and Rawlings and Mayne (2009), see also the references therein. Based on the above assumptions, it is then possible to state the following result on stability properties of the closed-loop system.

Theorem 2.1. *Let Assumptions 2.1–2.3 hold and let $\mathcal{X}_N$ denote the set of all states $x \in \mathcal{X}$ for which problem (2.3) admits a feasible solution. Then, for any $x(0) \in \mathcal{X}_N$, all future states $x = x(k)$ that are generated by the closed-loop system (2.4) will satisfy $x(k) \in \mathcal{X}_N$ for all $k \in \mathbb{N}$. Moreover, the resulting feedback $u(k) = u_0^*(x(k))$ asymptotically stabilizes the origin of system (2.4) with region of attraction $\mathcal{X}_N$ under satisfaction of the pointwise state and input constraints (2.2).*

A proof of this theorem can, *inter alia*, be found in (Mayne et al., 2000; Rawlings and Mayne, 2009; Grüne and Pannek, 2011). The key idea underlying the above stability result is to first exploit the properties of the terminal set together with the auxiliary controller in order to construct a feasible *candidate solution*, leading to recursive feasibility of problem (2.3), and then show that the MPC value function $J_N^*(x(k))$ can be used as a Lyapunov function for the closed-loop system (2.4). In fact, this idea is a recurring theme that is used in many MPC stability results and will also play a major role in the remainder of this thesis. In addition to the above terminal set based formulation, several researchers have presented alternative MPC approaches which do not rely on such an additional stabilizing constraint, see, e.g., (Grimm et al., 2005; Tuna et al., 2006; Grüne and Rantzer, 2008; Grüne et al., 2010). Beside a more natural formulation of the underlying MPC problem, which is in particular also often used in practice, the main advantage of these terminal set free MPC approaches is that they allow us to derive an estimate on the overall closed-loop performance in relation to the associated infinite-horizon problem. However, ensuring recursive feasibility and constraint satisfaction is, in particular in the presence of state constraints, not a simple task in this setting and seems still to be a topic of ongoing research. A unifying survey on these approaches and their systems theoretic properties has recently been presented by Grüne (2012).

Summarizing, there exists at least in the context of deterministic systems a solid theoretical foundation for handling the set-point stabilization problem within the MPC framework, comprising various well-understood methods for a constructive design of the underlying open-loop optimal control problem. On the other hand, many open problems remain (or open up along the way) and until today MPC continues to be a thriving and highly active research field, as for example illustrated by the survey paper (Mayne, 2014) and the huge number of recent advances and publications mentioned therein. In particular, although many powerful optimization algorithms based on sequential quadratic programming and interior-point methods exist, and predictive control concepts are by now more and more applied to challenging real-time applications, it is in many cases often not possible to rigorously guarantee stability and performance properties of the closed-loop system in situations in which the available sampling time or hardware resources do not allow the optimization algorithm to fully converge to the exact optimal solution. In addition, unknown external disturbances are often ignored in the MPC design process and there exists no general separation principle that allows for combining nominally stabilizing state feedback MPC schemes in a certainty equivalence output feedback fashion with suitable state estimation procedures. Apart from that, also the theoretically motivated concept of additional terminal set constraints is typically neither desired nor used in practice. As a consequence, there often is a gap between theoretical MPC schemes and concepts on the one hand and their actual algorithmic implementation on the other hand. Some, though unfortunately not all, of these problems may be alleviated in the context of MPC for linear systems, which will be the main focus of this thesis and which we therefore want to discuss in more detail in the following section.

Linear model predictive control

In linear model predictive control, the system to be controlled is assumed to be described by a linear model. Thus, the system dynamics (2.1) can be written as

$$x(k+1) = Ax(k) + Bu(k) \tag{2.5}$$

for given system matrices $A \in \mathbb{R}^{n\times n}$, $B \in \mathbb{R}^{n\times m}$, where we assume in the following that the pair (A, B) is stabilizable. Moreover, the state and input constraint sets $\mathcal{X}$ and $\mathcal{U}$ in (2.2) are typically given as polytopic sets of the form

$$\mathcal{X} = \{x \in \mathbb{R}^n : C_\mathrm{x} x \leq d_\mathrm{x}\}, \quad \mathcal{U} = \{u \in \mathbb{R}^m : C_\mathrm{u} u \leq d_\mathrm{u}\}, \tag{2.6}$$

where $C_\mathrm{x} \in \mathbb{R}^{q_\mathrm{x}\times n}$, $C_\mathrm{u} \in \mathbb{R}^{q_\mathrm{u}\times m}$ and $d_\mathrm{x} \in \mathbb{R}^{q_\mathrm{x}}_{++}$, $d_\mathrm{u} \in \mathbb{R}^{q_\mathrm{u}}_{++}$, with $q_x, q_u \in \mathbb{N}_+$ denoting the number of affine constraints in each case. Note that (2.6) includes in particular the case of simple box constraints and that both $\mathcal{X}$ and $\mathcal{U}$ inherently contain the origin in their interior due to the strict positivity of $d_\mathrm{x}, d_\mathrm{u}$. Obviously, the above problem setup immediately implies that Assumption 2.1 is satisfied. In accordance with Assumption 2.2, the stage and terminal costs are typically chosen to be quadratic functions of the form

$$\ell(x,u) = \|x\|_Q^2 + \|u\|_R^2, \qquad F(x) = \|x\|_P^2 \tag{2.7}$$

for suitable weighting matrices $Q \in \mathbb{S}^n_{++}$, $R \in \mathbb{S}^m_{++}$, and $P \in \mathbb{S}^n_{+}$. Under the assumption that the pair $(A, Q^{1/2})$ is detectable, positive-definiteness of Q may also be relaxed to the condition $Q \in \mathbb{S}^n_+$ (Mayne et al., 2000). While the stage cost matrices Q and R are typically chosen by the user, e.g., based on some energy-related performance criterion, the terminal cost matrix P needs to be computed in such a way that, together with a suitable terminal set, it satisfies the local CLF condition stated in Assumption 2.3. Constructive approaches for a suitable design of the pair $(\mathcal{X}_f, P)$ have, for example, been presented by Sznaier and Damborg (1987), Chmielewski and Manousiouthakis (1996), and Scokaert and Rawlings (1998). See also (Mayne et al., 2000; Rawlings and Mayne, 2009) for a comparison of these and other approaches. In particular, $F(x) = x^\top Px$ and an associated linear auxiliary controller $k_f(x) = Kx$ may be chosen as the value function and optimal controller for the associated unconstrained infinite-horizon linear quadratic regulation (LQR) problem, characterized by the discrete-time algebraic Riccati equation

$$K = -\left(R + B^\top PB\right)^{-1} B^\top PA\,, \tag{2.8a}$$

$$P = (A+BK)^\top P(A+BK) + K^\top RK + Q\,, \tag{2.8b}$$

while the terminal set X_f can be chosen as the maximal admissible set (Gilbert and Tan, 1991) for the system $x^+ = (A + BK)x$, defined as

$$\Omega_\infty := \left\{ x \in \mathbb{R}^n : (A + BK)^i x \in \mathcal{X},\ K(A + BK)^i x \in \mathcal{U}\ \ \forall\, i \in \mathbb{N} \right\} . \tag{2.9}$$

On the one hand, this approach allows to show that there does in fact exist a *finite* prediction horizon N such that the resulting MPC solution will actually be equivalent to that of the associated *infinite-horizon* constrained LQR problem. This result is possible since, in the linear case, the quadratic terminal cost function based on (2.8b) provides for all states in $\mathcal{X}_f = \Omega_\infty$ an *exact* estimate of the infinite-horizon cost-to-go – a property which does in general not hold for arbitrary nonlinear systems (Sznaier and Damborg, 1987; Chmielewski and Manousiouthakis, 1996). On the other hand, as $A + BK$ is stable and the sets $\mathcal{X}, \mathcal{U}$ are polytopes, the set Ω_∞ exists and will also be a polytope (Gilbert and Tan, 1991). Thus, based on (2.5)–(2.7) and an explicit representation $\Omega_\infty = \{x \in \mathbb{R}^n : C_\mathrm{f}\, x \leq d_\mathrm{f}\}$, the open-loop optimal control problem (2.3) can be reformulated as

$$\begin{aligned} J_N^*(x) = \min_{u}\ & \sum_{k=0}^{N-1} \|x_k\|_Q^2 + \|u_k\|_R^2 + \|x_N\|_P^2 && (2.10\text{a})\\ \text{s.t.}\ \ & x_{k+1} = Ax_k + Bu_k,\ x_0 = x\,, && (2.10\text{b})\\ & C_\mathrm{x}\, x_k \leq d_\mathrm{x},\ k = 0, \ldots, N-1,\ C_\mathrm{f}\, x_N \leq d_\mathrm{f}\,, && (2.10\text{c})\\ & C_\mathrm{u}\, u_k \leq d_\mathrm{u}\ k = 0, \ldots, N-1\,. && (2.10\text{d}) \end{aligned}$$

By introducing the stacked input vector $U = \begin{bmatrix} u_0^\top & u_1^\top & \ldots & u_{N-1}^\top \end{bmatrix}^\top \in \mathbb{R}^{Nm}$ and eliminating the linear system dynamics (2.10b), problem (2.10) can in this case be rewritten as an equivalent *condensed* quadratic program (QP) of the form

$$\begin{aligned} J_N^*(x) =& \min_{U} \frac{1}{2} U^\top H U + x^\top F U + \frac{1}{2} x^\top Y x && (2.11\text{a})\\ & \text{s.t.}\ \ GU \leq d + Ex\,, && (2.11\text{b}) \end{aligned}$$

which is obviously parametrized by the current system state. The matrices $H \in \mathbb{S}_{++}^{Nm}$, $F \in \mathbb{R}^{n \times Nm}$, $Y \in \mathbb{S}_{++}^n$, and $G \in \mathbb{R}^{q \times Nm}$, $d \in \mathbb{R}_{++}^q$, $E \in \mathbb{R}^{q \times n}$, with $q \in \mathbb{N}_+$ denoting the number of constraints, can be constructed from problem (2.10) by means of simple matrix operations (Bemporad et al., 2002). Note that positive-definiteness of H, and hence strict convexity of the QP (2.11) in the optimization variable U, follows from $R \in \mathbb{S}_{++}^m$, $Q \in \mathbb{S}_+^n$, and linearity of the underlying system dynamics. In addition to the above condensed formulation, it is also possible to write problem (2.10) as a non-condensed and typically sparse QP with linear equality constraints. The relation between condensed and sparse formulation as well as more details on the construction of the respective QP matrices will be discussed in Chapter 4 of this thesis. Interestingly, the above formulation is, with some slight modifications, also preserved when considering MPC formulations for tracking asymptotically constant references as well as in the context of tube-based robust MPC approaches that make use of a constraint tightening based on positive invariant sets.

Thus, when considering a linear system model, the MPC open-loop optimal control problem can in many cases be recast as a parametric quadratic program, solved at each sampling instant based on the current system state $x = x(k)$. As the resulting control input needs to be available in real-time, this immediately raises the question how this repeated optimization can be performed in an efficient and numerically stable manner. One particular approach is to apply multi-parametric quadratic programming techniques in order to precompute the parametrized optimal solution to (2.11) *offline* in form of a piecewise affine function defined over a polyhedral partition of the respective feasible set (Bemporad et al., 2002; Seron et al., 2003; Tøndel et al., 2003; Alessio and Bemporad, 2009; Gupta et al., 2011; Mönnigmann and Jost, 2012; Feller et al., 2013; Feller and Johansen, 2013). The resulting MPC feedback law is *explicit* in the sense that it provides an affine control law for each of the polyhedral regions in the partition, reducing the computation of the optimal control input at each sampling instant to the identification of the currently active critical region, followed by simple matrix-vector operations. Unfortunately, such explicit MPC approaches are in most cases only applicable to small systems (say, up to dimension six) as the complexity of the underlying polyhedral partition typically grows exponentially with the dimension of the state space. The more conventional approach with a much broader range of applicability is therefore to solve the underlying optimization problem *online* and in parallel to the evolving system dynamics. Two approaches that are often used in this context are active-set methods and interior-point methods, although there are also some interesting recent works on the use of fast (projected) gradient methods. For each of these approaches exist efficient optimization algorithms that are specifically tailored to the properties and characteristics of the MPC framework, see for example (Bartlett and Biegler, 2006; Pannocchia et al., 2007; Ferreau et al., 2008), (Rao et al., 1998; Wang and Boyd, 2010; Domahidi et al., 2012), as well as (Richter et al., 2009, 2012; Kögel and Findeisen, 2011a,b; Patrinos and Bemporad, 2014; Rubagotti et al., 2014; Giselsson, 2014). As the barrier function approach to constrained optimization and the related concept of interior-point methods will play a major role in the remainder of this thesis, some of the underlying main aspects are summarized briefly in the following section.

2.2 Barrier function methods for convex optimization

During the last three decades, the field of barrier function based optimization methods has seen a sheer outburst of research activity – reaching from Kamarkar's algorithm in 1984 over the theory of self-concordant barrier functions introduced by Nesterov and Nemirovskii in 1994 to the huge variety of significant results and efficient primal-dual interior-point algorithms that are available today, see for example Gondzio (2012). In the following, we are going to highlight some of the related main ideas and concepts, limiting ourselves, however, to those which are of immediate importance for the material presented in this thesis. More detailed and rigorous explanations as well as more general problem setups can be found in the relevant textbooks like, e.g., (Nesterov and Nemirovskii, 1994; Renegar, 2001; Boyd and Vandenberghe, 2004; Nesterov, 2004).

In order to simplify the discussion, we limit our considerations in the following to inequality constrained convex optimization problems of the form

$$\min_{x} \; f_0(x) \tag{2.12a}$$

$$\text{s.t.} \; f_i(x) \leq 0 \quad i = 1, \ldots, m, \tag{2.12b}$$

where $x \in \mathbb{R}^n$ is the optimization variable and $f_i : \mathbb{R}^n \to \mathbb{R}$, $i = 0, \ldots, m$, are continuously differentiable and convex functions. We assume that the problem is strictly feasible, that a well-defined (not necessarily unique) optimizer x^* exists, and we define $p^* := f_0(x^*) < \infty$. The principal idea of barrier function based optimization is to successively approximate the solution to the constrained problem (2.12) by the solutions to a suitably defined sequence of *unconstrained* problems. To this end, the inequality constrains in (2.12b) are incorporated into a modified, unconstrained version of the problem by means of so-called *barrier functions*. In accordance with Nesterov (2004), we introduce the following quite general definition of a barrier function for a closed set.

Definition 2.1. *Given a closed set $\mathcal{D}$ with nonempty interior, a continuous function $B : \mathcal{D}^\circ \to \mathbb{R}$ is called a* barrier function *for $\mathcal{D}$ if $B(x) \to \infty$ whenever x approaches the boundary of the set $\mathcal{D}$.*

A particular interesting class of barrier functions, both from a theoretical and a numerical point of view, is given by the class of *logarithmic* barrier functions, which we are going to discuss in more detail in the following and also in the remainder of this thesis. In particular, coming back to problem (2.12), a logarithmic barrier function for the associated feasible set may be defined as

$$B(x) = -\sum_{i=1}^{m} \ln\left(-f_i(x)\right), \tag{2.13}$$

where $\ln : \mathbb{R}_{++} \to \mathbb{R}$ denotes the natural logarithm defined over the positive reals. Obviously, the function $B : \mathcal{X}_F^\circ \to \mathbb{R}$ defined in (2.13) represents a barrier function for the feasible set $\mathcal{X}_F := \{x \in \mathbb{R}^n : f_i(x) \leq 0,\ i = 1, \ldots, m\}$ that has all the properties stated in Definition 2.1. Furthermore, it is smooth and convex on its domain of definition and possesses additional interesting properties that are beneficial for analyzing the convergence rate of the barrier function based optimization methods discussed in the following (in particular, the property of self-concordance (Nesterov and Nemirovskii, 1994)).
Aiming for an approximate solution to the constrained problem (2.12), we formulate the following barrier function based auxiliary problem

$$\min_{x} \; f_0(x) + \varepsilon B(x), \tag{2.14}$$

where $\varepsilon \in \mathbb{R}_{++}$ denotes a scalar barrier function weighting parameter that determines the impact of the barrier function $B(\cdot)$ on the overall cost function. As the considered logarithmic barrier function is convex, (2.14) is a convex and unconstrained optimization problem, which can be solved by means of efficient and well-understood optimization algorithms like the Newton method. Furthermore, as illustrated in Figure 2.2, the underlying weighted barrier function will approximate the indicator function for the underlying feasible set arbitrarily well for arbitrarily small values of ε. Thus, problem (2.14) can be interpreted as a smooth and more easily solvable approximation to problem (2.12) that gets more and more accurate as we decrease the barrier parameter ε. This directly motivates the so-called *barrier method* procedure, in which problem (2.14) is repeatedly solved for a decreasing sequence of barrier parameters, initializing each optimization with the solution from the previous step, cf. (Boyd and Vandenberghe, 2004; Nesterov, 2004).

Definition 2.2. *Given the above problem setup (2.12)–(2.14), the* barrier method *successively approximates the solution to problem (2.12) for $k \in \mathbb{N}$ according to the following iteration scheme:*

i) $k = 0$: *choose initialization* $\tilde{x}_0 \in \mathcal{X}_F^\circ$ *as well as* $\varepsilon_0 \in \mathbb{R}_{++}$.

ii) $k \geq 1$: *choose* $0 < \varepsilon_k < \varepsilon_{k-1}$, $\{\varepsilon_k\} \to 0$, *and determine* $\tilde{x}_k = \arg\min\{f_0(x) + \varepsilon_k B(x)\}$ *using* $\tilde{x}_{k-1}$ *as an initial guess. Define associated cost function value as* $\tilde{p}_k = f_0(\tilde{x}_k)$.

The resulting approximate optimizers $\tilde{x}_k$ all lie on the so-called *central path*, which is given by the set of optimal solutions $\tilde{x}^*(\varepsilon)$ to problem (2.14) for varying $\varepsilon \in \mathbb{R}_{++}$, see Figure 2.2. Based on the above arguments, intuition tells us that the sequence of solutions that is produced by the barrier method should converge to the optimal solution of the original, constrained problem. In fact, as shown by the following theorem, this claim is true for arbitrary barrier functions that comply with the conditions given in Definition 2.1.

Theorem 2.2. *Consider the problems (2.12) and (2.14) with* $B : \mathcal{X}_F^\circ \to \mathbb{R}$ *being a barrier function for the set* $\mathcal{X}_F$ *according to Definition 2.1. Furthermore, let* $\{\tilde{p}_k\}$ *be the sequence of solutions that is generated by the barrier method according to Definition 2.2. Then, it holds that* $\lim_{k\to\infty} \tilde{p}_k = p^*$.

A proof of Theorem 2.2, respectively of a slightly more general result that is applicable to not necessarily convex nonlinear programming problems of the form (2.12), can be found in (Nesterov, 2004, page 50). While Theorem 2.2 ensures that the solution of the barrier function based formulation will eventually converge to the solution of the original constrained problem [1], it does neither tell us how to actually update the barrier function weighting parameter at each step nor how good the resulting approximation will be for a given value of ε. However, in the case considered above, convexity allows for a more detailed interpretation that enables us to give answers to the aforementioned questions. In particular, due to convexity and strict feasibility of the problems (2.12) and (2.14), strong duality holds and the Karush-Kuhn-Tucker (KKT) conditions are necessary and sufficient conditions for optimality in both cases. Obviously, the KKT conditions for the unconstrained, barrier function based problem (2.14) are simply given by

$$\nabla f_0(\tilde{x}^*(\varepsilon)) + \sum_{i=1}^{m} \frac{\varepsilon}{-f_i(\tilde{x}^*(\varepsilon))} \nabla f_i(\tilde{x}^*(\varepsilon)) = 0. \tag{2.15}$$

[1] Note that $\lim_{k\to\infty} \tilde{p}_k = p^*$ implies that $\tilde{x}_k$ will converge to the associated set of optimizers, which is not necessarily a singleton. If the objective f_0 is strictly convex, the optimizer is unique and $\lim_{k\to\infty} \tilde{x}_k = x^*$.

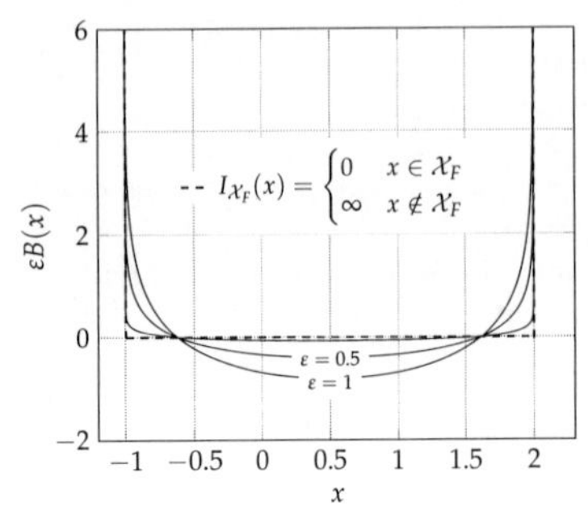

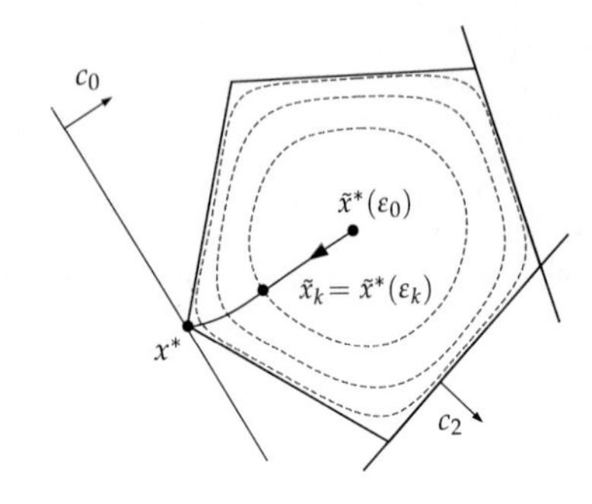

Figure 2.2. *Left:* Approximation of the indicator function $I_{\mathcal{X}_F}$ for the set $\mathcal{X}_F = \{x \in \mathbb{R} : -1 \leq x \leq 2\}$ by the associated weighted logarithmic barrier function for $\varepsilon \in \{1, 0.5, 0.1\}$. The approximation gets better for smaller ε. *Right:* Central path for a linear program with $n = 2$, $m = 5$ and $f_0(x) = c_0^\top x$, $f_i(x) = -c_i^\top x + d_i$. The dashed lines indicate level sets of the underlying logarithmic barrier function $B(x) = -\sum_{i=1}^{5} \ln(-c_i^\top x + d_i)$.

Introducing $\lambda_i^*(\varepsilon) := \varepsilon/(-f_i(\tilde{x}^*(\varepsilon)))$ for $i = 1, \ldots, m$, which implies $\lambda_i^*(\varepsilon) \geq 0$ for any feasible $\tilde{x}^*(\varepsilon)$, leads to the equivalent conditions

$$\nabla f_0(\tilde{x}^*(\varepsilon)) + \sum_{i=1}^{m} \lambda_i^*(\varepsilon) \nabla f_i(\tilde{x}^*(\varepsilon)) = 0\,, \tag{2.16a}$$

$$-\lambda_i^*(\varepsilon) f_i(\tilde{x}^*(\varepsilon)) = \varepsilon\,,\; i = 1, \ldots, m\,, \tag{2.16b}$$

$$\lambda_i^*(\varepsilon) \geq 0\,,\;\; f_i(\tilde{x}^*(\varepsilon)) \leq 0\,,\; i = 1, \ldots, m\,. \tag{2.16c}$$

However, these are nothing else than the KKT conditions for the original, constrained problem (2.12), with the only difference that the usual complementary slackness condition $\lambda_i f_i(x) = 0$ is relaxed to $\lambda_i f_i(x) = -\varepsilon$. Thus, together with $\lambda_i^*(\varepsilon)$ as defined above, the resulting optimal solution $\tilde{x}^*(\varepsilon)$ of (2.14) approximately satisfies for small values of the barrier parameter ε the KKT conditions for the original, constrained problem (2.12). In fact, based on the modified KKT conditions (2.16) and exploiting Lagrangian duality concepts, it is rather straightforward to show that the barrier function based solution will be no more than $m\varepsilon$-suboptimal in the sense that

$$f_0(\tilde{x}^*(\varepsilon)) - p^* \leq m\varepsilon \tag{2.17}$$

for any fixed $\varepsilon \in \mathbb{R}_{++}$ see (Boyd and Vandenberghe, 2004, p. 566). While this result suggests to simply choose the barrier function weighting parameter as $\varepsilon = \epsilon_{\text{tol}}/m$ in order to achieve a desired tolerance of $\epsilon_{\text{tol}} \in \mathbb{R}_{++}$, this approach does generally not work well in practice. The reason for this is that both gradient and Hessian of the function $f_0(x) + \varepsilon B(x)$ are typically rather sensitive near the boundary of the feasible set, which may cause numerical problems when applying first or second order methods directly with an arbitrary initial condition. In this light, the barrier method can be seen as a strategy that sequentially generates points on the central path and, by decreasing ε, eventually follows the central path to an arbitrarily small neighborhood of the true optimal solution. In addition to the barrier method discussed above, there also exist so-called *path-following* methods (sometimes also referred to as barrier methods), which compute at each iteration not the exact optimal solution for the respective value of ε_k but only an approximate solution in the vicinity of the central path, e.g., by performing only one Newton step for each ε_k and carefully choosing the update strategy for the barrier parameter. One particularly interesting example is the class of primal-dual interior-point algorithms, which update at each iteration both the primal and dual variables by successively applying Newton's method directly to the KKT conditions (2.16) (Mehrotra, 1992; Nesterov and Nemirovskii, 1994; Boyd and Vandenberghe, 2004). The primal-dual path-following approach will be discussed in more detail in Chapter 4 of this thesis. One of the main advantages of all the barrier function based optimization methods mentioned above is that, in contrast to other approaches like the simplex algorithm or active-set methods, they allow us to derive polynomial-time complexity estimates when applied in combination with Newton's method. As shown by Nesterov and Nemirovskii (1994) and Renegar (2001), a central role in the underlying complexity analysis is played by the concept of *self-concordant barrier functions*. We introduce and discuss this concept as well as some of the most important related results briefly in the following section.

Self-concordant barrier functions and complexity estimates

In the light of the above discussion, one might ask what is so special about the proposed class of logarithmic barrier functions and whether we could not make use of other barriers such as hyperbolic or exponential functions. In fact, one of the main results of the pioneering work of Nesterov and Nemirovskii (1994) was to show that logarithmic barrier functions enjoy a characteristic property that allows deriving important complexity estimates for the related class of optimization algorithms – namely the property of *self-concordance*. In particular, they defined a scalar function $f \in C^3 : \mathbb{R} \to \mathbb{R}$ to be self-concordant if it is convex and satisfies

$$|f'''(x)| \leq 2f''(x)^{\frac{3}{2}} \tag{2.18}$$

for all x in its domain of definition, where f''' and f'' denote the third and second derivative of the function f, respectively. Likewise, a multivariate function is said to be self-concordant if its restriction to every line in its domain of definition is self-concordant in the sense of (2.18). In this thesis, however, we will make use of the following slightly more general definition of self-concordance that is due to Renegar (2001)[2].

[2]Note that the original definition by Nesterov and Nemirovskii requires in the most general formulation only convexity, i.e., positive semi-definiteness instead of positive definiteness of the Hessian. However, in most cases (and also in this thesis), the considered barrier functions have positive definite Hessians by design, such that the latter condition is not really a restriction, cf. (Renegar, 2001). In this sense, Definition 2.3 is more general as it allows f to be only *twice* continuously differentiable, i.e., $f \in C^2$.

Definition 2.3. *Consider* $f \in \mathcal{C}^2 : \mathcal{D}_f \subseteq \mathbb{R}^n \to \mathbb{R}$ *satisfying* $\nabla^2 f(x) \in \mathbb{S}^n_{++}$ *for any* $x \in \mathcal{D}_f$ *and define* $\mathcal{B}(x,1) := \{y \in \mathbb{R}^n : \|y - x\|_{\nabla^2 f(x)} < 1\}$. *Then, the function* f *is said to be* self-concordant *if i) for any* $x \in \mathcal{D}_f$ *it holds that* $\mathcal{B}(x,1) \subseteq \mathcal{D}_f$*, and ii) whenever* $y \in \mathcal{B}(x,1)$*, then*

$$1 - \|y - x\|_{\nabla^2 f(x)} \leq \frac{\|v\|_{\nabla^2 f(y)}}{\|v\|_{\nabla^2 f(x)}} \leq \frac{1}{1 - \|y - x\|_{\nabla^2 f(x)}} \quad \text{for all } v \neq 0\,. \tag{2.19}$$

While Definition 2.3 might on the first glance appear to be quite different (and also more complicated) than the original definition by Nesterov and Nemirovskii (1994), Renegar proves that both definitions are in fact equivalent for the class of three times continuously differentiable functions with positive definite Hessians, see (Renegar, 2001, Section 2.5). In particular, both definitions capture the fact that a self-concordant function behaves locally, in a certain sense, like a quadratic function, which allows applying Newton's method in an efficient manner. As a result, Definition 2.3 allows one to derive essentially the same theoretical results as in the traditional framework, while only requiring the function f to be *twice* continuously differentiable (an advantage that we are going to exploit in Chapter 4 of this thesis). In accordance with the above, the class of self-concordant barrier functions is typically defined as follows.

Definition 2.4. *A function* $f \in \mathcal{C}^2 : \mathcal{D}_f \subseteq \mathbb{R}^n \to \mathbb{R}$ *is said to be a* ϑ-self-concordant barrier function *for the set* $\bar{\mathcal{D}}_f$ *if it is self-concordant and in addition satisfies*

$$\vartheta := \sup_{x \in \mathcal{D}_f} \left(\nabla f(x)^\top \left(\nabla^2 f(x)\right)^{-1} \nabla f(x) \right) < \infty\,. \tag{2.20}$$

Note in particular that although affine or quadratic functions are self-concordant, they are in general not self-concordant barrier functions according to Definition 2.4. In contrast to this, the primordial logarithmic barrier function $f(x) = -\sum_{i=1}^n \ln(x_i)$ for the nonnegative orthant $\mathbb{R}^n_+$ is a ϑ-self-concordant barrier function with $\vartheta = n$, see the example presented in Appendix B.1. The parameter $\vartheta \in \mathbb{R}_{++}$ is typically referred to as *"parameter of the barrier f"* (Nesterov and Nemirovskii, 1994; Nesterov, 2004) or *"complexity value of f"* (Renegar, 2001). As suggested by the latter name, the parameter ϑ plays a central role in deriving complexity bounds for iterative optimization algorithms that rely on self-concordant barrier functions. In particular, it can be shown that an ϵ_{tol}-suboptimal solution to both linear and quadratic optimization problems (and under certain assumptions also more general nonlinear programming problems) can be obtained in a maximal number of

$$\mathcal{O}\left(\sqrt{\vartheta}\ln\left(\frac{\vartheta\,\epsilon_0}{\epsilon_{\text{tol}}}\right)\right) \tag{2.21}$$

Newton steps when handling the underlying constraints by means of a ϑ-self-concordant barrier function, see for example Renegar (2001). Here, $\varepsilon_0 \in \mathbb{R}_{++}$ denotes the initial value of the barrier function weighting parameter, while $\epsilon_{\text{tol}} \in \mathbb{R}_{++}$ is the desired accuracy of the approximate solution (i.e., $f_0(\tilde{x}^*(\varepsilon)) - p^* \leq \epsilon_{\text{tol}}$). Note that the nominator in the logarithmic term in (2.21) can often be neglected in comparison with the typically very small denominator ϵ_{tol}, resulting in the complexity bound $\mathcal{O}(\sqrt{\vartheta}\ln(1/\epsilon_{\text{tol}}))$, cf. (Nesterov and Nemirovskii, 1994; Nesterov, 2004; Gondzio, 2012). A characteristic property that is central to the theory and the success of self-concordant barrier functions is that the complexity value ϑ, and hence the complexity estimate given in (2.21), is invariant under affine coordinate changes and does not necessarily depend on the problem dimension. Moreover, the complexity result remains also valid when considering in addition affine equality constraints of the form $Ax = b$ as these can be handled directly within the underlying Newton method. In order to illustrate the practical relevance of these observations, let us for example consider a quadratic programming problem of the form

$$\min_{x} \ \frac{1}{2}x^\top Qx + p^\top x \tag{2.22a}$$

$$\text{s.t.} \ \ Ax = b\,,\ Cx \leq d \tag{2.22b}$$

with optimization variable $x \in \mathbb{R}^n$, $Q \in \mathbb{S}^n_+$, $p \in \mathbb{R}^n$, and linear equality and inequality constraints defined by $A \in \mathbb{R}^{r\times n}$, $b \in \mathbb{R}^r$, $C \in \mathbb{R}^{m\times n}$, $d \in \mathbb{R}^m$ with $n, r, q \in \mathbb{N}_+$, $r < n$. As shown by Nesterov and Nemirovskii (1994), the associated logarithmic barrier function $B(x) = -\sum_{i=1}^{m}\ln(-C^i x + d^i)$ has (independently of the dimension of x) complexity value $\vartheta = m$, which reveals that the role of the complexity value is in this case played by the number of inequality constraints. In combination with (2.21), this shows that the complexity of solving problem (2.22) by means of a logarithmic barrier function based approach is growing only proportionally to $\sqrt{m}\ln(m)$. Obviously, this is a much better result than the exponential worst-case growth in the number of constraints that is for example inherent to active-set methods. An explicit formula for an upper bound on the number of required Newton steps for the considered problem class can for example be found in (Boyd and Vandenberghe, 2004, Section 11.5).

It should be noted that barrier and interior-point methods in practice usually require much less iterations than predicted by the corresponding polynomial-time complexity estimates, achieving convergence to a nearly optimal solution typically in an almost constant number of iterations. Moreover, the above discussion should be put into perspective by emphasizing that it is actually not clear whether self-concordant functions are *in general* more easily minimized by Newton's method than arbitrary convex functions (Boyd and Vandenberghe, 2004). Nevertheless, the concept of self-concordant barrier functions, of which we presented here only a brief and rather selective overview, provides a solid theoretical underpinning for the huge variety of barrier function based optimization algorithms that are available today. In particular, coming back to the main topic of this thesis, many of the available results, including the outlined polynomial-time complexity estimates, may be used in model predictive control approaches that either perform the required online optimization by means of barrier function based algorithms or which directly formulate the underlying open-loop optimal control problem based on logarithmic barrier functions. The latter approach, which has been introduced by Wills and Heath (2004) under the name of *barrier function based model predictive control*, will be briefly summarized in the following section. More details can be found in (Wills and Heath, 2002, 2004; Wills, 2003) as well as in Chapter 3 of this thesis.

2.3 Barrier function based model predictive control

Inspired by the barrier function based optimization methods discussed in the previous section, the principal idea of barrier function based model predictive control is to incorporate the underlying input and state constraints into the cost function by means of suitable barrier functions – reformulating the respective open-loop optimal control problem as

$$\tilde{J}_N^*(x) = \min_{u} \sum_{k=0}^{N-1} \tilde{\ell}(x_k, u_k) + \tilde{F}(x_N) \tag{2.23a}$$

$$\text{s.t. } x_{k+1} = f(x_k, u_k),\ x_0 = x\,, \tag{2.23b}$$

with barrier function based cost functions $\tilde{\ell} : \mathcal{X}^\circ \times \mathcal{U}^\circ \to \mathbb{R}$ and $\tilde{F} : \mathcal{X}_f^\circ \to \mathbb{R}$, respectively. In particular, the modified stage and terminal cost functions are chosen as

$$\tilde{\ell}(x,u) = \ell(x,u) + \varepsilon B_{\mathrm{x}}(x) + \varepsilon B_{\mathrm{u}}(u)\,, \quad \tilde{F}(x) = F(x) + \varepsilon B_{\mathrm{f}}(x)\,, \tag{2.24}$$

where $B_{\mathrm{x}} : \mathcal{X}^\circ \to \mathbb{R}$, $B_{\mathrm{u}} : \mathcal{U}^\circ \to \mathbb{R}$, and $B_{\mathrm{f}} : \mathcal{X}_f^\circ \to \mathbb{R}$ are logarithmic barrier functions for the sets $\mathcal{X} \subseteq \mathbb{R}^n$, $\mathcal{U} \subseteq \mathbb{R}^m$, and $\mathcal{X}_f \subseteq \mathbb{R}^n$, respectively, and $\varepsilon \in \mathbb{R}_{++}$ is the associated weighting parameter. Furthermore, $\ell : \mathbb{R}^n \times \mathbb{R}^m \to \mathbb{R}$ and $F : \mathbb{R}^n \to \mathbb{R}$ refer to the usual, e.g. quadratic, stage and terminal cost functions. As in the conventional interior-point framework, the barrier functions allow us to get rid of the inequality constraints in (2.3c,d) and lead to an equality constrained formulation which can then be tackled by efficient algorithms like the Newton method. The barrier function based reformulation may, thus, be seen as a first step towards an integrative MPC design approach that takes both systems theoretic and algorithmic aspects into account. Concerning the relation to the original MPC formulation, we know from the general convergence result of the barrier method, cf. Theorem 2.2, that the solution of problem (2.23) will converge to the solution of the original MPC problem (2.3) when decreasing the barrier function weighting parameter ε arbitrarily close to zero. However, one of the key ideas of the barrier function based MPC approach as proposed originally by Wills and Heath (2004) is to solve problem (2.23) for one *fixed* value $\varepsilon \in \mathbb{R}_{++}$ that does, in fact, not necessarily need to be close to zero. As a result, only *one* single equality constrained optimization problem needs to be solved at each sampling instant and no iteration of the barrier parameter is required. The resulting barrier function based MPC feedback is given by $u(k) = \tilde{u}_0^*(x(k))$, leading to the closed-loop system dynamics

$$x(k+1) = f\big(x(k), \tilde{u}_0^*(x(k))\big)\,, \tag{2.25}$$

with $\tilde{\boldsymbol{u}}^*(x) = \{\tilde{u}_0^*(x), \ldots, \tilde{u}_{N-1}^*(x)\}$ denoting the optimal input sequence associated to problem (2.23) for a given x. As pointed out in (Wills and Heath, 2004; Wills, 2003), another advantage of this approach is that the barrier function weighting parameter ε may be seen as a tuning parameter that allows us to influence how *"cautious"* the resulting control is close to the boundary of the feasible set. However, when considering a fixed barrier function weighting, stability of the resulting closed-loop system (2.25) will in general not be guaranteed. In particular, even when the original MPC problem satisfies all the sufficient stability conditions discussed in Section 2.1, the resulting stability properties will not necessarily be inherited by the associated barrier function based formulation (2.23). As a consequence, and as shown in the aforementioned references, both the underlying barrier functions and the design of terminal cost and terminal set need to be adjusted in a suitable way in order to ensure asymptotic stability of the closed-loop system. In the following, we briefly summarize the main ideas of the barrier function based MPC approach as proposed originally by Wills and Heath.

As in the conventional MPC stability analysis, the principal idea is to use the value function $\tilde{J}_N^*(x)$ associated to the open-loop optimal control problem (2.23) as Lyapunov function for the closed-loop system. On the one hand, this requires the value function to be positive definite with respect to the desired set-point, i.e., the origin, while, on the other hand, $\tilde{J}_N^*(x)$ needs to decrease under the closed-loop system dynamics. However, as illustrated in Figure 2.3 on page 27, any (logarithmic) barrier function will typically achieve its minimum at the analytic center of the respective constraint set, which will, in general, not be equal to the origin in the case of asymmetric constraints. As this inherently destroys the positive definiteness of the modified stage and terminal costs, and hence of the associated value function, the concept of *recentered barrier functions* has been proposed by Wills and Heath (2002, 2004). In particular, they introduce the concept of gradient recentered barrier functions as follows.

Definition 2.5. *Let $B : \mathcal{D} \to \mathbb{R}$ be a self-concordant barrier function on an open and non-empty convex set $\mathcal{D}$ with $0 \in \mathcal{D}$. Then, the function $\bar{B} : D \to \mathbb{R}_+$ defined as*

$$\bar{B}(z) = B(z) - B(0) - [\nabla B(0)]^\top z \tag{2.26}$$

is the gradient recentered self-concordant barrier function for the set $\overline{D}$ (around the origin).

Obviously, any gradient recentered barrier function is by design again self-concordant (we are only adding an affine term) and satisfies $\bar{B}(0) = 0$ as well as $\nabla \bar{B}(0) = 0$. In combination, this implies that $\bar{B}(x)$ attains its minimum at the origin as desired.

Remark 2.1. Note that the recentering can be performed for an arbitrary point $\bar{z} \in \mathcal{D}$ by simply substituting "0" in (2.26) with $\bar{z}$. In fact, the gradient recentered barrier function is nothing else than the *Bregman distance* or *Bregman divergence* $D_B(z, y) = B(z) - B(y) - [\nabla B(y)]^\top (z - y)$ in which the second argument is chosen to be the origin. Thus, the recentered barrier function measures the gap between the original barrier and its tangent at the origin, and desired properties like convexity and positive definiteness follow directly from the general properties of Bregman distances.

Example 2.1. Let us derive as an illustrating example the gradient recentered logarithmic barrier function for the set $\mathcal{X} = \{x \in \mathbb{R} : -1 \leq x \leq 2\}$. Without any recentering, the logarithmic barrier function for this set is given by $B(x) = -\ln(2 - x) - \ln(1 + x)$. As illustrated in Figure 2.3 on page 27, this function is not positive definite but attains its minimum at $x = 0.5$, the analytic center of the set $\mathcal{X}$. Following Definition 2.5, the associated gradient recentered barrier function is given by $\bar{B}(x) = -\ln(2-x) - \ln(1+x) + \ln(2) + \frac{1}{2}x$, where the additional terms follow from (2.26) and the fact that $B(0) = -\ln(2)$, while $\nabla B(0) = -1/2$. As desired, and as illustrated in Figure 2.3, the resulting recentered logarithmic barrier function is positive definite with respect to the origin.

In Chapter 3, we are going to introduce and discuss an alternative, weighting-based recentering concept for logarithmic barrier functions defined on polytopic sets, see also Appendix B.3. For the moment, we may simply see the recentering as a suitably defined transformation that ensures positive definiteness of the resulting barrier functions while preserving their main characteristics like convexity, self-concordance, and the barrier function property. For the remainder of this section, we make the following assumption concerning the barrier functions in (2.24).

Assumption 2.4. *The functions $B_x(\cdot)$, $B_u(\cdot)$, and $B_f(\cdot)$ are gradient recentered self-concordant barrier functions for the sets $\mathcal{X}, \mathcal{U}$, and $\mathcal{X}_f$, respectively.*

As the original stage and terminal cost are assumed to be continuous and positive definite, Assumption 2.4 implies that the modified cost functions $\tilde{\ell}(\cdot,\cdot)$ and $\tilde{F}(\cdot)$ are themselves continuous and positive definite. As a result, the associated value function $\tilde{J}_N^* : \mathcal{X}_N^\circ \to \mathbb{R}$, with $\mathcal{X}_N$ denoting the feasible set according to Theorem 2.1, is a positive definite function. A sufficient condition for ensuring that it will in addition also decrease under the closed-loop system dynamics is that the terminal set $\mathcal{X}_f$ and the modified terminal cost function $\tilde{F} : \mathcal{X}^\circ \to \mathbb{R}$ together satisfy the following conditions, which can in fact be seen as a slight modification of Assumption 2.3.

Assumption 2.5. *The terminal set $\mathcal{X}_f \subseteq \mathcal{X}$ is closed and convex and satisfies $0 \in \mathcal{X}_f^\circ$. Furthermore, there exists an auxiliary control law $u = k_f(x)$ such that the following holds for any $x \in \mathcal{X}_f^\circ$: i) $k_f(x) \in \mathcal{U}^\circ$, ii) $f(x, k_f(x)) \in \mathcal{X}_f^\circ$, and iii) $\tilde{F}(f(x,k_f(x))) - \tilde{F}(x) \leq -\tilde{\ell}(x, k_f(x))$.*

Theorem 2.3. *Let Assumptions 2.1, 2.2, 2.4, and 2.5 hold and let $\mathcal{X}_N$ denote the set of all states $x \in \mathcal{X}$ for which the original problem (2.3) admits a feasible solution. Then, for any $x(0) \in \mathcal{X}_N^\circ$, all future system states $x(k)$ that are generated by the closed-loop system (2.25) will satisfy $x(k) \in \mathcal{X}_N^\circ$ for all $k \in \mathbb{N}$. Moreover, the barrier function based feedback $u(k) = \tilde{u}_0^*(x(k))$ asymptotically stabilizes the origin of system (2.25) with region of attraction $\mathcal{X}_N^\circ$ under strict satisfaction of the pointwise state and input constraints (2.2).*

Theorem 2.3 and its proof presented in Appendix C.1 illustrate the fact that all the desired stability and constraint satisfaction properties of conventional MPC approaches can essentially be recovered within the barrier function based MPC framework. However, as also pointed out by Wills and Heath (2004), the main difficulty is to design the terminal set and the terminal cost such that Assumption 2.5 is satisfied. In particular, the underlying barrier functions need to be taken into account explicitly when aiming for a terminal cost function $\tilde{F}(\cdot)$ that satisfies the CLF condition *iii*). In the following, we briefly summarize the design approach presented in (Wills and Heath, 2004; Wills, 2003), which is based on upper bounding the stage cost barrier functions locally by a quadratic function and then using this quadratic upper bound for constructing a suitable terminal cost. As we will see, the property of self-concordance will play a central role also in this context. In order to simplify the discussion and with an eye on the focus of this thesis, we limit our considerations to the case of linear systems.

Barrier function based MPC of linear systems

Let us in the following consider the usual linear MPC problem setup with polytopic state and input constraints and quadratic stage and terminal cost functions, see (2.5)–(2.7). In this case, the corresponding barrier function based cost functions are given as

$$\tilde{\ell}(x,u) = \|x\|_Q^2 + \|u\|_R^2 + \varepsilon B_{\mathrm{x}}(x) + \varepsilon B_{\mathrm{u}}(u), \quad \tilde{F}(x) = \|x\|_P^2 + \varepsilon B_{\mathrm{f}}(x), \tag{2.27}$$

where $B_{\mathrm{x}} : \mathcal{X}^\circ \to \mathbb{R}_+$ and $B_{\mathrm{u}} : \mathcal{U}^\circ \to \mathbb{R}_+$ are gradient recentered self-concordant barrier functions for the polytopic sets $\mathcal{X}$ and $\mathcal{U}$, respectively. Furthermore, the ansatz that is proposed in the aforementioned references is to choose the terminal set as a sublevel set of the the quadratic terminal state penalty and to make use of an associated (recentered) logarithmic terminal set barrier function, i.e.,

$$\mathcal{X}_f = \{x \in \mathbb{R}^n : \|x\|_P^2 \leq \alpha^2\}, \quad B_{\mathrm{f}}(x) = \ln(\alpha^2) - \ln(\alpha^2 - \|x\|_P^2), \quad \alpha \in \mathbb{R}_{++}. \tag{2.28}$$

Obviously, Assumptions 2.1, 2.2, and 2.4 are inherently satisfied by the above problem formulation, and the stabilizing MPC design is reduced to choosing $P \in \mathbb{S}^n_{++}$ and the parameter $\alpha \in \mathbb{R}_{++}$ in such a way that Assumption 2.5 holds. Accounting for the linear system dynamics, we assume a linear auxiliary control law $k_f(x) = Kx$, where $K \in \mathbb{R}^{m \times n}$ is chosen such that system (2.5) is asymptotically stable under the feedback $u(k) = Kx(k)$. Consequently, conditions *i*) and *ii*) of Assumption 2.5 will be satisfied if the terminal set $\mathcal{X}_f$ is chosen as a positively invariant subset of the set

$$\mathcal{X}_K := \{x \in \mathcal{X} : Kx \in \mathcal{U}\}, \quad B_K(x) = B_{\mathrm{x}}(x) + B_{\mathrm{u}}(Kx), \tag{2.29}$$

where $B_K : \mathcal{X}_K^\circ \to \mathbb{R}_+$ denotes the barrier function for the set $\mathcal{X}_K$. The positive invariance condition on $\mathcal{X}_f^\circ$ leads to the requirement that $A_K x \in \mathcal{X}_f^\circ \; \forall \, x \in \mathcal{X}_f^\circ$, where $A_K := A + BK$. Obviously, positive invariance is ensured for any P satisfying $A_K^\top P A_K \preceq P$. Let us for the moment assume that this is the case and turn our attention to the CLF condition *iii*). With the above definitions, this condition is equivalent to

$$\|A_K x\|_P^2 - \|x\|_P^2 + \|x\|_Q^2 + \|Kx\|_R^2 + \varepsilon B_K(x) + \varepsilon \left(B_{\mathrm{f}}(A_K x) - B_{\mathrm{f}}(x)\right) \leq 0 \quad \forall x \in \mathcal{X}_f^\circ. \tag{2.30}$$

Note that (2.30) directly reveals how the barrier functions affect the design of the terminal cost matrix P. In particular, without the additional barrier function terms, inequality (2.30) would be equivalent to the usual linear MPC condition – leading to a discrete-time Lyapunov equation of the form (2.8b). In the current situation, however, the barrier functions have to be taken into account in addition to the usual quadratic cost function terms. The main observation of Wills and Heath was that under the assumption that $B_K : \mathcal{X}_K^\circ \to \mathbb{R}_+$ is a gradient recentered self-concordant barrier function, it can always be (locally) upper bounded by a suitably chosen quadratic function, which can then be used for computing a suitable matrix P. In particular, it holds that

$$B_K(x) \leq \frac{1}{2(1-r)^2} \|x\|^2_{\nabla^2 B_K(0)} \quad \forall\, x \in \mathcal{W}_{B_K}(0,r)\,, \tag{2.31}$$

where $\mathcal{W}_{B_K}(0,r)$ denotes the associated *Dikin ellipsoid* with radius $r \in (0,1)$ around the origin. The Dikin ellipsoid plays a central role in the theory of self-concordant functions and is typically defined as follows (Nesterov and Nemirovskii, 1994; Nesterov, 2004).

Definition 2.6. *Let $f : \mathcal{D}_f \subseteq \mathbb{R}^n \to \mathbb{R}$ be a self-concordant function on the open domain $\mathcal{D}_f \subset \mathbb{R}^n$. Then, the* Dikin ellipsoid *of f with radius $r \in [0,1)$ at point $\bar{x} \in \mathcal{D}_f$ is defined as*

$$\mathcal{W}_f(\bar{x},r) := \left\{ x \in \mathbb{R}^n : \|x - \bar{x}\|_{\nabla^2 f(\bar{x})} \leq r \right\}. \tag{2.32}$$

Note that the quadratic bound given in (2.31) follows in principle directly from Taylor's Theorem and the definition of self-concordance (Wills and Heath, 2004). For the sake of completeness, a proof of this result is presented in Appendix B.2.

In the following, we briefly summarize how the quadratic upper bound on the barrier function $B_K(\cdot)$ may be used for designing a suitable terminal cost matrix P. In particular, choosing P as positive definite solution to the (modified) discrete-time Lyapunov equation

$$P = (A+BK)^\top P(A+BK) + Q + K^\top RK + \varepsilon M \tag{2.33}$$

with $M := 1/(2(1-r)^2)\nabla^2 B_K(0) \in \mathbb{S}^n_{++}$ for a fixed $r \in (0,1)$ ensures satisfaction of (2.30) as, on the one hand, $\|A_K x\|_P^2 - \|x\|_P^2 + \|x\|_Q^2 + \|Kx\|_R^2 + \varepsilon B_K(x) \leq \|A_K x\|_P^2 - \|x\|_P^2 + \|x\|_Q^2 + \|Kx\|_R^2 + \varepsilon \|x\|_M^2 = 0$ and, on the other hand,

$$B_f(A_K x) - B_f(x) = \ln\left(\frac{\alpha^2 - \|x\|_P^2}{\alpha^2 - \|x\|^2_{A_K^\top P A_K}}\right) \leq \ln\left(\frac{\alpha^2 - \|x\|_P^2}{\alpha^2 - \|x\|_P^2}\right) = 0 \tag{2.34}$$

due to the fact that (2.33) implies $A_K^\top P A_K \preceq P$. In order to ensure that the above arguments are valid, the terminal set $\mathcal{X}_f$ in (2.28) needs to be a positively invariant subset of the Dikin ellipsoid, i.e., $\mathcal{X}_f \subseteq \mathcal{W}_{B_K}(0,r) \subseteq \mathcal{X}_K^\circ$, see Figure 2.3. As shown in (Wills and Heath, 2004), the maximal α^* that ensures the required set containment can be computed explicitly and is given by

$$\alpha^* = \sqrt{\frac{r^2}{\lambda_{\max}\left(P^{-1}\nabla^2 B_K(0)\right)}}\,. \tag{2.35}$$

Finally, note that (2.33) implies $\|A_K x\|_P \leq \|x\|_P$ for any $x \in \mathcal{X}_f$, which shows that the required invariance of the terminal set is implicitly ensured via the choice of P.

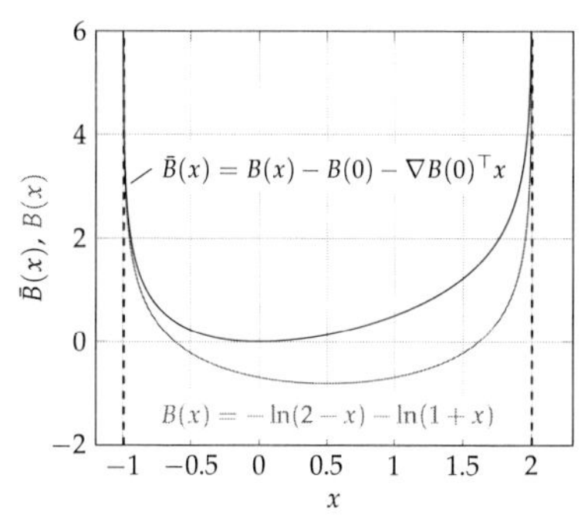

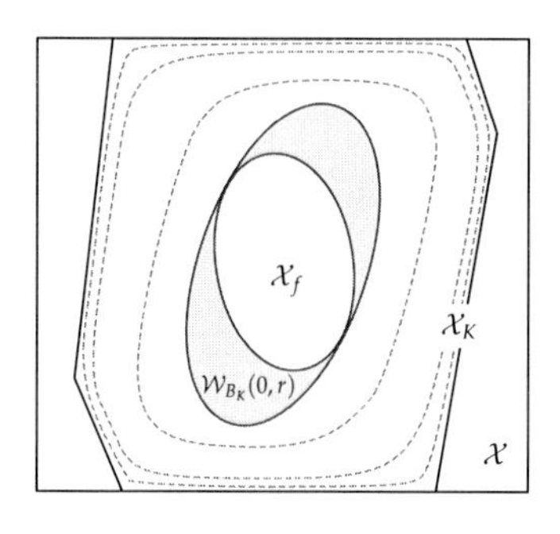

Figure 2.3. *Left:* Unlike the associated gradient recentered formulation, the logarithmic barrier function for the set $\mathcal{X} = \{x \in \mathbb{R} : -1 \leq x \leq 2\}$ attains its minimum not at the origin but at $x = 0.5$, the analytic center of the set $\mathcal{X}$. *Right:* Principal idea of choosing the terminal set $\mathcal{X}_f$ as a subset of the Dikin ellipsoid $\mathcal{W}_{B_K}(0,r) \subset \mathcal{X}_K \subseteq \mathcal{X} \subset \mathbb{R}^2$. Note that a larger $r \in (0,1)$ results in a larger terminal region at the cost of a larger terminal weight.

Thus, in summary, the above arguments provide a constructive approach for choosing the design parameters P and α in such a way that the terminal cost function defined in (2.27) and (2.28) satisfies, together with the underlying ellipsoidal terminal set, the sufficient stability conditions in Assumption 2.5. As also all other assumptions of Theorem 2.3 are satisfied, this directly implies asymptotic stability of the associated closed-loop system, which finally leads to the following result.

Corollary 2.1. *Consider the linear MPC problem setup according to (2.5)–(2.7) and let the corresponding barrier function based open-loop optimal control problem be given by*

$$\tilde{J}_N^*(x) = \min_u \sum_{k=0}^{N-1} \tilde{\ell}(x_k, u_k) + \tilde{F}(x_N) \tag{2.36a}$$

$$\text{s.t. } x_{k+1} = Ax_k + Bu_k,\ x_0 = x\,, \tag{2.36b}$$

where the cost functions $\tilde{\ell} : \mathcal{X}^\circ \times \mathcal{U}^\circ \to \mathbb{R}_+$ and $\tilde{F} : \mathcal{X}_f^\circ \to \mathbb{R}_+$ are chosen according to (2.27) and (2.28) for a fixed barrier function weighting $\varepsilon \in \mathbb{R}_{++}$ and with $P \in \mathbb{S}_{++}^n$ and $\alpha \leq \alpha^* \in \mathbb{R}_{++}$ chosen according to (2.33) and (2.35), respectively. Furthermore, let $\mathcal{X}_N^\circ$ denote the set of all states $x \in \mathcal{X}$ for which a well-defined solution to problem (2.36) exists. Then, for any $x(0) \in \mathcal{X}_N^\circ$, all future states $x(k)$ that are generated by the closed-loop system dynamics

$$x(k+1) = Ax(k) + B\tilde{u}_0^*(x(k)) \tag{2.37}$$

will satisfy $x(k) \in \mathcal{X}_N^\circ$ for all $k \in \mathbb{N}$. Moreover, the feedback $u(k) = \tilde{u}_0^*(x(k))$ asymptotically stabilizes the origin of the closed-loop system (2.37) with region of attraction $\mathcal{X}_N^\circ$ under strict satisfaction of the pointwise state and input constraints (2.2).

Remark 2.2. Note that, as also originally presented by Wills (2003) and Wills and Heath (2004), the above results hold for arbitrary (and not necessarily polytopic) convex constraint sets $\mathcal{X}, \mathcal{U}$. However, in compliance with the linear MPC problem setup that has been considered in Section 2.1 and which will also be the main focus of the remainder of this thesis, we restricted our discussion to the case of polytopic constraints.

Remark 2.3. The above design procedure is constructive in the sense that suitable parameters P and α are guaranteed to exist under the discussed, rather mild, assumptions. In particular, (2.33) admits a positive definite solution P with associated stabilizing control gain $K = -(R + B^\top PB)^{-1}B^\top PA$ whenever (A, B) is stabilizable and $Q \in \mathbb{S}^n_{++}$, $R \in \mathbb{S}^m_{++}$, where the latter condition may again be relaxed to $Q \in \mathbb{S}^n_{+}$ and $(A, Q^{1/2})$ being detectable. It is interesting to note that the original LQR solution given in (2.8) will be recovered when there are no state and input constraints (i.e., when $M = 0$) or whenever the barrier parameter ε approaches zero.

Remark 2.4. As part of the research that led to this thesis, we developed several alternative approaches for designing more general terminal sets as well as suitable associated terminal cost function formulations. In particular, this includes a novel, weighting based recentering procedure, a quadratic upper bound for logarithmic barrier functions that is valid on a polytopic set of adjustable size as well as the notion of barrier function based MPC schemes with *polytopic* terminal sets, see (Feller and Ebenbauer, 2013, 2014b, 2015a). A summary of these novel approaches, which may in many cases lead to an increased region of attraction of the closed-loop system and therefore represent an interesting contribution in their own right, is given in the Appendices B.3–B.5. In addition, they will in Chapter 3 also be discussed in the context of the relaxed logarithmic barrier function based MPC framework that constitutes the main theme of this thesis.

Remark 2.5. Analogously to Section 2.1, the linear system dynamics (2.36b) may be eliminated, leading to a condensed, unconstrained problem formulation of the form

$$\tilde{J}^*_N(x) = \min_{U} \frac{1}{2}U^\top HU + x^\top FU + \frac{1}{2}x^\top Yx + \varepsilon B_{\mathrm{xu}}(U, x), \tag{2.38}$$

where $B_{\mathrm{xu}} : \mathcal{D} \subseteq \mathcal{U}^{\circ N} \times \mathcal{X}^\circ_N \to \mathbb{R}_+$ refers to the overall barrier function for the feasible set defined by the state, input, and terminal set constraints. Due to $H \in \mathbb{S}^{Nm}_{++}$ and the properties of the underlying barrier functions, the overall cost function is twice continuously differentiable, strictly convex, and self-concordant in the optimization variable U.

As shown by Wills and Heath (2004), almost the same arguments can be applied in the context of nonlinear systems, leading to very similar stability results. In particular, the quadratic bound for the barrier function $B_K(\cdot)$ as well as the presented formulations of terminal set and terminal cost are still valid and remain unchanged. The main difference to the discussed linear MPC setup is given by the fact that the terminal cost matrix P now needs to be computed based on a linearization of the nonlinear plant dynamics around the origin. As a consequence, the terminal set $\mathcal{X}_f$ needs the be scaled in such a way that it is positively invariant under the *nonlinear* system dynamics $x^+ = f(x, Kx)$. While the resulting design approach is still constructive and suitable problem parameters are guaranteed to exist, the computation of a suitable scaling factor $\alpha \leq \alpha^*$ involves, similar to the situation in conventional terminal set based NMPC approaches, the global solution of a potentially non-convex optimization problem. More details can be found in the aforementioned references, in particular (Wills, 2003) and (Wills and Heath, 2004). A practical application of the discussed linear barrier function based MPC approach to the control of an edible oil refining process is discussed in (Wills and Heath, 2005).

In summary, the concept of barrier function based model predictive control allows us to deliberately integrate the numerically attractive interior-point approach from Section 2.2 into the conventional MPC framework introduced in Section 2.1. In particular, by fixing the respective barrier function weighting parameter to an arbitrary positive value, only one equality constrained (or even unconstrained) optimization problem needs to be solved at each sampling instant while still allowing to recover the important stability and constraint satisfaction properties of conventional MPC approaches by taking the underlying self-concordant barrier functions explicitly into account in the corresponding MPC design. In this light, the discussed results may be seen as a first step towards a novel class of MPC schemes that consider and exploit characteristics of the applied optimization algorithm explicitly within the MPC design and deliberately dissolve the oftentimes tacitly assumed separation of MPC analysis and design on the one and the actual realization of the underlying optimization algorithm on the other hand. Herein, as well as in the fact that it allows for a twice continuously differentiable formulation of the resulting cost function, the barrier function based MPC approach conceptually differs from so-called "soft constrained" MPC schemes, which are based on the idea of eliminating the respective inequality constraints by means of additional slack variables in conjunction with suitably defined (exact) penalty functions, see for example (Zheng and Morari, 1995; Scokaert and Rawlings, 1999; Kerrigan and Maciejowski, 2000; Zeilinger et al., 2010, 2014b).

One might of course argue that applying a barrier function based optimization algorithm directly to a given stabilizing MPC formulation will typically work well in practice for small values of the barrier function parameter. However, when aiming for rigorous stability guarantees, the influence of the barrier functions needs to be taken into account. A further interesting aspect of the discussed barrier function based reformulation is given by the fact that it results in a *smoothed* approximation of the original, constrained problem, which makes it possible to asymptotically track the corresponding optimal solution by means of suitably defined continuous-time tracking algorithms, see (Ohtsuka, 2004; DeHaan and Guay, 2007) as well as Chapter 4 of this thesis. Seeing that the outlined stability results as well as related ideas like the concept of recentered self-concordant barrier functions can in addition be readily combined with interior-point based optimization algorithms that have been specifically tailored for MPC applications, see for example (Wright, 1997a; Rao et al., 1998; Wang and Boyd, 2010; Domahidi et al., 2012), it seems almost a bit surprising that the framework of barrier function based MPC has not been studied more extensively during the last decade.

2.4 Chapter reflection

In this introductory chapter, we briefly recalled some of the main concepts and results from different strands of research that are forming the background for the material presented in this thesis – namely, model predictive control, barrier function methods for solving convex optimization problems, and barrier function based model predictive control. In Section 2.1, we introduced the principal idea of model predictive control and discussed constructive approaches for designing the underlying open-loop optimal control problem in such a way that it is possible to guarantee asymptotic stability of the closed-loop system. In the case of linear system dynamics that are subject to polytopic state and input constraints, this may be achieved by constructing a suitable quadratic terminal cost as well as a polytopic terminal set based on the corresponding unconstrained infinite-horizon LQR solution. In Section 2.2, we gave a brief overview on barrier function and interior-point methods for convex optimization and showed how the solution to a given inequality constrained problem can be approximated arbitrarily close by successively solving a sequence of barrier function based unconstrained problems. In particular, we introduced the concept of self-concordant barrier functions and discussed how their characteristic properties can be exploited in order to derive polynomial-time complexity estimates for the related class of optimization algorithms. Finally, in Section 2.3, we introduced and summarized the concept of barrier function based model predictive control, in which the principal idea of barrier function based optimization algorithms is integrated directly into the design of the MPC open-loop optimal control problem. Based on a recentering of the underlying barrier functions and a suitable design of both the terminal cost and the associated terminal set, all the desired stability and constraint satisfaction properties of conventional MPC approaches could be recovered, while at the same time reducing the respective open-loop optimal control problem to an unconstrained and convex optimization problem that allows for an efficient algorithmic implementation. Interestingly, the concept of self-concordant barrier functions played a central role also in the discussed stabilizing MPC design procedure. With the exception of Theorem 2.3 and the corresponding Corollary 2.1, most of the results presented in this section are based on the works by Wills (2003) and Wills and Heath (2004).

In the following chapter, which also constitutes the first main part of the thesis, we will generalize and significantly extend some of the ideas presented above by introducing the novel concept of *relaxed barrier function based model predictive control*. As we will see, the relaxed barrier function approach will allow us to eliminate the disadvantages of the conventional logarithmic barrier function based formulation discussed above, e.g., that the resulting feedback law is only defined on the interior of the associated feasible set and that the required local quadratic bound on the input and state constraint barrier function inherently necessitates the use of a suitably chosen compact terminal set. In addition, the concept of relaxed logarithmic barrier functions will enable us to derive important inherent robustness properties of the resulting closed-loop system as well as numerically efficient MPC algorithms with both systems theoretic and algorithmic guarantees.

Chapter 3

Relaxed Barrier Function Based MPC Theory

In this chapter, we introduce a novel framework for the design and analysis of model predictive control schemes that are based on the concept of so-called relaxed logarithmic barrier functions. We present different constructive approaches for choosing the ingredients of the associated open-loop optimal control problem (i.e., the terminal set and/or terminal cost) in a suitable way and discuss related systems theoretical aspects such as stability of the resulting closed-loop system or the satisfaction of input and state constraints. As we will see, the resulting relaxed barrier function based MPC framework is more than a straightforward extension of the conventional, nonrelaxed barrier function based MPC approach discussed in the previous chapter. In particular, it enables the formulation of conceptually novel and simple linear MPC schemes that are not necessarily based on an additional terminal set constraint and that nevertheless possess important stability, constraint satisfaction, and inherent robustness properties.
The results presented in this chapter are based on (Feller and Ebenbauer, 2013, 2014a,b, 2015a,b, 2017a) and (Feller et al., 2016). Suitable tailored iteration schemes and optimization algorithms that allow for a numerically efficient and stability-preserving algorithmic implementation of the resulting MPC schemes will be discussed in Chapter 4.

3.1 Problem setup and chapter outline

The major part of this chapter will be concerned with the case of linear discrete-time systems subject to polytopic input and state constraints. In particular, we consider the prototypic linear MPC problem setup that we already encountered in Chapter 2 and which is in the following briefly recalled for the sake of convenience. Some preliminary results and an outlook on relaxed barrier function based MPC for nonlinear discrete-time systems will be discussed briefly in Section 3.7 at the end of this chapter.

Let us in the following assume that the system dynamics of the plant are described by a discrete-time linear time-invariant system of the form

$$x(k+1) = Ax(k) + Bu(k)\,, \tag{3.1}$$

where $x(k) \in \mathbb{R}^n$ and $u(k) \in \mathbb{R}^m$ denote the vectors of system states and inputs, both at time $k \in \mathbb{N}$. Furthermore, $A \in \mathbb{R}^{n\times n}$ and $B \in \mathbb{R}^{n\times m}$ denote given system matrices, where we assume (if not stated otherwise) that the pair (A, B) is stabilizable.

At each sampling instant $k \in \mathbb{N}$, the states and inputs are required to satisfy pointwise polytopic state and input constraints of the form

$$x(k) \in \mathcal{X} := \{x \in \mathbb{R}^n : C_{\mathrm{x}} x \leq d_{\mathrm{x}}\}, \tag{3.2a}$$
$$u(k) \in \mathcal{U} := \{u \in \mathbb{R}^m : C_{\mathrm{u}} u \leq d_{\mathrm{u}}\}, \tag{3.2b}$$

where $C_{\mathrm{x}} \in \mathbb{R}^{q_{\mathrm{x}} \times n}$, $C_{\mathrm{u}} \in \mathbb{R}^{q_{\mathrm{u}} \times m}$ and $d_{\mathrm{x}} \in \mathbb{R}^{q_{\mathrm{x}}}_{++}$, $d_{\mathrm{u}} \in \mathbb{R}^{q_{\mathrm{u}}}_{++}$, with $q_{\mathrm{x}}, q_{\mathrm{u}} \in \mathbb{N}_+$ denoting the number of affine constraints in each case. Note that both $\mathcal{X}$ and $\mathcal{U}$ inherently contain the origin in their interior due to the fact that the elements of $d_{\mathrm{x}}, d_{\mathrm{u}}$ are strictly positive. As in the previous discussions, we consider quadratic stage and terminal cost function terms of the form

$$\ell(x,u) = \|x\|_Q^2 + \|u\|_R^2, \qquad F(x) = \|x\|_P^2, \tag{3.3}$$

where $Q \in \mathbb{S}^n_{++}$, $R \in \mathbb{S}^m_{++}$, and $P \in \mathbb{S}^n_{+}$ are suitable weighting matrices that may be chosen by the user and can thus be considered as tuning parameters.

Remark 3.1. Note that for most of the results in this chapter, the positive definiteness of Q may also be relaxed to $Q \in \mathbb{S}^n_+$ under the condition that the pair $(A, Q^{1/2})$ is detectable, cf. Mayne et al. (2000). Furthermore, the presented relaxed barrier function based MPC framework also allows one to handle in a straightforward way more general problem setups involving mixed state and input constraints or mixed cost function terms. However, for the sake of simplicity, we restrict ourselves to the problem setup specified above.

In total, the considered problem setup is identical to the one that we discussed in Chapter 2, which would in principle allow us to employ either the conventional linear MPC design summarized in Section 2.1 or the barrier function based MPC approach presented in Section 2.3. However, both approaches have the already discussed disadvantages that the respective open-loop optimal control problems are only defined on the associated feasible set $\mathcal{X}_N$ and that they require an additional terminal set constraint, which is not desired and typically also not used in practice. Especially when considering the barrier function based approach, the fact that the underlying barrier functions are only defined in the interior of the respective constraint sets may be problematic both from a practical and from a conceptual point of view. On the one hand, violations of the state, input, and terminal set constraints are not tolerated at all, which might cause severe problems in the presence of uncertainties, disturbances, noise, observer errors, or sensor outliers. On the other hand, the existing stability results outlined in Section 2.3 inherently require the use of a compact terminal set constraint as they are based on upper bounding the logarithmic barrier for the combined state and input constraints by a quadratic function, which is of course only possible locally in a region around the origin. These considerations motivate in a direct and intuitive way the formulation of barrier function based MPC schemes in which the underlying barrier functions are *relaxed*, for example by smoothly extending them by a globally defined penalty function when coming close to the boundary of the respective constraint set. While such a relaxed problem formulation immediately implies that the associated open-loop optimal control problem, and hence also the associated barrier function based MPC feedback, will be defined globally, it also raises several important questions concerning the stability and constraint satisfaction properties of the resulting closed-loop system. In particular, it is not immediately clear in which way some of the results discussed above, for example the concept of barrier function recentering or the terminal set based MPC design, may carry over to the relaxed barrier function case. Furthermore, as the handling of constraints is one of the main assets of both conventional and nonrelaxed barrier function based MPC, it needs to be investigated to which extent the proposed relaxation may lead to a violation of the underlying polytopic state and input constraints. Finally, we might ask whether we can say anything about the robustness properties of the resulting closed-loop system that goes beyond the fact that a well-defined control input will exist for every realization of the system state. The purpose of this chapter is to give answers to these and similar fundamental questions, developing thereby different classes of relaxed barrier function MPC schemes as well as a systems theoretical framework for their analysis. The chapter is structured as follows.

In Section 3.2, we introduce the concept of relaxed logarithmic barrier functions, including different relaxation procedures as well as a novel, weighting based recentering method. In Section 3.3, we state the relaxed barrier function based MPC problem in a quite general form, before discussing in Section 3.4 different constructive design approaches (both with and without terminal sets) that allow to guarantee important stability properties of the closed-loop system. Methods that allow us to estimate and control the maximally possible constraint violation are discussed in Section 3.5, while important robust stability properties as well as their implications for the case of certainty equivalence output feedback are discussed in Section 3.6. In Section 3.7, we briefly outline how some of the presented results may carry over to the case of nonlinear systems, before concluding the chapter with a summarizing discussion in Section 3.8.

3.2 Relaxed logarithmic barrier functions

The basic idea of relaxed logarithmic barrier functions is to smoothly extend a given conventional logarithmic barrier function by a suitably defined penalizing term – the so-called relaxing function, see Figure 3.1. We begin our studies by introducing the following more rigorous definition.

Definition 3.1 (Relaxed logarithmic barrier function). *Consider a scalar parameter $\delta \in \mathbb{R}_{++}$ and let $\beta(\cdot\,;\delta) : \mathbb{R} \to \mathbb{R}$ be a continuously differentiable function that satisfies $\beta(\delta;\delta) = -\ln(\delta)$ as well as $\lim_{z\to-\infty} \beta(z;\delta) = \infty$ and is strictly monotone for all $z \in (-\infty, \delta]$. Moreover, let $\beta(z;\delta)$ for any $z \in \mathbb{R}$ be continuous with respect to δ. Then, we call $\hat{B} : \mathbb{R} \to \mathbb{R}$ defined as*

$$\hat{B}(z) := \begin{cases} -\ln(z) & z > \delta \\ \beta(z;\delta) & z \leq \delta \end{cases} \tag{3.4}$$

the relaxed logarithmic barrier function for the set $\mathbb{R}_+$. Moreover, we call the parameter $\delta \in \mathbb{R}_{++}$ the relaxation parameter and refer to the function $\beta(\cdot;\delta)$ as the relaxing function.

A graphical illustration of the basic idea is given in Figure 3.1 on the next page. Based on this definition, relaxed logarithmic barrier functions for more general sets of the form $\mathcal{X} = \{x \in \mathbb{R}^n : f_i(x) \leq 0\}$ may quite naturally be defined as $\hat{B}_\mathrm{x} = \sum_i \hat{B}(-f_i(x))$, see also the discussion below. Note that we do not indicate the explicit dependence of the relaxed barrier function on the relaxation parameter δ for the sake of notational simplicity. However, we will subsequently use $\hat{B}(\cdot)$ to denote the relaxed version of a barrier function based expression $B(\cdot)$. With an eye on the regularity and convexity properties of the resulting relaxed barrier functions, it is advisable to choose the relaxing function $\beta(\cdot;\delta)$ as a strictly convex and twice continuously differentiable function that extends the natural logarithm at $z = \delta$ with a certain degree of smoothness, i.e.,

$$\frac{\partial}{\partial z}\beta(z;\delta)\big|_{z=\delta} = -\frac{1}{\delta}, \qquad \frac{\partial^2}{\partial z^2}\beta(z;\delta)\big|_{z=\delta} = \frac{1}{\delta^2}. \tag{3.5}$$

In this case, the primordial relaxed barrier $\hat{B} : \mathbb{R} \to \mathbb{R}$ in (3.4) will itself be a strictly convex and twice continuously differentiable function, which makes it accessible to standard nonlinear programming techniques like the Newton method, see also Chapter 4.
Some of the first ideas on relaxed (or approximate) logarithmic barrier functions were proposed by Ben Tal et al. (1992) and Nash et al. (1994), respectively, by making use of a quadratic relaxing function. Hauser and Saccon (2006) extended the concept to more general polynomial penalty terms and applied it in the context of continuous-time trajectory optimization. In particular, they proposed the following class of relaxing functions:

$$\beta_k(z;\delta) = \frac{k-1}{k}\left[\left(\frac{z-k\delta}{(k-1)\delta}\right)^k - 1\right] - \ln(\delta), \quad k = 2,4,\dots, \tag{3.6}$$

where $k > 1$ is an even integer representing the degree of the resulting polynomial. It can be easily verified that $\beta_k(\cdot;\delta)$ satisfies both Definition 3.1 and the regularity conditions (3.5), resulting for any given relaxation parameter $\delta \in \mathbb{R}_{++}$ and any choice of k in a smooth (i.e., twice continuously differentiable) relaxation of the natural logarithm. Note that for $k = 2$, the resulting relaxing function is quadratic and very similar to the ones discussed by Ben Tal et al. (1992) and Nash et al. (1994). On the other hand, for $k \to \infty$, (3.6) reduces to the exponential relaxing function

$$\beta_\mathrm{e}(z;\delta) = \exp\left(1 - \frac{z}{\delta}\right) - 1 - \ln(\delta), \tag{3.7}$$

which has been introduced in (Feller and Ebenbauer, 2014a) as a limiting case of the above polynomial approach. In fact, using the respective derivatives and the limit representation of the exponential function, it can be shown that $\beta_k(z;\delta) \leq \beta_{k+2}(z;\delta) \leq \beta_\mathrm{e}(z;\delta)$ for any $k = 2,4,\dots$ and that $\lim_{k\to\infty}\beta_k(z;\delta) = \beta_\mathrm{e}(z;\delta)$ for any $z \leq \delta$. Furthermore, a combined "barrier-penalty function" which extends the natural logarithm with a *linear* penalty function term (prohibiting therefore twice continuous differentiability) has been proposed by Srinivasan et al. (2008). Note that while being in principle interesting from a conceptual point of view, the exponential relaxation (3.7) is in many cases not really suited

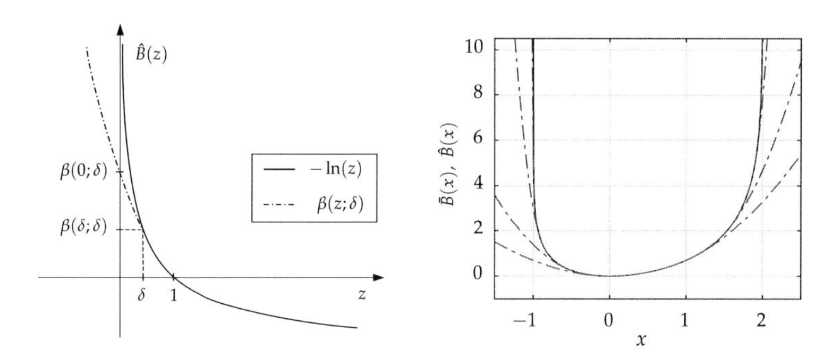

Figure 3.1. *Left:* Principal idea of a relaxed logarithmic barrier function for the set $\mathbb{R}_+$. *Right:* Weight recentered nonrelaxed and weight recentered relaxed logarithmic barrier function for the set $\mathcal{X} = \{x \in \mathbb{R} : -1 \leq x \leq 2\}$ from Example 3.1 below when using the quadratic relaxation given in (3.8) for decreasing values of the relaxation parameter δ. The depicted relaxed barrier functions correspond to $\delta \in \{1, 0.5, 0.1, 0.01\}$.

for practical implementation. The main reason for this is that the underlying exponential function increases very rapidly for arguments outside of the feasible region (imagine for example a scenario in which $\delta = 10^{-3}$ and $\bar{z} = -0.1$, resulting in $\beta_e(\bar{z};\delta) \approx 7.31 \times 10^{43}$). In this thesis, we will henceforth focus on a simple quadratic relaxation of the form

$$\beta(z;\delta) = \frac{1}{2}\left[\left(\frac{z-2\delta}{\delta}\right)^2 - 1\right] - \ln(\delta)\,, \tag{3.8}$$

which is nothing else than the above polynomial approach when choosing $k = 2$, cf. (3.6). Beside its simplicity, our main reasons for choosing a quadratic relaxing function are, on the one hand, that it will later allow us to derive a *global* quadratic upper bound for the resulting relaxed logarithmic barrier function and, on the other hand, that quadratic relaxations typically also work very well in practice, see also (Hauser and Saccon, 2006). In fact, as will be discussed in Chapter 4, the quadratic relaxation preserves also the self-concordance of the underlying logarithmic barrier functions, which allows us to derive the same algorithmic complexity estimates as in the conventional, nonrelaxed case. Nevertheless, many of the results discussed in the following do, in general, also hold when considering relaxing functions of higher order, as for example given by (3.6) and (3.7).

As illustrated in Figure 3.1, the parameter δ determines the steepness of the relaxing function and allows one to control how close the original logarithmic barrier is approximated by the respective relaxation. In this context, it is worth mentioning that the values of original and relaxed logarithmic barrier function will be identical for a given $\bar{z} \in \mathbb{R}_{++}$

whenever we choose $\delta \in (0, \bar{z}]$. In the limit, this implies that $\lim_{\delta \to 0} \hat{B}(z) \to B(z)$ for any strictly feasible $z \in \mathbb{R}_{++}$, which shows that the nonrelaxed formulation can always be recovered by choosing the relaxation parameter arbitrarily small. These arguments will also play a major role in the remainder of this chapter, in particular in the context of ensuring constraint satisfaction properties of relaxed logarithmic barrier function MPC.

Recentered relaxed logarithmic barrier functions

Before we turn our attention to the formulation and design of MPC schemes that make use of the above class of relaxed logarithmic barrier functions, we first want to revisit an important concept from the previous chapter, namely the concept of *recentered* barrier functions. In accordance with our problem setup, we consider a polytopic set of the form $\mathcal{P} = \{z \in \mathbb{R}^{n_z} : Cz \leq d\}$ with $C \in \mathbb{R}^{q \times n_z}$ and $d \in \mathbb{R}^q_{++}$, cf. Equation (3.2). Using the above definition, the relaxed logarithmic barrier function for this set may be formulated as

$$\hat{B}_z(z) = \sum_{i=1}^{q} \hat{B}(-C^i z + d^i), \tag{3.9}$$

where the primordial relaxed logarithmic barrier function $\hat{B}(\cdot)$ is given by (3.4) and we assume a quadratic relaxation based on (3.8). While this function is twice continuously differentiable and convex, it will in general not attain its minimum at the origin, i.e., it is not *recentered* in the sense of Section 2.3. However, as will be discussed in the following, recentering of a relaxed logarithm barrier functions can be achieved in a rather straightforward way by simply recentering the underlying nonrelaxed logarithmic barrier function and then ensuring via the choice of the relaxation parameter that the relaxation has no effect at the origin. In particular, for the considered problem setup this can always be achieved by choosing $0 < \delta \leq \min(d)$. Using the concept of gradient recentered barrier functions due to Wills and Heath (2004), this leads to the following definition, cf. Section 2.3.

Definition 3.2 (Gradient recentered relaxed logarithmic barrier function). *Let $\mathcal{P} = \{z \in \mathbb{R}^n : Cz \leq d\}$ with $C \in \mathbb{R}^{q \times n_z}$, $d \in \mathbb{R}^q_{++}$, be a polytopic set containing the origin. Assume a given relaxation parameter $0 < \delta \leq \min(d)$ and consider the primordial relaxed logarithmic barrier function $\hat{B} : \mathbb{R} \to \mathbb{R}$ for the set $\mathbb{R}_+$ as given in (3.4). Then, the function $\hat{B}_{z,G} : \mathbb{R}^{n_z} \to \mathbb{R}_+$ defined as*

$$\hat{B}_{z,G}(z) = \sum_{i=1}^{q} \hat{B}(-C^i z + d^i) + \ln(d^i) - \frac{C^i z}{d^i} \tag{3.10}$$

is called the gradient recentered relaxed logarithmic barrier function for the set $\mathcal{P}$.

Note that the choice of δ ensures that $\hat{B}_{z,G}(0) = 0$ as well as $\nabla \hat{B}_{z,G}(0) = 0$. As the recentering preserves in addition the convexity of the underlying relaxed barrier function, this implies $\hat{B}_{z,G}(z) \geq 0 \; \forall z \in \mathbb{R}^{n_z}$, which shows that $\hat{B}_{z,G}(\cdot)$ attains its global minimum at the origin. As an alternative to the gradient based recentering approach, the concept of so-called *weight recentered* logarithmic barrier functions has been presented in (Feller and

Ebenbauer, 2015a), see also Appendix B.3. Applied in the context of relaxed logarithmic barrier functions, this leads to the following definition.

Definition 3.3 (Weight recentered relaxed logarithmic barrier function). *Let $\mathcal{P} = \{z \in \mathbb{R}^n : Cz \leq d\}$ with $C \in \mathbb{R}^{q \times n_z}$, $d \in \mathbb{R}^q_{++}$, be a polytopic set containing the origin. Assume a given relaxation parameter $0 < \delta \leq \min(d)$ and consider the primordial relaxed logarithmic barrier function $\hat{B} : \mathbb{R} \to \mathbb{R}$ for the set $\mathbb{R}_+$ as given in (3.4). Furthermore, let $w_z \in \mathbb{R}^q_{++}$ be a given weighting vector with nonzero elements, satisfying $\bar{C}w_z = 0$, where $\bar{C} := C^\top \mathrm{diag}(\frac{1}{d^1}, \ldots, \frac{1}{d^q})$. Then, the function $\hat{B}_{z,W} : \mathbb{R}^{n_z} \to \mathbb{R}_+$ defined as*

$$\hat{B}_{z,W}(z) = \sum_{i=1}^{q} w_z^i \left(\hat{B}(-C^i z + d^i) + \ln(d^i) \right) \tag{3.11}$$

is called the weight recentered relaxed logarithmic barrier function for the set $\mathcal{P}$.

Note that $\hat{B}_{z,W}(0) = \sum_{i=1}^q w_z^i \left(\hat{B}(d^i) + \ln(d^i)\right)$ and $\nabla \hat{B}_{z,W}(0) = -\sum_{i=1}^q w_z^i {C^i}^\top \nabla \hat{B}(d^i)$, which implies for $\delta \leq \min(d)$ nothing else than

$$\hat{B}_{z,W}(0) = 0, \quad \nabla \hat{B}_{z,W}(0) = \sum_{i=1}^{q} w_z^i \frac{{C^i}^\top}{d^i} = C^\top \mathrm{diag}\left(\frac{1}{d^1}, \ldots, \frac{1}{d^q}\right) w_z = \bar{C} w_z = 0. \tag{3.12}$$

Thus, the above choice of the weighting vector w_z ensures that the gradient of (3.11) vanishes at the origin. As $\hat{B}_{z,W}(\cdot)$ is a convex function, this implies that $\hat{B}_{z,W}(z) \geq 0 \; \forall z \in \mathbb{R}^{n_z}$. As discussed in Appendix B.3, a suitable weighting vector satisfying $\bar{C}w_z = 0$ always exists and can be computed in a straightforward and constructive fashion. In fact, from a geometric point of view, the above procedure is equivalent to ensuring that the so-called *weighted analytic center* of the set $\mathcal{P}$ is located at the origin. By choosing $\delta \leq \min(d)$, the resulting recentering is then also inherited by the associated relaxed barrier function.

Example 3.1. Let us revisit as an illustrating example the set $\mathcal{X} = \{x \in \mathbb{R} : -1 \leq x \leq 2\}$, which we discussed in Section 2.3 already in the nonrelaxed context. Obviously, $\mathcal{X}$ is of the considered polytopic form with $C_x = [1, -1]^\top$, $d_x = [2, 1]^\top$. Without any recentering, the relaxed logarithmic barrier function for this set is given by

$$\hat{B}_x(x) = \hat{B}(2-x) + \hat{B}(1+x) = \begin{cases} -\ln(2-x) + \beta(1+x;\delta) & x < -1+\delta \\ -\ln(2-x) - \ln(1+x) & -1+\delta \leq x \leq 2-\delta \\ \beta(2-x;\delta) - \ln(1+x) & x > 2-\delta\,, \end{cases} \tag{3.13}$$

where the relaxing function $\beta(\cdot;\delta)$ is defined in (3.8) and $0 < \delta \leq \min(d_x) = 1$ is a given relaxation parameter. According to the above discussion, recentering of (3.13) can be achieved by simply recentering the underlying nonrelaxed barrier function. In particular, following Definition 3.2, the associated gradient recentered relaxed logarithmic barrier function can for example be constructed as (cf. also Example 2.1)

$$\hat{B}_{x,G}(x) = \hat{B}_x(x) + \ln(2) + \frac{1}{2}x\,. \tag{3.14}$$

Concerning the weighting-based recentering according to Definition 3.3, the condition for the weighting vector w_{x} reduces to $\frac{1}{2}w_{\mathrm{x}}^1 - w_{\mathrm{x}}^2 = 0$. Consequently, a suitable weighting is given by $w_{\mathrm{x}} = [2,\, 1]^\top$, which results in

$$\hat{B}_{\mathrm{x,W}}(x) = \begin{cases} 2\big(-\ln(2-x) + \ln(2)\big) + \beta(1+x;\delta) & x < -1+\delta \\ 2\big(-\ln(2-x) + \ln(2)\big) - \ln(1+x) & -1+\delta \le x \le 2-\delta \\ 2\big(\beta(2-x;\delta) - \ln(2)\big) - \ln(1+x) & x > 2-\delta\,. \end{cases} \tag{3.15}$$

Using the above definitions, the considered example can be easily generalized to higher dimensional polytopic constraints. A plot of the weight recentered relaxed barrier function (3.15) for different values of the relaxation parameter δ can be found in Fig. 3.1. As desired, the result is a positive definite and (strictly) convex function which approximates the original logarithmic barrier function arbitrarily close by virtue of the parameter δ.

Remark 3.2. In principle, all the results that will be discussed in the remainder of the thesis hold independently of the underlying recentering method. However, for the sake of simplicity, we will in the following mainly focus on the case of weight recentered barrier functions. Note also that the weighting based recentering approach has the advantage that it does not introduce an additional term related to the gradient, which might be beneficial when recentering certain classes of barrier functions for polytopic terminal sets (Feller and Ebenbauer, 2015a).

3.3 The relaxed barrier function based MPC problem

Based on the problem setup stated in Section 3.1 and the concept of relaxed logarithmic barrier functions introduced in Section 3.2, we are now in a position to formulate the relaxed logarithmic barrier function based MPC problem. In particular, we consider the following general relaxed barrier function based MPC formulation

$$\hat{J}_N^*(x) = \min_{u} \sum_{k=0}^{N-1} \hat{\ell}(x_k, u_k) + \hat{F}(x_N) \tag{3.16a}$$

$$\text{s.t. } x_{k+1} = Ax_k + Bu_k,\; x_0 = x\,, \tag{3.16b}$$

where $N \in \mathbb{N}_+$ is again the finite prediction horizon, $x = x(k) \in \mathbb{R}^n$ refers to the current system state, and (A, B) are real matrices describing the dynamics of system (3.1). Furthermore, $\hat{\ell} : \mathbb{R}^n \times \mathbb{R}^m \to \mathbb{R}_+$ denotes a modified stage cost function, defined as

$$\hat{\ell}(x,u) := \|x\|_Q^2 + \|u\|_R^2 + \varepsilon\hat{B}_{\mathrm{x}}(x) + \varepsilon\hat{B}_{\mathrm{u}}(u)\,, \quad \varepsilon \in \mathbb{R}_{++}\,, \tag{3.17}$$

where $\hat{B}_{\mathrm{x}}(\cdot)$ and $\hat{B}_{\mathrm{u}}(\cdot)$ are recentered relaxed logarithmic barrier functions for the sets $\mathcal{X}$ and $\mathcal{U}$, respectively, based on a suitably chosen relaxation parameter $\delta \in \mathbb{R}_{++}$. Moreover, $\varepsilon \in \mathbb{R}_{++}$ denotes the associated barrier function weighting parameter. We assume the following to hold throughout the remainder of this thesis.

Assumption 3.1. *The functions $\hat{B}_{\mathrm{x}} : \mathbb{R}^n \to \mathbb{R}_+$ and $\hat{B}_{\mathrm{u}} : \mathbb{R}^m \to \mathbb{R}_+$ in (3.17) are weight recentered relaxed logarithmic barrier functions for the polytopic sets $\mathcal{X}$ and $\mathcal{U}$, respectively.*

Recalling Definition 3.3 and using the polytopic representation of the constraint sets given in (3.2), the respective barrier functions may be written as

$$\hat{B}_{\mathrm{x}}(x) = \sum_{i=1}^{q_{\mathrm{x}}} w_{\mathrm{x}}^i \left(\hat{B}(-C_{\mathrm{x}}^i x + d_{\mathrm{x}}^i) + \ln(d_{\mathrm{x}}^i) \right), \quad w_{\mathrm{x}} \in \mathbb{R}_{++}^{q_{\mathrm{x}}}, \tag{3.18a}$$

$$\hat{B}_{\mathrm{u}}(u) = \sum_{i=1}^{q_{\mathrm{u}}} w_{\mathrm{u}}^i \left(\hat{B}(-C_{\mathrm{u}}^i u + d_{\mathrm{u}}^i) + \ln(d_{\mathrm{u}}^i) \right), \quad w_{\mathrm{u}} \in \mathbb{R}_{++}^{q_{\mathrm{u}}}, \tag{3.18b}$$

where $\hat{B} : \mathbb{R} \to \mathbb{R}$ refers to the primordial relaxed logarithmic barrier function given in (3.4), with $\delta \in \mathbb{R}_{++}$ satisfying $\delta \leq \min(d_{\mathrm{x}}^\top, d_{\mathrm{u}}^\top)$. Furthermore, w_{x} and w_{u} denote suitable weighting vectors for a recentering according to Definition 3.3. Finally, the term $\hat{F} : \mathbb{R}^n \to \mathbb{R}_+$ denotes a suitably defined (relaxed) terminal cost function, which we do, however, not specify for the moment. In fact, as we will see in the following section, the choice of the terminal cost function term is crucial for ensuring stability properties of the closed-loop system and can be seen as one of the main design parameters in the context of relaxed barrier function based MPC.

Remark 3.3. In principle, the input and state constraint barrier functions may make use of different relaxation parameters $\delta_{\mathrm{x}} \leq \min(d_{\mathrm{x}})$ and $\delta_{\mathrm{u}} \leq \min(d_{\mathrm{u}})$, cf. Definition 3.3. However, for the sake of simplicity we will in the following assume that $\delta_{\mathrm{x}} = \delta_{\mathrm{u}} = \delta$ with $\delta \leq \min(d_{\mathrm{x}}^\top, d_{\mathrm{u}}^\top)$. Nevertheless, it should be noted that individual values for δ_{x} and δ_{u} might in some practical application be useful in order to enforce prioritized satisfaction of certain constraints, e.g., in the presence of physically motivated hard input constraints. This issue will be discussed in more detail in Section 3.5 of this chapter.

As usual in MPC, the resulting feedback law is obtained by solving at each sampling instant the relaxed barrier function based open-loop optimal control problem (3.16) based on the current system state $x = x(k)$ and then applying only the first element of the associated optimal input sequence $\hat{\boldsymbol{u}}^*(x) = \{\hat{u}_0^*(x), \ldots \hat{u}_{N-1}^*(x)\}$ to the plant. The resulting closed-loop system is therefore given by

$$x(k+1) = Ax(k) + B\hat{u}_0^*(x(k)). \tag{3.19}$$

The rest of this chapter will be concerned with the question of what can be said about the systems theoretic properties of system (3.19). In particular, we are going to discuss in the following Section 3.4 suitable choices for the terminal cost function $\hat{F}(\cdot)$ that ensure (global) asymptotic stability of the resulting closed-loop system. Further properties like the influence of the relaxation parameter on the satisfaction of state and input constraints or the robustness of the closed loop with respect to additive disturbances will be discussed in Sections 3.5 and 3.6, respectively. Note that we assume throughout the chapter that the exact optimal control input is applied to the plant and that no state or control input saturations are considered, see also Remark 3.6 below.

3.4 Stability of the closed-loop system

As in the context of nonrelaxed barrier function based MPC we in the following want to investigate under which conditions the origin of system (3.19) will be asymptotically stable with a certain (possibly maximal) region of attraction. More precisely, we are interested in constructive design approaches and recipes that allow us to choose the characteristic ingredients of problem (3.16) such that this goal is achieved. The results that we are going to present in the following can be structured in two conceptually distinct parts, which will be discussed separately in Sections 3.4.1 and 3.4.2. While the first part is closely related to the nonrelaxed barrier function based MPC approach discussed in Section 2.3, making use of a suitably defined compact terminal set, the second part follows a different route and elaborates on the design of relaxed barrier function based MPC schemes that do not necessarily rely on such an additional stabilizing terminal set constraint.

The results presented in the following are mainly based on (Feller and Ebenbauer, 2013, 2014a,b, 2015a, 2017a). Where indicated, some related auxiliary results are discussed in more detail in Appendix B. All the technical proofs can be found in Appendix C.

3.4.1 Stabilizing approaches based on terminal sets

As summarized in Section 2.3, the stability and constraint satisfaction properties of conventional MPC schemes may in the context of barrier function based MPC approaches be recovered by taking the underlying barrier functions explicitly into account in the design of the terminal set and associated terminal cost function formulation. In particular, the key idea of Wills and Heath (2004) was to choose the terminal set as a positively invariant ellipsoidal subset of a region in the state space in which the effect of the underlying barrier functions can be upper bounded by a quadratic function. A suitable quadratic terminal cost function term is then chosen such that it compensates for this effect, allowing to ensure asymptotic stability of the closed-loop system. In the following, we are going to apply this approach in the context of the above *relaxed* barrier function based problem setup, widening the scope in addition to more general (i.e., not necessarily ellipsoidal) terminal set formulations. In particular, we consider terminal sets of the form

$$\mathcal{X}_f = \{x \in \mathbb{R}^n : \varphi_f(x) \leq 1\}, \tag{3.20}$$

where $\varphi_f : \mathbb{R}^n \to \mathbb{R}_+$ is a suitably defined positive definite, convex, and radially unbounded function. Obviously, $\mathcal{X}_f$ will in this case be a convex and compact set that contains the origin in its interior. Based on (3.20), we consider the following nonrelaxed and relaxed logarithmic barrier function formulations for the terminal set

$$B_f(x) = -\ln(1 - \varphi_f(x)), \qquad \hat{B}_f(x) = \hat{B}(1 - \varphi_f(x)), \tag{3.21}$$

in which $\hat{B} : \mathbb{R} \to \mathbb{R}_+$ denotes for a given relaxation parameter $\delta \leq 1$ the primordial relaxed logarithmic barrier function according to Definition 3.1. Note that both terminal set barrier functions are by design positive definite with respect to the origin, which reveals that no additional recentering is needed. As will be discussed below, the function $\varphi_f(\cdot)$ may in the case of an ellipsoidal terminal set simply be chosen as $\varphi_f(x) = x^\top P_f x$, $P_f \in \mathbb{S}^n_{++}$, whereas more involved formulations are needed in the context of polytopic terminal set constraints, see also Appendix B.5.

Based on this quite general problem setup, we are going to present in the following two different approaches that ensure stability of the closed-loop system based on a suitable terminal set of the form (3.20). The first approach (*A*) makes use of a *relaxed* terminal set barrier function formulation and can thus be seen as a full relaxation of the underlying nonrelaxed MPC problem. The proposed terminal cost function term in this case is

$$\hat{F}(x) = x^\top P x + \varepsilon \hat{B}_f(x)\,, \tag{3.22}$$

where $P \in \mathbb{S}^n_{++}$ is a suitably chosen terminal weight matrix, $\varepsilon \in \mathbb{R}_{++}$, and $\hat{B}_f : \mathbb{R}^n \to \mathbb{R}_+$ denotes the globally defined relaxed terminal set barrier function given in (3.21). In contrast to this, the second approach (*B*) makes use of a *nonrelaxed* terminal set barrier function, resulting in the modified terminal cost function term

$$\hat{F}(x) = x^\top P x + \varepsilon B_f(x)\,. \tag{3.23}$$

In both approaches, the key task is to choose the pair $(\mathcal{X}_f, P)$, respectively (φ_f, P), in such a way that important stability properties of the closed-loop can be guaranteed. The results presented in the subsequent discussion can be summarized as follows:

- Under rather mild assumptions on the function $\varphi_f(\cdot)$ and the matrix P, the approach based on the relaxed terminal cost function (3.22) allows to guarantee strict constraint satisfaction as well as asymptotic stability of the closed-loop system for a region of attraction whose size depends on the choice of the relaxation parameter δ. In addition, the constraint satisfaction and stability properties of the corresponding nonrelaxed MPC formulation can always be recovered for sufficiently small δ.
- Under the same conditions on $\varphi_f(\cdot)$ and P, and under the additional assumption that the considered plant model is controllable instead of merely stabilizable, the second approach based on the nonrelaxed terminal cost function (3.23) allows one to guarantee *global* asymptotic stability of the closed-loop system. A detailed analysis of the resulting constraint satisfaction properties will be given in Section 3.5.

Note that the above generalized problem setup as well as many of the results presented in the following can also be applied directly in the context of conventional, nonrelaxed barrier function based MPC approaches. For the sake of completeness, a discussion of the nonrelaxed case, including in particular a novel and generalized stability theorem, is presented in Appendix B.4. In addition, different constructive approaches for computing suitable choices for $\varphi_f(\cdot)$ and P for the cases of ellipsoidal and polytopic terminal sets are presented and discussed in detail in Appendix B.5, see also Remark 2.4 on page 28.

Main results

In the following, we present our main results on the stability properties of relaxed barrier function based MPC schemes that make use of an additional terminal set constraint. As outlined above, we will discuss two conceptually different approaches, which mainly differ in whether the underlying logarithmic barrier function for the terminal set is relaxed or not. All the results will be presented for the rather general problem formulation introduced above, allowing thereby in principle for general semi-algebraic terminal sets.

We begin our discussion with extending some of the ideas of Wills and Heath (2004) to the above general problem setup. In particular, we make the following assumptions on the terminal set and the (now relaxed) logarithmic barrier functions.

Assumption 3.2. *For (A, B) stabilizable, given $Q \in \mathbb{S}^n_{++}$, $R \in \mathbb{S}^m_{++}$, $\varepsilon \in \mathbb{R}_{++}$, and a stabilizing local control gain $K \in \mathbb{R}^{m \times n}$ with $A_K := A + BK$, $|\lambda_i(A_K)| < 1$, the following holds:*

A1: The barrier functions $\hat{B}_{\mathrm{x}} : \mathbb{R}^n \to \mathbb{R}_+$ and $\hat{B}_{\mathrm{u}} : \mathbb{R}^m \to \mathbb{R}_+$ are suitably recentered relaxed logarithmic barrier functions for the sets $\mathcal{X}$ and $\mathcal{U}$, that is, Assumption 3.1 holds.

A2: There exist a compact set $\mathcal{N}_K \subseteq \mathcal{X}_K := \{x \in \mathcal{X} : Kx \in \mathcal{U}\}$ with $0 \in \mathcal{N}_K^\circ$ and an associated $M \in \mathbb{S}^n_+$ such that the quadratic upper bound $\hat{B}_K(x) \leq x^\top M x$ holds for any $x \in \mathcal{N}_K$, where the relaxed barrier function $\hat{B}_K : \mathbb{R}^n \to \mathbb{R}_+$ is defined as $\hat{B}_K = \hat{B}_{\mathrm{x}}(x) + \hat{B}_{\mathrm{u}}(Kx)$.

A3: The terminal set $\mathcal{X}_f$ is given by (3.20), where $\varphi_{\mathrm{f}} : \mathbb{R}^n \to \mathbb{R}_+$ is a positive definite, convex, and radially unbounded function that satisfies $\varphi_{\mathrm{f}}(A_K x) \leq \varphi_{\mathrm{f}}(x)$ for any $x \in \mathcal{X}_f$. Moreover, $\varphi_{\mathrm{f}}(\cdot)$ is chosen in such a way that $\mathcal{X}_f \subseteq \mathcal{N}_K$ for the set $\mathcal{N}_K$ is given in A2.

A4: The terminal cost matrix $P \in \mathbb{S}^n_{++}$ is chosen as the positive definite solution to the Lyapunov equation $P = A_K^\top P A_K + K^\top R K + Q + \varepsilon M$, where the matrix $M \in \mathbb{S}^n_+$ is given in A2.

Note that Assumption 3.2 is in fact only a slight modification of the assumptions that are required in the nonrelaxed case, see Appendix B.4. In particular, as the relaxed barrier function $\hat{B}_K(\cdot)$ is on the set $\mathcal{X}_K$ always upper bounded by its nonrelaxed counterpart $B_K(\cdot)$, the same quadratic upper bound as in the nonrelaxed case can be applied. As a consequence, one possible way to ensure satisfaction of Assumption 3.2 in a constructive fashion is given by the original approach of Wills and Heath (2004), in which the role of the set $\mathcal{N}_K$ is played by the Dikin ellipsoid $\mathcal{W}_{B_K}(0, r)$ with adjustable radius $r \in (0, 1)$ and the terminal set is chosen as a scaled sublevel set of the quadratic terminal cost, that is, $\varphi_{\mathrm{f}}(x) = \frac{1}{\alpha} x^\top P x$ with suitably chosen $\alpha \in \mathbb{R}_{++}$, see Section 2.3. Moreover, several alternative constructive design approaches for choosing the set $\mathcal{N}_K$, the terminal set $\mathcal{X}_f$, and the terminal cost matrix P have been presented in (Feller and Ebenbauer, 2013, 2014b, 2015a). Besides introducing the novel concept of barrier function based MPC with *polytopic* terminal sets, these approaches lead in many cases to an increased region of attraction of the closed-loop system and may therefore be considered as an interesting contribution in their own right. However, for the sake of a simplified and streamlined discussion, which in particular focuses on the different stability results as the main topic of this section, these alternative approaches are summarized and discussed in Appendix B.5.

A) Stabilization with guaranteed constraint satisfaction

Let us return to the relaxed logarithmic barrier function based MPC formulation (3.16), where we in the following assume that the terminal cost function is given as

$$\hat{F}(x) = x^\top P x + \varepsilon \hat{B}_\mathrm{f}(x)\,, \tag{3.24}$$

with $\hat{B}_\mathrm{f} : \mathbb{R}^n \to \mathbb{R}_+$ denoting the relaxed logarithmic terminal set barrier function given in (3.21). In particular, using the definition of the primordial relaxed logarithmic barrier function, the function $B_\mathrm{f}(\cdot)$ can more explicitly be written as

$$\hat{B}_\mathrm{f}(x) = \begin{cases} -\ln(1-\varphi_\mathrm{f}(x)) & 1-\varphi_\mathrm{f}(x) > \delta \\ \beta(1-\varphi_\mathrm{f}(x);\delta) & 1-\varphi_\mathrm{f}(x) \le \delta\,, \end{cases} \tag{3.25}$$

where $\delta \in \mathbb{R}_{++}$ satisfying $\delta \le 1$ is the corresponding relaxation parameter and $\varphi_\mathrm{f}(\cdot)$ is the function that defines the terminal set according to (3.20).

Assumption 3.3. *The relaxed terminal set barrier function $\hat{B}_\mathrm{f}(\cdot)$ in (3.24) is given by (3.25), where $\beta(\cdot;\delta)$ is the quadratic relaxing function according to (3.8) and $0 < \delta \le \min(d_\mathrm{x}^\top, d_\mathrm{u}^\top, 1)$ is the underlying overall relaxation parameter.*

As all the state, input, and terminal set constraints are relaxed, problem (3.16) can be seen as a full relaxation of the associated nonrelaxed problem discussed in Appendix B.4. It is now a quite natural question whether and to which extent the stability results presented in Appendix B.4 for the nonrelaxed barrier function based MPC case will actually carry over to the relaxed MPC formulation introduced above. Concerning this point, the following results show that asymptotic stability of the closed-loop system and even satisfaction of state and input constraints can still be guaranteed for a certain region of attraction whose size depends on the choice of the relaxation parameter δ. Furthermore, the region of attraction of the original, nonrelaxed barrier function based MPC approach can always be recovered by making the relaxation parameter arbitrarily small.

Definition 3.4. *Consider the above problem setup and let Assumptions 3.1 and 3.3 hold for a given relaxation parameter $\delta \in \mathbb{R}_{++}$. Then, we define the scalar $\bar{\beta}(\delta) \in \mathbb{R}_{++}$ as*

$$\bar{\beta}(\delta) := \min\left\{\bar{\beta}_\mathrm{x}(\delta), \bar{\beta}_\mathrm{u}(\delta), \bar{\beta}_\mathrm{f}(\delta)\right\}\,, \tag{3.26}$$

where

$$\bar{\beta}_\mathrm{x}(\delta) := \min_{i,\xi}\{\hat{B}_\mathrm{x}(\xi) |\, C_\mathrm{x}\xi \le d_\mathrm{x},\, C_\mathrm{x}^i \xi = d_\mathrm{x}^i\}, \quad i = 1,\dots,q_\mathrm{x}\,, \tag{3.27a}$$

$$\bar{\beta}_\mathrm{u}(\delta) := \min_{j,v}\{\hat{B}_\mathrm{u}(v) |\, C_\mathrm{u} v \le d_\mathrm{u},\, C_\mathrm{u}^j v = d_\mathrm{u}^j\}, \quad j = 1,\dots,q_\mathrm{u}\,, \tag{3.27b}$$

$$\bar{\beta}_\mathrm{f}(\delta) := \min_{\xi}\{\hat{B}_\mathrm{f}(\xi) | \varphi_\mathrm{f}(\xi) = 1\} = \beta(0;\delta)\,. \tag{3.27c}$$

For a given relaxation parameter δ, the scalar $\bar{\beta}(\delta)$ represents a lower bound for the minimal value that can be attained by the relaxed barrier functions $\hat{B}_{\mathrm{x}}(\cdot)$, $\hat{B}_{\mathrm{u}}(\cdot)$, and $\hat{B}_{\mathrm{f}}(\cdot)$ when evaluated on the borders of the respective constraint sets $\mathcal{X}$, $\mathcal{U}$, and $\mathcal{X}_f$. Note that the values $\bar{\beta}_{\mathrm{x}}(\delta)$, $\bar{\beta}_{\mathrm{u}}(\delta)$, and hence also $\bar{\beta}(\delta)$, can be computed easily for a given $\delta \in \mathbb{R}_{++}$ as the optimization problems in (3.27a) and (3.27b) are convex due to convexity of the underlying relaxed barrier functions. The following Lemma reveals that the sublevel sets of the relaxed barrier functions $\hat{B}_{\mathrm{x}}(\cdot)$, $\hat{B}_{\mathrm{u}}(\cdot)$, and $\hat{B}_{\mathrm{f}}(\cdot)$ related to the values $\bar{\beta}_{\mathrm{x}}(\delta)$, $\bar{\beta}_{\mathrm{u}}(\delta)$, and $\bar{\beta}_{\mathrm{f}}(\delta)$ will always be contained within the sets $\mathcal{X}, \mathcal{U}$, and $\mathcal{X}_f$, respectively.

Lemma 3.1. *Let Assumptions 3.1 and 3.3 hold and let the values $\bar{\beta}_{\mathrm{x}}(\delta)$, $\bar{\beta}_{\mathrm{u}}(\delta)$, and $\bar{\beta}_{\mathrm{f}}(\delta)$ be defined according to Definition 3.4. Then, the associated barrier function sublevel sets satisfy*

$$\mathcal{S}_{\hat{B}_{\mathrm{x}}}(\delta) := \{x \in \mathbb{R}^n \,|\, \hat{B}_{\mathrm{x}}(x) \leq \bar{\beta}_{\mathrm{x}}(\delta)\} \subseteq \mathcal{X}\,, \tag{3.28a}$$

$$\mathcal{S}_{\hat{B}_{\mathrm{u}}}(\delta) := \{u \in \mathbb{R}^m \,|\, \hat{B}_{\mathrm{u}}(u) \leq \bar{\beta}_{\mathrm{u}}(\delta)\} \subseteq \mathcal{U}\,, \tag{3.28b}$$

$$\mathcal{S}_{\hat{B}_{\mathrm{f}}}(\delta) := \{x \in \mathbb{R}^n \,|\, \hat{B}_{\mathrm{f}}(x) \leq \bar{\beta}_{\mathrm{f}}(\delta)\} \subseteq \mathcal{X}_f\,. \tag{3.28c}$$

A proof of this quite intuitive results is given in Appendix C.2. Note that a key insight that is provided by Lemma 3.1 is that satisfaction of all state, input, and terminal set constraints will be guaranteed as long as the values of the corresponding relaxed barrier functions do not exceed the threshold $\bar{\beta}(\delta)$. Based on this idea, we may now define the following set of initial conditions for which we can guarantee not only strict satisfaction of all input and state constraints under the closed-loop system dynamics but, as stated in Theorem 3.1 below, also asymptotic stability of the origin.

Definition 3.5. *Given a relaxation parameter $\delta \in \mathbb{R}_{++}$ and a corresponding $\bar{\beta}(\delta) \in \mathbb{R}_{++}$ according to Definition 3.4, let the set $\hat{\mathcal{X}}_N(\delta)$ be defined as $\hat{\mathcal{X}}_N(\delta) := \{x \in \mathbb{R}^n \,|\, \hat{J}_N^*(x) \leq \varepsilon\bar{\beta}(\delta)\}$, where $\hat{J}_N^* : \mathbb{R}^n \to \mathbb{R}_+$ refers to the value function of problem (3.16).*

Theorem 3.1. *Consider the relaxed barrier function based problem (3.16) with stage cost (3.17) and terminal cost (3.24), and let Assumptions 3.1, 3.2, and 3.3 hold true. Moreover, let the set $\hat{\mathcal{X}}_N(\delta)$ be defined according to Definition 3.5 with $\delta \in \mathbb{R}_{++}$ denoting the underlying relaxation parameter. Then, the origin of the resulting closed-loop system (3.19) is asymptotically stable with region of attraction $\hat{\mathcal{X}}_N(\delta)$. Furthermore, the relaxed barrier function based feedback $u(k) = \hat{u}_0^*(x(k))$ ensures strict satisfaction of the pointwise state and input constraints (3.2) for all $k \in \mathbb{N}$.*

A proof of this first main result is given in Appendix C.3. A quite natural question that arises from the above results is whether we can actually characterize the dependence of the resulting region of attraction on the choice of the relaxation parameter δ in a more precise way. To this end, let us recall the region of attraction of the associated conventional or nonrelaxed MPC formulation, given by the feasible set

$$\mathcal{X}_N := \left\{x \in \mathcal{X} : \exists\, \boldsymbol{u} = \{u_0, \ldots, u_{N-1}\} \text{ s. t. } \; u_k \in \mathcal{U},\, x_k(\boldsymbol{u}, x) \in \mathcal{X},\, x_N(\boldsymbol{u}, x) \in \mathcal{X}_f\right\}. \tag{3.29}$$

The following result states some useful properties of the set $\hat{\mathcal{X}}_N(\delta)$, and reveals that the constraint satisfaction and stability results from the nonrelaxed barrier function based problem formulation can always be recovered by making the relaxation parameter arbitrarily small. A proof of Lemma 3.2 can be found in Appendix C.4.

Lemma 3.2. *Consider $\mathcal{X}_N$ given in (3.29). Furthermore, let the assumptions of Theorem 3.1 hold and let the associated region of attraction $\hat{\mathcal{X}}_N(\delta)$ be defined according to Definition 3.5. Then, $\hat{\mathcal{X}}_N(\delta)$ is a nonempty, compact, and convex set, satisfying $\hat{\mathcal{X}}_N(\delta) \subseteq \mathcal{X}_N^\circ$ for any $\delta \in \mathbb{R}_{++}$. Moreover, for any compact set $\mathcal{X}_0 \subseteq \mathcal{X}_N^\circ$ there exists a well-defined relaxation parameter $\bar{\delta}_0 \in \mathbb{R}_{++}$ such that $\mathcal{X}_0 \subseteq \hat{\mathcal{X}}_N(\delta)$ for any $\delta \leq \bar{\delta}_0$.*

A direct consequence of Lemma 3.2 is that $\hat{\mathcal{X}}_N(\delta) \to \mathcal{X}_N$ for $\delta \to 0$, which shows that the original feasible region given in (3.29) may always be recovered by making the relaxation parameter arbitrarily small. Thus, apart from the immediate stability and constraint satisfaction guarantees, the above results reveal a direct link between the choice of the relaxation parameter δ and important systems theoretical properties of the closed-loop system, such as the size of the resulting (guaranteed) region of attraction. By combining Theorem 3.1 and Lemma 3.2, we can state the following result.

Corollary 3.1. *Consider the relaxed barrier function based problem (3.16) with stage cost (3.17) and terminal cost (3.24) and let Assumptions 3.1, 3.2, and 3.3 hold true. Furthermore, let $\mathcal{X}_N$ denote the feasible set given in (3.29). Then, for any compact set $\mathcal{X}_0 \subseteq \mathcal{X}_N^\circ$ there exists a $\bar{\delta}_0 \in \mathbb{R}_{++}$ such that for any relaxation parameter $\delta \leq \bar{\delta}_0$, the origin of the resulting closed-loop system (3.19) is asymptotically stable with region of attraction $\mathcal{X}_0$. Furthermore, for any $x(0) \in \mathcal{X}_0$, the associated relaxed barrier function based feedback $u(k) = \hat{u}_0^*(x(k))$ ensures strict satisfaction of the pointwise state and input constraints (3.2) for any $k \in \mathbb{N}$.*

Remark 3.4. It has to be noted that the above conditions for closed-loop stability and constraint satisfaction are of course only sufficient and may be rather conservative. In particular, for a given relaxation parameter $\delta \in \mathbb{R}_{++}$, the actual region of attraction of the closed-loop system may be considerably larger than the set $\hat{\mathcal{X}}_N(\delta)$ introduced in Theorem 3.1. Likewise, a very small δ may be needed to achieve $\mathcal{X}_0 \subseteq \hat{\mathcal{X}}_N(\delta)$, especially when $\mathcal{X}_0$ approaches $\mathcal{X}_N^\circ$. However, despite possible practical limitations, the presented results provide from a conceptual point of view some first interesting insights as well as a theoretical justification for the use of relaxed barrier functions in the context of MPC.

Remark 3.5. Note that the crucial role for ensuring asymptotic stability in the proof of Theorem 3.1 is played by the terminal set condition $x_N \in \mathcal{X}_f$. Consequently, following the above arguments, a less conservative estimate for the actual region of attraction is given by $\hat{\mathcal{X}}'_N(\delta) := \{x \in \mathbb{R}^n \mid \hat{J}_N^*(x) \leq \varepsilon\beta_f(\delta)\}$. However, in this case it is not necessarily guaranteed that the state and input constraints will also be satisfied.

B) Global stabilization with nonrelaxed terminal set constraint

As discussed in the proof of Theorem 3.1 as well as by Remark 3.5, satisfaction of the underlying terminal set constraint was the most important requirement for being able to prove asymptotic stability of the closed-loop system. In particular, it was shown in part *A)* that even when relaxing the respective terminal set barrier function, this condition can still be ensured for initial conditions within a suitably chosen sublevel set of the value function. Under the slightly stronger assumption of controllability of system (3.1), we will in the following present a second approach that allows us to guarantee asymptotic stability for *any* initial condition. This is achieved by simply relaxing only the barrier functions of the state and input constraints while strictly enforcing the stabilizing terminal set constraint by means of a nonrelaxed terminal set barrier function. In particular, we consider the following, nonrelaxed terminal cost formulation

$$\hat{F}(x) = x^\top P x + \varepsilon B_\mathrm{f}(x)\,, \qquad B_\mathrm{f}(x) = -\ln(1 - \varphi_\mathrm{f}(x))\,, \tag{3.30}$$

with $P \in \mathbb{S}^n_{++}$ and the function $\varphi_\mathrm{f} : \mathbb{R}^n \to \mathbb{R}_+$ describing the terminal set chosen according to Assumption 3.2. Assuming now controllability of the underlying system dynamics and exploiting again the properties ensured by Assumption 3.2, we can state the following *global* stability result, for which a proof can be found in Appendix C.5 .

Assumption 3.4. *The pair (A, B) describing the system dynamics (3.1) is controllable and the matrix $\begin{bmatrix} A^{N-1}B & \cdots & AB & B \end{bmatrix}$ has full row rank n.*

Theorem 3.2. *Consider the relaxed barrier function based problem (3.16) with stage cost (3.17) and terminal cost (3.30) and let Assumptions 3.1, 3.2, and 3.4 hold true. Then, independently of the relaxation parameter $\delta \in \mathbb{R}_{++}$, the origin of the resulting closed-loop system (3.19) is globally asymptotically stable.*

Remark 3.6. Note that the above global stability results are inherently bought at the cost of potential input and state constraint violations. However, as we will see in Section 3.5 of this thesis, the maximally possible violation of input and state constraint will be bounded and can in fact be computed and controlled a priori via the relaxation parameter δ.

Discussion

In summary, the results presented above show that, on the one hand, the desirable stability and constraint satisfaction properties of nonrelaxed barrier function based MPC can always be recovered when considering a full relaxation of both state/input and terminal set constraints. In particular the resulting region of attraction depends directly on the relaxation parameter δ and approximates the original feasible set whenever δ approaches zero. On the other hand, we were able to prove *global* asymptotic stability when assuming controllability of the underlying dynamical system and making use of a nonrelaxed logarithmic terminal set constraint barrier function.

As outlined above, different constructive MPC design approaches for choosing the terminal set and terminal cost in such a way that Assumption 3.2 holds are summarized in Appendix B.5. Aiming for an enlargement of the resulting region of attraction, these approaches not only allow one to choose the terminal set in principle as the maximal volume ellipsoid that can be inscribed within the set $\mathcal{X}_K$ but they also constitute the novel concept of (relaxed) barrier function based MPC based on *polytopic* terminal sets.

3.4.2 Stabilizing approaches without terminal sets

In the previous subsection, we showed how asymptotic stability and even strict constraint satisfaction can in the context of relaxed barrier function based MPC be guaranteed by making use of suitable terminal set formulations. On the one hand, the terminal set has been used for ensuring the existence of a feasible local control law that can be appended at the end of the prediction horizon and, thus, leads to recursive feasibility of the corresponding open-loop optimal control problem. On the other hand, only restricting the terminal state to a compact set around the origin allowed us to derive a quadratic upper bound for the barrier functions $B_K(\cdot)$ and $\hat{B}_K(\cdot)$, respectively, see Assumption 3.2 as well as the proofs of Theorems 3.1 and 3.2. In the following, we show that the use of relaxed logarithmic barrier functions in fact allows us to circumvent both of these problems, enabling us to design novel and conceptually simple MPC schemes that ensure important stability properties of the resulting closed-loop system *without* the need for an explicit terminal set constraint. In particular, we present three different approaches that can be distinguished by the respective choice of the terminal cost function $\hat{F} : \mathbb{R}^n \to \mathbb{R}_+$. While the first two approaches are based on the idea of overestimating the exact infinite-horizon cost-to-go by a simple quadratic penalty term or a suitably chosen tail-sequence of finite length, we also present a third approach which makes no use of a terminal cost at all. All presented approaches allow to guarantee global asymptotic stability of the closed loop and are constructive in the sense that the parameters of the associated MPC formulation can be computed in a constructive and numerically straightforward manner. Important constraint satisfaction and robustness properties of the resulting relaxed barrier function based MPC schemes will be discussed in the Sections 3.5 and 3.6, respectively.

Quadratic terminal cost

We begin our considerations with a rather intuitive approach that is in particular also often used in practice, namely that of choosing the terminal cost as a purely quadratic penalty term. More precisely, we propose to make use of

$$\hat{F}(x) = x^\top P x, \tag{3.31}$$

where $P \in \mathbb{S}^n_{++}$ is a suitably chosen terminal cost matrix. The key insight underlying the following results is that when making use of (quadratically) relaxed logarithmic barrier functions as introduced above, one can actually derive *global* quadratic upper bounds for the respective state and input constraint barrier functions, which then allows to overestimate the infinite-horizon cost-to-go *globally* and without making use of an additional terminal set constraint. More precisely, we can state the following general auxiliary result, of which we will make use multiple times throughout the remainder of this thesis. A proof of Lemma 3.3 is presented in Appendix C.6.

Lemma 3.3. *Let $\hat{B}_z : \mathbb{R}^n \to \mathbb{R}_+$ be a weight recentered relaxed logarithmic barrier function for a polytopic set of the form $\mathcal{P} = \{z \in \mathbb{R}^{n_z} : Cz \leq d\}$ with $C \in \mathbb{R}^{q \times n_z}$ and $d \in \mathbb{R}^q_{++}$, and* *let $w_z \in \mathbb{R}^{n_z}_{++}$ denote the corresponding weighting vector. Moreover, let the underlying relaxing function be quadratic and chosen according to (3.8) for a suitable relaxation parameter $\delta \in \mathbb{R}_{++}$. Then, it holds that*

$$\hat{B}_z(z) \leq z^\top M z \quad \forall\, z \in \mathbb{R}^{n_z}, \tag{3.32}$$

where the square matrix $M \in \mathbb{S}^{n_z}_{++}$ is defined as $M := \frac{1}{2\delta^2} C^\top \mathrm{diag}(w_z) C$.

Returning to our problem setup and the required upper bound for the combined state and input constraint barrier function $\hat{B}_K(\cdot)$, this directly leads to the following result, which is an immediate consequence of applying Lemma 3.3 to $\hat{B}_x(x)$ and $\hat{B}_u(u)$ and then inserting $u = Kx$ into the latter.

Lemma 3.4. *Consider the weight recentered relaxed logarithmic barrier functions $\hat{B}_x : \mathbb{R}^n \to \mathbb{R}_+$ and $\hat{B}_u : \mathbb{R}^m \to \mathbb{R}_+$ given in (3.18), with weighting vectors $w_x \in \mathbb{R}^n_{++}$ and $w_u \in \mathbb{R}^m_{++}$, relaxation parameter $\delta \in \mathbb{R}_{++}$, and making use of a quadratic relaxation according to (3.8). Then, for any control gain $K \in \mathbb{R}^{m \times n}$, the associated barrier function $\hat{B}_K(x) := \hat{B}_x(x) + \hat{B}_u(Kx)$ satisfies*

$$\hat{B}_K(x) \leq x^\top \left(M_x + K^\top M_u K \right) x \quad \forall\, x \in \mathbb{R}^n, \tag{3.33}$$

where the matrices $M_x \in \mathbb{S}^n_{++}$ and $M_u \in \mathbb{S}^m_{++}$ are defined as $M_x := \frac{1}{2\delta^2} C_x^\top \mathrm{diag}(w_x)\, C_x$ and $M_u := \frac{1}{2\delta^2} C_u^\top \mathrm{diag}(w_u)\, C_u$, respectively.

Based on Lemma 3.4 and inspired by our choice for the terminal cost in Section 3.4.1, we now propose to choose the control gain K and the terminal cost matrix $P \in \mathbb{S}^n_{++}$ as solutions to the following discrete-time algebraic Riccati equation

$$K = -\left(R + B^\top P B + \varepsilon M_u \right)^{-1} B^\top P A \tag{3.34a}$$

$$P = (A + BK)^\top P (A + BK) + K^\top (R + \varepsilon M_u) K + Q + \varepsilon M_x\,, \tag{3.34b}$$

where $M_x \in \mathbb{S}^n_+$ and $M_u \in \mathbb{S}^m_+$ are defined according to Lemma 3.4. Note that the controller gain K is in principle arbitrary as long as P is chosen according to (3.34b). However, the above choice results in a minimal value of the terminal cost function and ensures that for $\varepsilon \to 0$ or in the absence of constraints, K and P will reduce to the solution of the unconstrained LQR problem. Note that a suitable solution (K, P) to (3.34) always exists as Q and R are positive definite and (A, B) is assumed to be stabilizable.

Concerning the stability properties of the resulting terminal set free MPC scheme, we can then state the following theorem, for which a proof of is given in Appendix C.7.

Theorem 3.3. *Consider the relaxed barrier function based problem (3.16) with stage cost (3.17) and let Assumption 3.1 hold, assuming a quadratic relaxation according to (3.8). Moreover, let the terminal cost $\hat{F} : \mathbb{R}^n \to \mathbb{R}_+$ be given by (3.31), where the associated terminal cost weighting matrix $P \in \mathbb{S}^n_{++}$ is chosen according to (3.34). Then, the origin of the resulting closed-loop system (3.19) is globally asymptotically stable.*

Thus, by exploiting characteristic properties of the underlying quadratic relaxation, we can derive a *global* quadratic upper bound for the influence of the involved relaxed barrier functions and, thereby, prove important stability properties of the closed-loop system without the need for an additional terminal set constraint. However, one disadvantage of the presented approach is given by the fact that the employed quadratic upper bound may be overly conservative and scales with $1/\delta^2$, which may especially for small values of the relaxation parameter δ result in conservative and suboptimal choices for the terminal cost matrix P given in (3.34), see also the numerical example in Section 3.4.4. An alternative approach which circumvents this problem is presented in the following subsection.

Finite-tail terminal cost

It is a well-known result that closed-loop stability of both linear and nonlinear MPC schemes may be ensured by choosing the terminal cost as a suitable control Lyapunov function (CLF) that is an upper bound for the infinite-horizon cost-to-go, see for example (Jadbabaie and Hauser, 2001). In the presence of input and state constraints, deriving such a function in global form is generally not possible, which directly motivates the use of a local CLF in combination with a corresponding terminal set constraint. However, when considering the relaxed problem formulation (3.16), we may exploit the fact that *any* input sequence which steers the state to the origin in a finite number of steps can actually be used to derive an upper bound on the infinite-horizon cost-to-go. In the following, we discuss this idea in more detail and then introduce the related concept of *tail-sequence* based terminal cost functions.

Assumption 3.5. *Consider system (3.1) and let $\boldsymbol{v}(x) := \{v_0(x), \ldots, v_{T-1}(x)\}$ be a parametrized input sequence that steers the respective system state from any initial condition $x(0) = x \in \mathbb{R}^n$ to the origin in a finite number of $T \geq n$ steps. Let $\boldsymbol{z}(x) := \{z_0(x), \ldots, z_{T-1}(x)\}$ with $z_0(x) = x$, $z_{l+1}(x) = Az_l(x) + Bv_l(x)$, $l = 0, \ldots, T-1$, and $z_T(x) = 0$ be the associated state sequence. Furthermore, assume that $v_l(x) = 0\ \forall\, l = 0, \ldots, T-1 \Leftrightarrow x = 0$.*

As discussed below, one possible approach to obtain a suitable parametrized input sequence $\boldsymbol{v}(x)$ in a constructive fashion is to make use of a linear dead-beat controller. Based on Assumption 3.5 and the associated input and state tail-sequences $\boldsymbol{v}(x)$ and $\boldsymbol{z}(x)$, we then propose to choose the terminal cost function as

$$\hat{F}(x) = \sum_{l=0}^{T-1} \hat{\ell}(z_l(x), v_l(x)), \tag{3.35}$$

where $\hat{\ell} : \mathbb{R}^n \times \mathbb{R}^m \to \mathbb{R}_+$ simply refers to the relaxed barrier function based stage cost defined above in (3.16). Due to the relaxation, $\boldsymbol{v}(x)$ is a well-defined but possibly suboptimal input sequence that steers the state from $z_0(x) = x$ for any $x \in \mathbb{R}^n$ to the origin in a finite number of steps. Consequently, $\hat{F}(\cdot)$ given in (3.35) represents a global upper bound for the infinite-horizon cost-to-go, which allows us to state the following stability result. A proof of Theorem 3.4 can be found in Appendix C.8.

Theorem 3.4. *Consider the relaxed barrier function based problem (3.16) with stage cost (3.17) and tail-sequence based terminal cost according to (3.35), and let Assumptions 3.1 and 3.5 hold. Then, the origin of the resulting closed-loop system (3.19) is globally asymptotically stable.*

Remark 3.7. As in the previous approach based on a purely quadratic terminal cost, the key to the presented stability result is to derive a *global* upper bound for the infinite-horizon cost-to-go. In particular, by assuming the existence of suitable input sequences that steer the system to the origin in finite time, this can be achieved here without making use of the potentially conservative barrier function bound given in Lemma 3.4.

Remark 3.8. Note that in order to ensure convexity of the terminal cost function $\hat{F} : \mathbb{R}^n \to \mathbb{R}_+$ given in (3.35), the elements of the parametrized tail input sequence $v(x)$ should be affine in the argument x. In combination with the condition that $v_l(x) = 0 \ \forall l = 1, \dots, T-1 \Leftrightarrow x = 0$, this in fact limits $v(\cdot)$ to contain a sequence of linear state feedback laws, that is, $v(x) = \{K_0 x, \dots, K_{T-1} x\}$. Different constructive design approaches that meet this requirement, and thus allow us to guarantee stability of the closed-loop system as well as convexity of the resulting overall cost function, will be discussed in the following.

Probably the easiest way to design suitable parametrized tail-sequences $v(\cdot)$ and $z(\cdot)$ that satisfy the conditions stated in Assumption 3.5 and Remark 3.8 is by making use of a linear dead-beat controller. To this end, we may choose a static control gain $K \in \mathbb{R}^{m \times n}$ in such a way that the matrix $A_K = A + BK$ is nilpotent, i.e., that it satisfies $A_K^r = 0$ for some $r \leq n$. Under the assumption that (A, B) is controllable, this can for example be achieved by a suitable pole placement procedure which ensures that all eigenvalues of the matrix A_K are located at the origin of the complex plane. Based on these ideas, we may set the tail-sequence horizon T equal to the state dimension, i.e., $T = n$, and choose the input tail-sequence as

$$v(x) = \left\{ Kx, KA_K x, \dots, KA_K{}^{n-1} x \right\}, \tag{3.36}$$

which results in the corresponding tail-sequence $z(x) = \{x, \dots, A_K^{n-1} x\}$ for the predicted system state. Due to the design of the matrix K, it obviously holds that $z_n = A_K^n x = 0$. Thus, $v(\cdot)$ and $z(\cdot)$ satisfy the conditions in Assumption 3.5, implying that Theorem 3.4 can be used to conclude asymptotic stability of the closed-loop system.

While the above approach based on a static dead-beat controller allows for a rather simple design of the tail-sequences $v(\cdot)$ and $z(\cdot)$, the implicit requirement that the predicted terminal state is steered to the origin in at most n steps might be restrictive, leading to suboptimal or even aggressive behavior of the overall closed-loop system. In the following, we therefore present a second approach that eliminates this restriction by allowing for tail-sequences with $T \geq n$ elements. In particular, we propose to choose the parametrized input sequence $v(\cdot)$ as the solution to the associated finite-horizon LQR problem with zero terminal state constraint, that is,

$$v(x) = \arg\min_{v} \sum_{l=0}^{T-1} \ell(z_l, v_l) \tag{3.37a}$$

$$\text{s.t. } z_{l+1} = Az_l + Bv_l, \quad l = 0, \dots, T-1, \tag{3.37b}$$

$$z_T = 0, \; z_0 = x, \tag{3.37c}$$

where $\ell(z,v) = \|z\|_Q^2 + \|v\|_R^2$ refers to the conventional quadratic stage cost with $Q \in \mathbb{S}^n_{++}$ and $R \in \mathbb{S}^m_{++}$ as defined above. When assuming that the underlying system dynamics (A,B) are controllable, it can be shown that the solution $v^*(x)$ to problem (3.37) can be expressed as a sequence of static linear state feedbacks of the form $v_l^*(x) = K_l x$ with $l = 0, \dots, T-1$. In particular, explicit expressions for the optimal control gains K_l and the associated state trajectories were derived by Ntogramatzidis (2003) as well as Ferrante and Ntogramatzidis (2005). Furthermore, as discussed in Appendix B.6 of this thesis, the optimal solution to problem (3.37) may also be computed directly in vector form as

$$V^*(x) = \begin{bmatrix} v_0^{*\top}(x) & \cdots & v_{T-1}^{*\top}(x) \end{bmatrix}^\top = K_V x, \tag{3.38}$$

where $K_V \in \mathbb{R}^{Tm \times n}$ denotes a suitably defined static control gain matrix. As the resulting parametrized input sequence satisfies by design the conditions of Assumption 3.5, global asymptotic stability of the closed-loop system can again be concluded from Theorem 3.4. However, the possibility to choose an arbitrary horizon within the tail-sequence based terminal cost typically leads to an improved overall closed-loop performance. Especially in scenarios in which the open-loop state and input tail-sequences stay away from the boundaries of the constraint sets, the terminal cost function $\hat{F}(\cdot)$ based on (3.35) and (3.37) may actually give a quite good approximation of the exact infinite-horizon cost-to-go.

Remark 3.9. Note that by trying to force the predicted state to the origin in $N+T$ steps, the presented approach is conceptually similar to MPC schemes that are based on an explicit zero terminal state constraint of the form $x_N = 0$, see for example Keerthi and Gilbert (1988). However, by using the proposed relaxed barrier function based terminal cost function, this behavior is enforced implicitly, i.e., without introducing an explicit equality constraint within the optimization problem. Furthermore, stability can be guaranteed for a global region of attraction and the tail-sequence horizon T may be made arbitrary large without increasing the number of optimization variables.

Remark 3.10. Note that controllability of the system dynamics (A,B) is in general a sufficient but not a necessary condition for the existence of suitable tail-sequences $v(x)$ and $z(x)$. In particular, parametrized input sequences that lead to finite-time convergence of the system state to the origin may also exist for stabilizable systems whose uncontrollable modes exhibit a dead-beat characteristic. Nevertheless, Assumption 3.5 is a stronger requirement than the stabilizability assumption that is required by the purely quadratic terminal approach presented in the previous section.

Summarizing, we conclude that by making use of suitably defined *tail-sequences*, asymptotic stability of the closed-loop system can again be guaranteed without the help of an additional explicit terminal set constraint. In contrast to the previous approach with purely quadratic terminal cost, the assumed finite length of the tails thereby allowed us to circumvent the use of the conservative barrier function bound given in Lemma 3.4. However, the price to pay is given by the slightly stronger Assumption 3.5 and a more complicated expression for the terminal cost. In the following section, we will present yet another approach, which demonstrates that we do actually not need *any* terminal cost as long as the prediction horizon N is chosen large enough.

Zero terminal cost

While the vast majority of stability proofs in the MPC literature make in one way or another use of a suitably chosen terminal cost function term in combination with an explicit or implicit terminal set constraint, various recent result have shown that imposing such a terminal set constraint is in general not necessary for ensuring stability of the closed-loop system, see for example (Grüne and Pannek, 2011) as well as (Grüne, 2012) and references therein. In fact, these works show that not even a specifically chosen terminal cost function term is needed as long as the underlying open-loop optimal control problem satisfies certain controllability conditions and the prediction horizon N is chosen long enough. Following similar ideas, we will now present a relaxed barrier function based MPC scheme that ensures asymptotic stability of the closed loop system based on a zero terminal cost, i.e.,

$$\hat{F}(x) = 0\,. \tag{3.39}$$

The associated open-loop optimal control problem (3.16) therefore reduces to the following simplified problem, which we recall here for the sake of convenience:

$$\hat{J}_N^*(x) = \min_u \sum_{k=0}^{N-1} \hat{\ell}(x_k, u_k) \tag{3.40a}$$

$$\text{s.t. } x_{k+1} = Ax_k + Bu_k,\ x_0 = x\,. \tag{3.40b}$$

Here, $N \in \mathbb{N}_+$ is again the prediction horizon and $\hat{\ell} : \mathbb{R}^n \times \mathbb{R}^m \to \mathbb{R}_+$ refers to the relaxed barrier function based stage cost according to (3.17), which we assume to comply with Assumption 3.1. Related to this problem setup, we subsequently show the following two main results: first, that asymptotic stability of the resulting closed-loop system can be guaranteed for any $N \geq \bar{N}$, where the critical horizon $\bar{N} < \infty$ can be computed in a constructive fashion, and, second, that an estimate for the resulting closed-loop performance can be given in terms of the associated infinite-horizon problem

$$\hat{J}_\infty^*(x) = \min_u \sum_{k=0}^{\infty} \hat{\ell}(x_k, u_k) \tag{3.41a}$$

$$\text{s.t. } x_{k+1} = Ax_k + Bu_k,\ x_0 = x\,. \tag{3.41b}$$

To this end, we define

$$\hat{J}^{\text{cl}}_{\infty}(x) := \sum_{k=0}^{\infty} \hat{\ell}(x(k), \hat{u}^*_0(x(k))), \quad x(0) = x, \tag{3.42}$$

which represents the cumulated closed-loop stage cost under the barrier function based feedback $u(k) = \hat{u}^*_0(x(k))$. Note that $\hat{J}^{\text{cl}}_{\infty}(x) \geq \hat{J}^*_{\infty}(x)$ by definition.

The main idea underlying the following stability and performance results relies again on the fact that the presented quadratic barrier function bounds allow us to derive a global upper bound for the infinite-horizon cost – a result that we basically already exploited in the proof of Theorem 3.3. In fact, in the case of a zero terminal cost, this directly implies a uniform upper bound on the value function $\hat{J}^*_N : \mathbb{R}^n \to \mathbb{R}_+$ that is valid for any $N \in \mathbb{N}_+$.

Lemma 3.5. *Consider the relaxed barrier function based open-loop optimal control problem (3.40) with (A, B) stabilizable and let $\hat{J}^*_N : \mathbb{R}^n \to \mathbb{R}_+$ denote the associated value function. Furthermore, let $P \in \mathbb{S}^n_{++}$ be the positive definite solution to the discrete-time algebraic Riccati equation (3.34). Then, for any $x \in \mathbb{R}^n$ and any prediction horizon $N \in \mathbb{N}_+$, it holds that*

$$\hat{J}^*_N(x) \leq \hat{J}^*_{\infty}(x) \leq x^\top P x. \tag{3.43}$$

The proof of Lemma 3.5 mainly exploits the quadratic barrier function bounds provided by the Lemmas 3.3 and 3.4, respectively, and can be found in Appendix C.9. Based on Lemma 3.5, we can then state the following main result.

Theorem 3.5. *Consider the relaxed barrier function based problem (3.16) with stage cost (3.17) and zero terminal cost, leading to the simplified problem formulation (3.40). Furthermore, define $\gamma \in (1, \infty)$ as $\gamma := \lambda_{\max}(PQ^{-1})$, where $Q \in \mathbb{S}^n_{++}$ refers to the weighting matrix of the quadratic cost function term and $P \in \mathbb{S}^n_{++}$ is given according to Lemma 3.5. Then, the origin of the resulting closed-loop system (3.19) is globally asymptotically stable if the prediction horizon N satisfies*

$$N \geq \bar{N} := \left\lceil 1 + \frac{2\ln(\gamma)}{\ln(\gamma) - \ln(\gamma - 1)} \right\rceil. \tag{3.44}$$

In addition, for any such N, the cumulated stage cost given in (3.42) satisfies for any $x \in \mathbb{R}^n$ the infinite-horizon performance estimate

$$\hat{J}^{\text{cl}}_{\infty}(x) \leq \hat{J}^*_{\infty}(x)/\alpha_N, \tag{3.45}$$

where $\alpha_N \in (0, 1)$ is defined as $\alpha_N := 1 - (\gamma - 1)^N/\gamma^{N-2}$. The performance index α_N increases monotonically with the horizon N and satisfies $\lim_{N\to\infty} \alpha_N = 1$.

A proof of this result is given in Appendix C.10. Note that Theorem 3.5 has two interesting main implications. First, it provides an alternative MPC design approach that allows us to guarantee asymptotic stability of the resulting closed-loop system solely based on the underlying prediction horizon. In contrast to the two previously discussed

approaches, this is not only achieved without enforcing an additional terminal set constraint but also without making use of any terminal cost function term. The price to pay is given by the fact that the prediction horizon needs to be chosen larger than some (potentially conservative) critical horizon $\bar{N}$. Second, Theorem 3.5 reveals that as soon as the underlying horizon is long enough, any further increase in the prediction horizon will also result in an increased cumulated closed-loop performance – allowing to recover for $N \to \infty$ in the limit the closed-loop performance of the infinite-horizon problem.
However, one disadvantage of the presented approach is that it relies again, like the purely quadratic terminal cost function approach discussed above, on the potentially conservative quadratic upper bounds given in Lemma 3.3 and Lemma 3.5, respectively. As a consequence, the critical horizon $\bar{N}$ that is sufficient for ensuring the desired stability properties, and which depends via γ and P directly on the choice of the relaxation parameter δ, may also be conservative. In particular, $\bar{N}$ may become quite large for small values of δ, see also the numerical example in Section 3.4.4. Nevertheless, the presented approach is interesting from a conceptual point of view as it links the proposed relaxed barrier function based MPC framework to other recent works on MPC without terminal sets, e.g., the works of Tuna et al. (2006), Grüne and Rantzer (2008), and Grüne (2012).

Remark 3.11. Note that it may in principle be possible to obtain better and less conservative estimates for the critical horizon $\bar{N}$. This, may for example be achieved by deriving less conservative upper bounds for the infinite-horizon value function, by including a suitably chosen nonzero terminal cost function term, or by exploiting the fact that the underlying system dynamics are linear and hence Lipschitz continuous (Grüne, 2012). However, as we are in this thesis mainly interested in the alternative conceptual approach as such, we limit ourselves to the considered simple problem formulation.

Discussion

In summary, we presented three conceptually different design approaches that allow guaranteeing (global) asymptotic stability of the resulting closed-loop system without making use of any additional relaxed or nonrelaxed terminal set constraint. Being all based on the general relaxed barrier function based MPC problem formulation (3.16) introduced at the beginning of this chapter, the different approaches are mainly characterized by the choice of the respective terminal cost function term $\hat{F} : \mathbb{R}^n \to \mathbb{R}_+$.
First, we presented a stabilizing MPC approach that makes use of a purely quadratic terminal cost, exploiting thereby the fact that the effect of the underlying relaxed logarithmic barrier functions can be upper bounded by a quadratic function. While allowing for an intuitive and simple overall problem formulation, the main disadvantage of this approach is that the required terminal cost function weight tends to be conservative and may grow quite fast for small values of the relaxation parameter.
Second, we presented an alternative approach which exploits the fact that any (possibly suboptimal) input sequence that steers the system state to the origin in a finite number of steps can due to the underlying relaxation be used to construct an explicit upper bound

for the infinite-horizon cost-to-go. As a consequence, global asymptotic stability of the resulting closed-loop system can be ensured by making use of a terminal cost function that consists of suitably chosen *tail-sequences* of finite length. This approach circumvents the potentially conservative quadratic upper bound for the barrier functions, but results in a slightly more complex terminal cost function term and, in general, requires to assume some form of controllability of the underlying system dynamics.
Third, by exploiting again the upper bound on the relaxed barrier functions, we showed that the same stability properties can in fact be guaranteed without *any* terminal cost function term as long as the prediction horizon is chosen long enough. Beside the fact that it results in a natural and simple problem formulation and allows for important infinite-horizon performance estimates, this approach establishes in addition an interesting link to already existing works on terminal set free MPC approaches. However, due to the involved conservativeness, it may suffer from similar shortcomings as the first approach.

3.4.3 Summary

In the following, we briefly summarize our main results on closed-loop stability and related constructive design approaches in the context of model predictive control based on relaxed barrier functions. The discussion deliberately focuses on the main conceptual aspects, while some more implementation-oriented advantages and disadvantages of the respective approaches will be discussed in the following numerical example.

Starting out from the definition of a relaxed logarithmic barrier function and the associated relaxed barrier function based MPC formulation, our first main result was that the (local) stability and constraint satisfaction properties of the corresponding *nonrelaxed* formulation may always be recovered by a sufficiently small choice of the relaxation parameter (Theorem 3.1). This underlines the character of the proposed relaxed barrier function based MPC framework as a meaningful and coherent extension of existing barrier function based MPC concepts. Our second main result was that the relaxed problem formulation not only leads to a globally defined reformulation of the associated open-loop optimal control problem but in fact allows to guarantee *global* asymptotic stability of the resulting closed-loop system if the underlying cost function terms are chosen in a suitable way. Related to this finding, our third main contribution consisted in the presentation and discussion of different stabilizing design approaches, which in particular revealed that in principle neither an additional terminal set constraint nor a specifically chosen terminal cost function term are necessary in order to prove global asymptotic stability of the origin. The different approaches presented in this section are summarized in Table 3.1 on the next page, together with some comparing information concerning the necessity of a terminal set, the required assumptions on the underlying system dynamics and prediction horizon, and the resulting region of attraction (ROA). At this point we want to emphasize again that all the required ingredients – from the terminal sets underlying the approaches in Section 3.4.1 to the critical horizon that is sufficient for ensuring stability in the case of zero terminal cost – can be computed in a systematic and constructive way, and that we presented in each case constructive approaches for doing so.

Table 3.1. Summary of the presented relaxed barrier function based MPC schemes.

Section	Terminal cost $\hat{F}(x)$	$\mathcal{X}_f$	(A,B)	Horizon	ROA	Theorem (page)
3.4.1	$x^\top Px + \varepsilon\hat{B}_f(x)$	yes	stabilizable	$N \in \mathbb{N}_+$	$\hat{\mathcal{X}}_N(\delta)$	Thm. 3.1 (p. 44)
3.4.1	$x^\top Px + \varepsilon B_f(x)$	yes	controllable	$N \in \mathbb{N}_+$	$\mathbb{R}^n$	Thm. 3.2 (p. 46)
3.4.2	$x^\top Px$	no	stabilizable	$N \in \mathbb{N}_+$	$\mathbb{R}^n$	Thm. 3.3 (p. 48)
3.4.2	$\sum_{l=0}^{T-1} \hat{\ell}(z_l(x), v_l(x))$	no	controllable	$N \in \mathbb{N}_+$	$\mathbb{R}^n$	Thm. 3.4 (p. 50)
3.4.2	0	no	stabilizable	$N \geq \bar{N}$	$\mathbb{R}^n$	Thm. 3.5 (p. 53)

As outlined above, one of the main advantages of relaxed barrier function based MPC formulations is given by the fact that the stabilizing control input can be characterized as the minimizer of a globally defined and strongly convex cost function. In fact, after elimination of the linear system dynamics, the associated open-loop optimal control problem can be formulated as unconstrained minimization of a cost function of the form

$$\hat{J}_N(U,x) = \frac{1}{2}U^\top HU + x^\top FU + \frac{1}{2}x^\top Yx + \varepsilon\hat{B}_{\mathrm{xu}}(U,x) \tag{3.46}$$

with optimization variable $U := \begin{bmatrix} u_0^\top & \cdots & u_{N-1}^\top \end{bmatrix}^\top \in \mathbb{R}^{Nm}$. Furthermore, $\hat{B}_{\mathrm{xu}} : \mathbb{R}^{Nm} \times \mathbb{R}^n \to \mathbb{R}_+$ is a positive definite, convex, and twice continuously differentiable relaxed logarithmic barrier function for polytopic constraints of the form $GU \leq d + Ex$. The matrices $H \in \mathbb{S}_{++}^{Nm}$, $F \in \mathbb{R}^{n\times Nm}$, $Y \in \mathbb{S}_{++}^n$, $G \in \mathbb{R}^{q\times Nm}$, $d \in \mathbb{R}_{++}^q$, and $E \in \mathbb{R}^{q\times n}$ can be constructed from (3.16) and the corresponding constraints by means of simple matrix operations, see Appendix B.9. Due to the design of the relaxed barrier functions, the above cost function is twice continuously differentiable, strongly convex in U and (at least in the case of a full relaxation) globally defined. More details and further beneficial properties of the associated optimization problem will be discussed in Chapter 4 of this thesis.

Thus, in total we can conclude that the use of relaxed logarithmic barrier functions not only leads to an intuitive and numerically attractive formulation of the underlying open-loop optimal control problem but also allows us to guarantee in a constructive fashion important stability properties of the resulting closed-loop system. As this in particular involves conceptually novel and terminal set free design approaches, the concept of relaxed barrier function based MPC can in fact be considered to be far more than a mere extension of existing approaches based on nonrelaxed logarithmic barrier functions. As we will see in the remainder of this thesis, this claim is also supported by beneficial robustness and algorithmic properties that are implied by the underlying relaxation and the globally stabilizing MPC design, see in particular Section 3.6 and Chapter 4. However, as already discussed above, both the relaxation of the barrier functions and the presented global stability results rely on the assumption that the underlying input and state constraints may in principle be violated in closed-loop operation. As the handling of constraints is typically one of the main assets of conventional MPC approaches, this necessitates to investigate whether we can despite the relaxation still give guarantees on the resulting maximal constrain violation and whether and how we may control this maximal violation in a constructive way via the relaxation parameter δ. This issue will be analyzed in more detail in the following Section 3.5.

3.4.4 Numerical example

In the following, we want to briefly discuss and illustrate some of our findings from above by means of an academic numerical example. In particular, we consider a discrete-time double integrator system with the linear system dynamics

$$x(k+1) = \begin{bmatrix} 1 & T_s \\ 0 & 1 \end{bmatrix} x(k) + \begin{bmatrix} T_s^2 \\ T_s \end{bmatrix} u(k), \qquad T_s = 0.1. \tag{3.47}$$

The input and state constraints are assumed to be given by box constraints of the form $\mathcal{U} = \{u \in \mathbb{R} \mid -1 \leq u \leq 1\}$ and $\mathcal{X} = \{x \in \mathbb{R}^2 \mid -2 \leq x_1 \leq 3,\ -1 \leq x_2 \leq 1\}$. In the language of our problem setup above, this translates into

$$C_\mathrm{u} = \begin{bmatrix} 1 \\ -1 \end{bmatrix}, \; d_\mathrm{u} = \begin{bmatrix} 1 \\ 1 \end{bmatrix}, \qquad C_\mathrm{x} = \begin{bmatrix} 1 & 0 \\ 0 & 1 \\ -1 & 0 \\ 0 & -1 \end{bmatrix}, \; d_\mathrm{x} = \begin{bmatrix} 3 \\ 1 \\ 2 \\ 1 \end{bmatrix}. \tag{3.48}$$

The goal is now to steer the state of system (3.47) to the origin while satisfying the polytopic input and state constraints that are induced by (3.48). Note that the above dynamical system is controllable, which implies that we can apply any of the barrier function based MPC approaches introduced above.

The MPC problem setup

In order to illustrate again the design process and highlight possible advantages and disadvantages, we now implement and test each of the relaxed barrier function based MPC approaches summarized in Table 3.1.

The basic MPC problem parameters, i.e., the prediction horizon and the quadratic stage cost weighting matrices, are chosen as $N = 15$, $Q = \mathrm{diag}(1,\ 0.1)$, and $R = 0.1$. The barrier function weighting and relaxation parameters are chosen as $\varepsilon = \delta = 10^{-3}$. The weighting vectors for recentering the resulting barrier functions are constructed by making use of the procedure in Appendix B.3, leading to

$$w_\mathrm{u} = \begin{bmatrix} 1.0000 & 1.0000 \end{bmatrix}^\top, \qquad w_\mathrm{x} = \begin{bmatrix} 1.0000 & 1.0000 & 1.5000 & 1.0000 \end{bmatrix}^\top. \tag{3.49}$$

Based on the constructive procedure explained in Appendix B.5, we can compute a suitable terminal set as well as a corresponding terminal cost matrix $P \in \mathbb{S}^n_{++}$ that allows us to ensure stability of the closed-loop system for the terminal set based approaches presented in Section 3.4.1. In particular, choosing $\gamma = 200$, this results in

$$K = \begin{bmatrix} -1.5584 & -1.9051 \end{bmatrix}, \qquad P = \begin{bmatrix} 18.3374 & 7.7918 \\ 7.7918 & 10.0254 \end{bmatrix} \tag{3.50}$$

as well as in a contractive polytopic terminal set of the form

$$\mathcal{X}_f = \{x \in \mathbb{R}^n \mid C_f x \leq \mathbb{1}\}, \qquad C_f \in \mathbb{R}^{14\times 2}. \tag{3.51}$$

Making use of a smooth approximation of the Minkowski functional (see again Appendix B.5), a smoothed polytopic terminal set approximation with an associated smooth relaxed logarithmic terminal set barrier function can be computed as

$$\mathcal{X}_f^p = \{x \in \mathbb{R}^n \mid \varphi_f(x) \leq 1\}, \qquad \hat{B}_f(x) = \hat{B}(1 - \varphi_f(x)). \tag{3.52}$$

In particular, $\varphi_f : \mathbb{R}^n \to \mathbb{R}_+$ is given by (B.43) with (in this case) $p = 46$. For the approach based on a nonrelaxed terminal set, the primordial relaxed barrier function $\hat{B}(\cdot)$ in (3.52) is simply replaced by the natural logarithm.

In contrast to this, all that we have to do for the alternative, terminal set free approach based on a purely quadratic terminal cost is to compute a solution to the modified algebraic Riccati equation (3.34), resulting in

$$K = \begin{bmatrix} -1.0222 & -1.6466 \end{bmatrix}, \qquad P = \begin{bmatrix} 2.0152 & 1.0223 \\ 1.0223 & 1.7468 \end{bmatrix} \times 10^4 \tag{3.53}$$

In contrast to the one given in (3.50) above, this choice allows ensuring global asymptotic stability without the need for an additional terminal set constraint. However, we also see that the norm of the matrix P is quite large, leading to a relatively heavy penalty on the terminal state. We will discuss this point in more detail after we have introduced also the remaining MPC approaches from Table 3.1.

Turning our attention to the next terminal set free approach, we note that suitable tail-sequences with $T = n = 2$ elements may for example be computed based on a linear dead-beat controller of the form

$$v(x) = \{Kx, K(A+BK)x\}, \quad z(x) = \{x, (A+BK)x\}, \quad K = \begin{bmatrix} -100 & -10 \end{bmatrix}. \tag{3.54}$$

Furthermore, more general tail-sequences with $T \geq n$ elements can be constructed based on parametrized solutions to the associated LQR problem with zero terminal state constraint, see Section 3.4.2 as well as Appendix B.6. For the following simulation experiments, we choose $T = N = 15$. We do not explicitly state the resulting parametrized input and state sequences here for the sake of simplicity.

Finally, concerning the approach based on a zero terminal cost, we obtain based on the definitions in Theorem 3.5

$$\gamma = \lambda_{\max}(PQ^{-1}) = 1.8117 \times 10^5 , \qquad \bar{N} = 4.3869 \times 10^6 , \tag{3.55}$$

where $P \in \mathbb{S}^n_{++}$ refers to the matrix given in (3.53) and $\bar{N}$ denotes the critical horizon that is required to ensure asymptotic stability of the closed-loop system. Obviously, $\bar{N}$ is far too large for any meaningful numerical or practical implementation. The main reason for this prohibitively large prediction horizon is that both the global quadratic upper bounds that are exploited in the computation of the matrix P and the lower bound for the critical horizon tend do be highly conservative. As a result, both P and $\bar{N}$ will typically grow quite fast for small values of the relaxation parameter, see Figure 3.2 on the next page. While this is from a conceptually point of view not really a problem for the approach based on a purely quadratic terminal cost, it may in certain cases prevent the implementation of the approach based on a zero terminal cost (at least with the guaranteed global stability properties as derived above). However, as will be discussed below, a quite good closed-loop behavior can for the considered double integrator example in fact already be observed for a horizon of $N = 15$.

In the following, we consider three different scenarios for studying the behavior of the closed-loop system for each of the above relaxed barrier function based MPC approaches. We focus deliberately on the main conceptual aspects outlined above, and do not discuss details concerning the implementation or the computational efficiency of the respective MPC schemes. The considered scenarios are briefly discussed in the following paragraphs. The related simulation results are depicted in Figure 3.3. All computations and simulations where performed using MATLAB/Simulink R2015b. Furthermore, the Multi-Parametric Toolbox (Kvasnica et al., 2004) has been used for constructing the required contractive polytopic terminal set.

Scenario 1

We consider a feasible initial condition of $x(0) = x_{0,1} = [2.5, -0.65]^\top$ and analyze the behavior of the resulting closed-loop system when making use of the globally stabilizing approach based on a nonrelaxed terminal set constraint. In particular, we vary the value for the relaxation parameter δ and observe how this affects the resulting constraint satisfaction properties. For the presented simulation results, we chose five different values according to $\delta \in \{10^{-1}, 2.5 \times 10^{-2}, 10^{-2}, 10^{-3}, 10^{-5}\}$. As can be seen from the subplots (a) and (b) of Figure 3.3, there is a quite severe violation of state and input constraints if δ is chosen too large. On the other hand, the resulting closed-loop trajectories seem to satisfy the constraints as soon as the relaxation parameter is chosen small enough (here as soon as $\delta \leq 10^{-3}$). This observed constraint satisfaction behavior will be discussed (and in fact theoretically confirmed) in Section 3.5 of this thesis. Furthermore, it can be seen that the system state, as ensured by Theorem 3.2, converges to the origin in all cases.

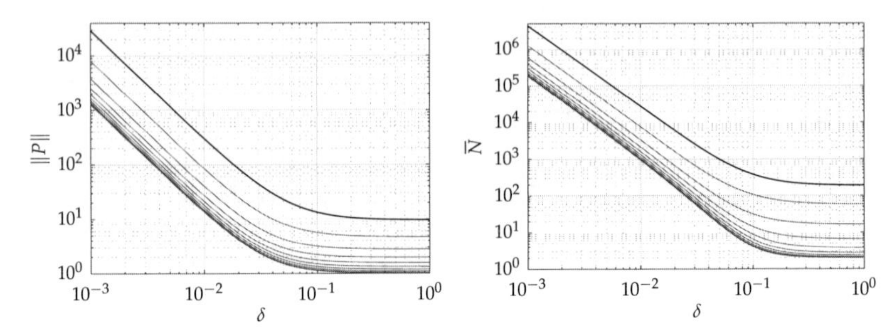

Figure 3.2. Evolution of the terminal cost matrix P and the critical horizon $\bar{N}$ for decreasing values of the relaxation parameter δ. As can be seen, both the norm of P and the critical horizon $\bar{N}$ remain almost constant as long as $\delta \geq 0.1$, whereas they start to increase exponentially as soon as the relaxation parameter is decreased below this threshold. The gray lines indicate the respective behavior when stabilizing the dynamic matrix of the double integrator artificially by means of a multiplicative scaling factor $\kappa = 0.9, 0.8, \ldots, 0.1$. It can be seen that the evolution does in principle depend on the system dynamics, but that the qualitative behavior stays the same.

Scenario 2

As a second scenario, we consider the *infeasible* initial condition $x_{0,2} = [-1.7, -1.25]^\top$ and compare the resulting closed-loop behavior when applying the different relaxed barrier function based MPC schemes that we presented above. As can be seen from Figure 3.3 (and as guaranteed by the respective stability theorems), asymptotic convergence to the origin is achieved independently of the underlying MPC approach. In fact, this even holds for the approach based on a *relaxed* terminal set constraint barrier function, for which stability of the closed-loop system can according to Theorem 3.1 only be guaranteed for all initial conditions within the (apparently conservative) set $\mathcal{X}_N(\delta)$. Note that the input constraints are at the beginning violated by all approaches due to the fact the initial condition is infeasible. However, as soon as the state trajectory has entered the feasible set, the respective controllers tend to achieve satisfaction of the underlying input and state constraints for all future time steps. An exception is given by the approaches based on a purely quadratic terminal cost and a finite-tail terminal cost relying on a linear dead-beat controller. For these approaches, which also show the most severe input constraint violations at initialization, the upper bounds on x_2 are also slightly violated after having already entered the set of feasible states, see subplot (a). One possible explanation for this behavior is that both approaches actually tend to force the predicted terminal state x_N very close to the origin, accepting potentially necessary constraint violations to achieve this goal. In the first case, this is due to the heavy influence of the terminal cost matrix P, see (3.53), whereas in the second case the system state is implicitly forced to the origin in $N + 2$ steps. Concerning the latter approach, a prolonged tail-sequence based on a parametrized LQR solution with zero terminal state already leads to a less aggressive closed-loop behavior with considerably improved constraints satisfaction properties, see Figure 3.3. Moreover, as will be shown in Section 3.5 of this thesis, also the behavior of the approach with purely quadratic terminal cost may be improved significantly by increasing the prediction horizon N. Note furthermore that all approaches result qualitatively in the same control input as soon as the system state gets close to the origin. This illustrates nicely that the influence of the terminal cost – and thus the main difference between the proposed approaches – is more and more diminished as the closed-loop system state approaches the desired set point.

Scenario 3

We consider again an infeasible initial condition, given by $x_{0,3} = [1.25,\ 1.25]^\top$, for which we now want to compare the resulting closed-loop behavior for the approach based on a zero terminal cost when choosing different values for the underlying prediction horizon. In particular, for the presented simulation results, we chose five different horizon lengths according to $N \in \{10, 15, 20, 30, 100\}$. As can be seen from subplots (a) and (d) of Figure 3.3, the resulting closed-loop behavior is almost the same for different lengths of the prediction horizon. In fact, the behavior converges quite fast to that of the associated infinite-horizon problem. A qualitative change can only be observed when decreasing the horizon as far as $N = 5$, for which the resulting input trajectory is depicted in gray in subplot (d). Thus, at least for the considered example and the above initial condition, the computed choice for the critical horizon $\bar{N}$ seems to be highly conservative. However, note that $\bar{N}$ was designed in such a way that it ensures asymptotic stability for *any* initial condition and that it is exactly this global aspect which necessitates the underlying conservative global quadratic upper bound on the value function. In addition, the stability result in given in Theorem 3.5 provides a theoretical underpinning and justification for the applicability of the corresponding MPC approach as such. For practical applications, a more meaningful and implementable choice for the prediction horizon may – as it has also been done in this example – be found based on numerical simulations.

Summarizing, we can conclude that the presented numerical example illustrates and confirms many of our theoretical results, revealing also some potential advantages and disadvantages of the different MPC approaches. Concerning the numerical performance, it should be noted that the computation times for the terminal set free approaches with quadratic or zero terminal cost were observed to be far lower than those for the more complicated terminal set based formulations (roughly up to a factor of 30 on average). From the perspective of real-time implementation, this illustrates an additional potential advantage of the proposed terminal set free MPC approaches. A more detailed complexity analysis and numerical benchmark results will be presented in Chapter 4 of this thesis.

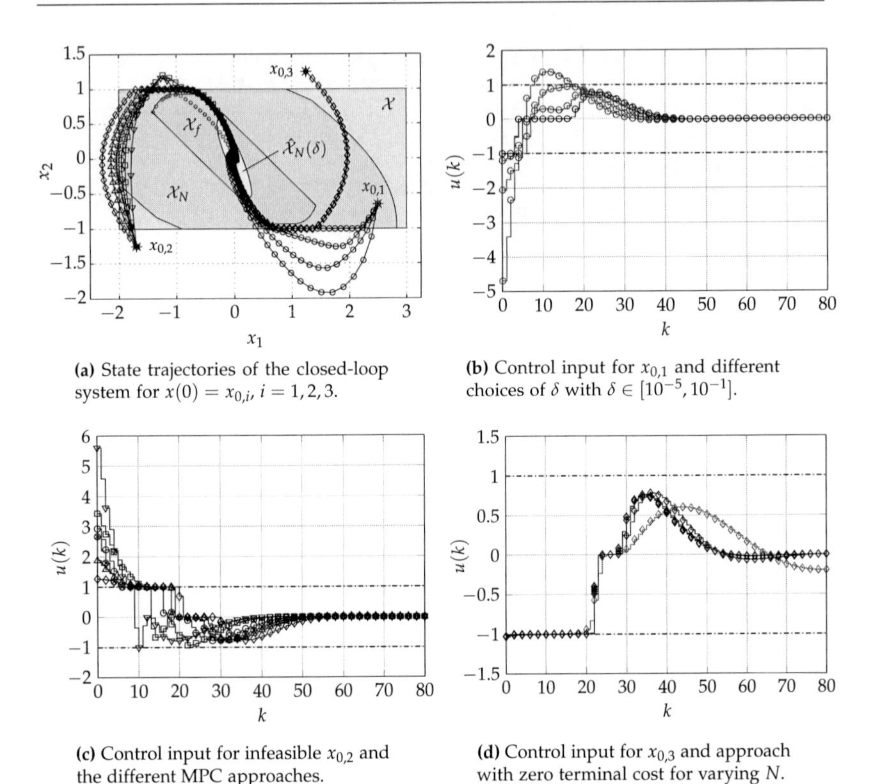

(a) State trajectories of the closed-loop system for $x(0) = x_{0,i}$, $i = 1, 2, 3$.

(b) Control input for $x_{0,1}$ and different choices of δ with $\delta \in [10^{-5}, 10^{-1}]$.

(c) Control input for infeasible $x_{0,2}$ and the different MPC approaches.

(d) Control input for $x_{0,3}$ and approach with zero terminal cost for varying N.

Figure 3.3. Results for the double integrator system when applying the different relaxed barrier function based MPC schemes. Depicted are the closed-loop state and input trajectories for the approaches based on a relaxed (⊛) and nonrelaxed (○) terminal set constraint, a purely quadratic terminal cost (□), a tail-sequence based terminal cost using a linear dead-beat controller (▽), a finite-tail terminal cost with a length of $T = N = 15$ (△), and a zero terminal cost with $N << \bar{N}$ (◇). Asymptotic convergence to the origin can be observed for all scenarios, while the resulting constraint satisfaction properties depend on the initial condition and the choice of the relaxation parameter δ. Depicted in (a) are also the state constraint set $\mathcal{X}$, the polytopic terminal set $\mathcal{X}_f$, and the resulting feasible set $\mathcal{X}_N$. Furthermore, $\hat{\mathcal{X}}_N(\delta)$ denotes for the given value of δ the set of initial conditions for which asymptotic stability and strict satisfaction of all input and state constraints can be guaranteed for the approach based on a relaxed terminal set constraint, cf. Theorem 3.1.

3.5 Satisfaction of input and state constraints

As outlined in the introduction of this thesis, one of the main advantages of model predictive control is given by its ability to handle constraints on the input and state variables in a constructive way. While we in the previous section tacitly assumed that the underlying input and state constraints may in the context of the proposed relaxed barrier function based MPC approaches in principle be violated, we are of course nevertheless interested in the constraint satisfaction properties of the resulting closed-loop system as well as in constructive procedures that allow us to guarantee different levels of approximate constraint satisfaction based on a suitable choice of the relaxation parameter. Concerning these aspects, we are going to show in this section that *i*) the maximal constraint violation that may occur in closed-loop operation can always be bounded and computed a priori and that *ii*) the underlying relaxed barrier functions can always be designed in such a way that they allow one to structurally guarantee (for a certain set of initial conditions) the satisfaction of *any* prescribed constraint violation tolerance – including the case of exact input and state constraint satisfaction, see Theorems 3.6 and 3.7 below.

3.5.1 Problem setup

In order to simplify the discussion, we in the following mainly focus on the relaxed barrier function based MPC formulation that makes use of a purely quadratic terminal cost as introduced in Section 3.4.2 above. In particular, we consider the MPC problem

$$\hat{J}_N^*(x) = \min_{U} \sum_{k=0}^{N-1} \hat{\ell}(x_k, u_k) + x_N^\top P x_N \tag{3.56a}$$

$$\text{s.t. } x_{k+1} = Ax_k + Bu_k, \; x_0 = x, \tag{3.56b}$$

where

$$\hat{\ell}(x,u) = \|x\|_Q^2 + \|u\|_R^2 + \varepsilon \hat{B}_\mathrm{x}(x) + \varepsilon \hat{B}_\mathrm{u}(u), \quad \varepsilon \in \mathbb{R}_{++}, \tag{3.57}$$

denotes the globally defined stage cost with $\hat{B}_\mathrm{x} : \mathbb{R}^n \to \mathbb{R}_+$ and $\hat{B}_\mathrm{u} : \mathbb{R}^m \to \mathbb{R}_+$ being recentered relaxed logarithmic barrier functions for the sets $\mathcal{X}$ and $\mathcal{U}$, respectively. As above, we assume that $Q \in \mathbb{S}^n_{++}$, $R \in \mathbb{S}^m_{++}$. Note that we make use of the input vector notation based on $U = [u_0^\top, \dots, u_{N-1}^\top]^\top \in \mathbb{R}^{Nm}$. Furthermore, following our results from Section 3.4.2, the terminal cost matrix $P \in \mathbb{S}^n_{++}$ is chosen based on

$$K = -\left(R + B^\top PB + \varepsilon M_\mathrm{u}\right)^{-1} B^\top PA \tag{3.58a}$$

$$P = (A+BK)^\top P(A+BK) + K^\top (R + \varepsilon M_\mathrm{u})K + Q + \varepsilon M_\mathrm{x}\,, \tag{3.58b}$$

where $M_\mathrm{x} = \frac{1}{2\delta^2} C_\mathrm{x}^\top \operatorname{diag}(w_\mathrm{x})\, C_\mathrm{x}$ and $M_\mathrm{u} = \frac{1}{2\delta^2} C_\mathrm{u}^\top \operatorname{diag}(w_\mathrm{u})\, C_\mathrm{u}$ are defined according to Lemma 3.4, with $\delta \in \mathbb{R}_{++}$ referring to the underlying relaxation parameter and $w_\mathrm{x} \in \mathbb{R}^{q_\mathrm{x}}_{++}$, $w_\mathrm{u} \in \mathbb{R}^{q_\mathrm{u}}_{++}$ to the respective recentering vectors. The resulting closed-loop system dynamics are then given by (3.19), which we repeat here for the sake of convenience:

$$x(k+1) = Ax(k) + B\hat{u}_0^*(x(k)). \tag{3.59}$$

From Theorem 3.3 in Section 3.4.2 we know that the origin of system (3.59) is globally asymptotically stable and that the associated value function satisfies for any $x(0) \in \mathbb{R}^n$

$$\hat{J}_N^*(x(k+1)) - \hat{J}_N^*(x(k)) \leq -\hat{\ell}(x(k), \hat{u}_0^*(x(k))) \;\; \forall\, k \in \mathbb{N}, \tag{3.60}$$

Note that beside (3.60) also all other properties that we are going to exploit in the following do in principle hold for any of the globally stabilizing MPC approaches presented in Section 3.4. Consequently, the following results on the constraint satisfaction properties of system (3.59) apply to these alternative design approaches as well. Our main reason for focusing on the above quadratic terminal cost approach is that, while resulting in a simple and intuitive MPC formulation with beneficial numerical properties, it makes the theoretical analysis more challenging (and thus more interesting) due to the fact that the underlying terminal cost depends via P (respectively $M_{\mathrm{x}}, M_{\mathrm{u}}$) directly on the choice of the relaxation parameter δ. However, where appropriate, we will point out relations and implications for the alternative MPC design approaches presented above, and we will briefly revisit and discuss all of them in the numerical example at the end of this section.

3.5.2 Main results

We begin our analysis with the following theorem, which provides a time-varying upper bound on the maximally possible closed-loop constraint violation that is, in addition, shown to be monotonically decreasing over time.

Theorem 3.6. *Consider the relaxed barrier function based problem setup based on (3.56)–(3.58) and let $\boldsymbol{x}_{cl} = \{x(0), x(1), \ldots\}$ and $\boldsymbol{u}_{cl} = \{u(0), u(1), \ldots\}$ with $u(k) = \hat{u}_0^*(x(k))$ denote the resulting state and input sequences associated to the closed-loop system (3.59). Then, for any initial condition $x_0 = x(0) \in \mathbb{R}^n$ and any $k \in \mathbb{N}$, it holds that*

$$C_{\mathrm{x}}\, x(k) \leq d_{\mathrm{x}} + \hat{z}_{\mathrm{x}}(k), \quad C_{\mathrm{u}}\, u(k) \leq d_{\mathrm{u}} + \hat{z}_{\mathrm{u}}(k), \tag{3.61}$$

where the elements of the maximal constraint violation vectors $\hat{z}_{\mathrm{x}}(k) \in \mathbb{R}^{q_{\mathrm{x}}}$ and $\hat{z}_{\mathrm{u}}(k) \in \mathbb{R}^{q_{\mathrm{u}}}$ decrease strictly monotonically over time. Furthermore, for each $k \in \mathbb{N}$, the respective elements can be computed explicitly by solving a convex optimization problem, and there exists a finite $k_0 \in \mathbb{N}$ such that $\hat{z}_{\mathrm{x}}(k) \leq 0$, $\hat{z}_{\mathrm{u}}(k) \leq 0$ for any $k \geq k_0$.

A proof of this result as well as explicit characterizations for the upper bounds are given in Appendix C.11. Theorem 3.6 implies that the maximally possible violation of the state and input constraints will for all time instants $k \in \mathbb{N}$ be upper bounded by $\hat{z}_{\mathrm{x}}(0)$ and $\hat{z}_{\mathrm{u}}(0)$, which can for any given initial state $x(0) \in \mathbb{R}^n$ be computed by solving the convex optimization problems stated in (C.39) for $\hat{\alpha}(0) := \hat{J}_N^*(x(0)) - x(0)^\top P_{\mathrm{uc}}^*\, x(0)$. Here, $P_{\mathrm{uc}}^* \in \mathbb{S}_{++}^n$ denotes solution to the associated infinite-horizon LQR problem, see the proof of Theorem 3.6. This shows that an upper bound on the maximal constraint violation in closed-loop operation can be computed a priori by solving a finite number of convex optimization problems. Furthermore, note that when making use of gradient recentered barrier functions, it is in fact possible to derive explicit formulas for the upper bounds $\hat{z}_{\mathrm{x}}(0)$ and $\hat{z}_{\mathrm{u}}(0)$, see Remark 3.13 below. In both cases, the obtained upper bounds for the maximal constraint violations may in principle also be negative, in which case the corresponding constraints will not be violated at all but rather satisfied with a respective safety margin. In fact, an important consequence of Theorem 3.6 is that there always exists a nonempty set of initial conditions for which the associated closed-loop trajectories will satisfy the underlying input and state constraints without any constraint violation.

Corollary 3.2. *Consider the relaxed barrier function based problem setup based on (3.56)–(3.58) and let $\boldsymbol{x}_{cl} = \{x(0), x(1), \ldots\}$ and $\boldsymbol{u}_{cl} = \{u(0), u(1), \ldots\}$ with $u(k) = \hat{u}_0^*(x(k))$ denote the resulting state and input sequences associated to the closed-loop system (3.59). Furthermore, for given $\delta \in \mathbb{R}_{++}$ let the set $\hat{\mathcal{X}}_N(\delta)$ be defined as*

$$\hat{\mathcal{X}}_N(\delta) := \left\{ x \in \mathcal{X} \mid \hat{J}_N^*(x) - x^\top P_{\mathrm{uc}}^* x \leq \varepsilon \bar{\beta}(\delta) \right\}, \tag{3.62}$$

where $\bar{\beta}(\delta) := \min\{\bar{\beta}_{\mathrm{x}}(\delta), \bar{\beta}_{\mathrm{u}}(\delta)\}$, with $\bar{\beta}_{\mathrm{x}}(\delta)$ and $\bar{\beta}_{\mathrm{u}}(\delta)$ being defined according to (3.27) in Definition 3.4. Then, for any initial condition $x(0) \in \hat{\mathcal{X}}_N(\delta)$ and any $k \in \mathbb{N}$ it holds that

$$C_{\mathrm{x}}\, x(k) \leq d_{\mathrm{x}}, \qquad C_{\mathrm{u}}\, u(k) \leq d_{\mathrm{u}}. \tag{3.63}$$

This result follows directly from the fact that $x(0) \in \hat{\mathcal{X}}_N(\delta)$ implies together with (C.36) that $\hat{B}_{\mathrm{x}}(x(k)) \leq \bar{\beta}_{\mathrm{x}}(\delta)$ and $\hat{B}_{\mathrm{u}}(u(k)) \leq \bar{\beta}_{\mathrm{u}}(\delta)$ for any $k \in \mathbb{N}$. Applying Lemma 3.1 then immediately reveals that $x(k) \in \mathcal{X}$ and $u(k) \in \mathcal{U}$, by which (3.63) holds. A more detailed explanation is also given in the proof of Theorem 3.7 below.
One may now ask how the size of the set $\hat{\mathcal{X}}_N(\delta)$ changes depending on the choice of the relaxation parameter δ and whether and how exact or approximate constraint satisfaction may be guaranteed for a *given* set of initial conditions. An answer to this question is provided by the following theorem, which essentially reveals that the input and state constraints can be satisfied with any desired tolerance – including the case of exact constraint satisfaction – if the relaxation parameter δ is chosen small enough and the initial state lies in the set $\mathcal{X}_N$ as defined in (3.64). A proof of Theorem 3.7 is given in Appendix C.12.

Theorem 3.7. *Consider the relaxed barrier function based problem setup based on (3.56)–(3.58) and let $\boldsymbol{x}_{cl} = \{x(0), x(1), \ldots\}$ and $\boldsymbol{u}_{cl} = \{u(0), u(1), \ldots\}$ with $u(k) = \hat{u}_0^*(x(k))$ denote the resulting state and input sequences associated to the closed-loop system (3.59). Furthermore, assume that $\begin{bmatrix} A^{N-1}B & \cdots & AB & B \end{bmatrix}$ has full row rank n and let the set $\mathcal{X}_N \subseteq \mathcal{X}$ be defined as*

$$\mathcal{X}_N := \left\{ x \in \mathcal{X} \mid \exists U \in \mathbb{R}^{Nm} \text{ s.t. } u_k \in \mathcal{U}, x_k(U, x) \in \mathcal{X}, x_N(U, x) = 0 \right\}, \tag{3.64}$$

where $x_k(U, x)$ is for $k = 1, \ldots, N$ given according to (3.56b). Then, for any compact set $\mathcal{X}_0 \subseteq \mathcal{X}_N^\circ$ and any $\hat{z}_{\mathrm{x,tol}} \in \mathbb{R}_+^{q_x}$, $\hat{z}_{\mathrm{u,tol}} \in \mathbb{R}_+^{q_u}$, there exists $\bar{\delta}_0 \in \mathbb{R}_{++}$ such that for any relaxation parameter $0 < \delta \leq \bar{\delta}_0$, any initial condition $x(0) \in \mathcal{X}_0$, and any $k \in \mathbb{N}$, it holds that

$$C_{\mathrm{x}}\, x(k) \leq d_{\mathrm{x}} + \hat{z}_{\mathrm{x,tol}}, \quad C_{\mathrm{u}}\, u(k) \leq d_{\mathrm{u}} + \hat{z}_{\mathrm{u,tol}}. \tag{3.65}$$

3.5.3 Discussion

Thus, summarizing, we can conclude that when applying the relaxed logarithmic barrier function based MPC approaches presented in the previous section, we can despite the underlying "constraint softening" still give rigorous guarantees on the constraint satisfaction properties of the resulting closed-loop system. In particular, Theorem 3.6 and Corollary 3.2 reveal that an upper bound for the maximally possible constraint violation can always be computed a priori and that exact satisfaction of all input and state constraints can for any choice of the relaxation parameter be guaranteed for a certain set of initial conditions. Furthermore, Theorem 3.7 showed in addition that any desired level of constraint satisfaction may in principle be achieved for a *given* set of initial conditions by choosing the relaxation parameter sufficiently small. Based on the above results, we may in fact formulate an iterative algorithm that gradually decreases the relaxation parameter δ and computes for each realization of δ the resulting guaranteed maximal constraint violation. This procedure can be performed offline and returns at termination a suitable choice for the relaxation parameter which guarantees that the resulting constraint violation lies below a desired tolerance. For the sake of completeness, a possible realization of such an algorithm is given in Appendix B.8 of this thesis.

While the aforementioned results have in this section be shown for the case of a purely quadratic terminal cost, we want to emphasize again that they in general also hold for the other globally stabilizing MPC approaches presented in Section 3.4, see also Remark 3.14 below. However, it should be noted that the constraint violation bounds provided by Theorem 3.6 as well as the restriction of x_0 to the set $\mathcal{X}_N^{\circ}$ that is required by Theorem 3.7 are more of a conceptual nature and tend to be rather conservative when compared to the actual closed-loop behavior, see in particular also the numerical example in the following section. Thus, while the presented results are constructive in the sense that they allow us to compute an upper bound for the largest possible constraint violation in a constructive way, they may also be conservative, and more appropriate choices or selection regimes for the relaxation parameter δ may in the context of practical applications be found based on numerical simulations. The facts that the maximal possible constraint violation is monotonically decaying and that any desired level of constraint satisfaction can in principle be achieved by a suitable choice of the underlying problem parameters nevertheless reveal and constitute important and favorable properties, which, in particular, also provide a theoretical justification for typical constraint satisfaction behavior that is usually observed in numerical simulations as well as in practical applications.

Remark 3.12. A quite common scenario in practical MPC applications is the existence of hard input constraints which may be impossible to violate due to physical reasons (e.g., actuator saturation). In order to prevent the controller from steering the plant to a configuration from where it can not recover, the underlying input constraints may in such a situation be prioritized and enforced by making use of different relaxation parameters for the input and state constraints and choosing $\delta_{\mathrm{u}} \ll \delta_{\mathrm{x}}$. Further approaches might be to adapt the relaxation parameter δ_{u} online or to a priori tighten the input constraints with a suitable safety margin based on the constraint violation bounds presented above. In this light, the consideration of hard input constraints and control input saturations leads to many interesting research questions and opens up various directions for possible future work that are, however, beyond the scope of the research presented in this thesis.

Remark 3.13. As outlined above, the presented constraint satisfaction results apply as well when making use of gradient recentered relaxed logarithmic barrier functions instead of the weighting-based formulation on which we focus throughout this thesis. In fact, concerning the maximal constraint violations specified in Theorem 3.6, it is in this case possible to exploit the fact that the recentering is performed for each of the linear constraints separately. As a consequence, $\hat{B}_{\mathrm{x}}(\cdot)$ and $\hat{B}_{\mathrm{u}}(\cdot)$ in (C.36) consist of separable positive definite terms, which allows one to derive explicit expressions for upper bounds on the maximal constraint violations given in (C.39). In particular,

$$\hat{z}^i_{\mathrm{x}}(k) \leq \hat{z}^i_{\mathrm{x}}(0) \leq \max\left\{-\delta\left(p_{\mathrm{x},i} - \sqrt{p^2_{\mathrm{x},i} - r_{\mathrm{x},i}}\right), 0\right\}, \quad i = 1, \ldots, q_{\mathrm{x}}, \tag{3.66a}$$

$$\hat{z}^j_{\mathrm{u}}(k) \leq \hat{z}^j_{\mathrm{u}}(0) \leq \max\left\{-\delta\left(p_{\mathrm{u},j} - \sqrt{p^2_{\mathrm{u},j} - r_{\mathrm{u},j}}\right), 0\right\}, \quad j = 1, \ldots, q_{\mathrm{u}}, \tag{3.66b}$$

where $p_{\mathrm{x},i} = 2 - \delta/d^i_{\mathrm{x}}$, $p_{\mathrm{u},j} = 2 - \delta/d^j_{\mathrm{u}}$, $r_{\mathrm{x},i} = 1 + 2\ln(d^i_{\mathrm{x}}/\delta) - \frac{2}{\varepsilon}\hat{\alpha}(x(0);\delta)$, and $r_{\mathrm{u},j} = 1 + 2\ln(d^j_{\mathrm{u}}/\delta) - \frac{2}{\varepsilon}\hat{\alpha}(x(0);\delta)$ with $\hat{\alpha}(\cdot;\delta)$ defined according to (C.43). For more details, see Theorem 14 and Corollary 17 in (Feller and Ebenbauer, 2014a).

Remark 3.14. By virtue of the cost function decrease (3.60), the results of Theorem 3.6 and Corollary 3.2 directly apply also to all the other globally stabilizing MPC approaches presented in Section 3.4. Furthermore, Theorem 3.7 holds as well if the set $\mathcal{X}_N$ on which it is possible to guarantee exact constraint satisfaction is modified to

$$\mathcal{X}_N = \{x \in \mathcal{X} \mid \exists U \in \mathbb{R}^{Nm} \text{ s.t. } u_k \in \mathcal{U}, x_k(U,x) \in \mathcal{X}, x_N(U,x) \in \mathcal{X}_f\} \tag{3.67}$$

for the approach based on a nonrelaxed terminal set constraint (Theorem 3.2) and to

$$\mathcal{X}_N = \{x \in \mathcal{X} \mid \exists U \in \mathbb{R}^{Nm} \text{ s.t. } u_k \in \mathcal{U}, x_k(U,x) \in \mathcal{X}, z_l(x_N(U,x)) \in \mathcal{X}, v_l(x_N(U,x)) \in \mathcal{U}\} \tag{3.68}$$

for the approach based on auxiliary tail-sequences (Theorem 3.4), where in both cases $k = 0, \ldots, N-1$. For the approach based on a zero terminal cost (Theorem 3.5), the problem occurs that the required prediction horizon depends on the choice of the relaxation parameter. However, the conditions required in the proof of Theorem 3.7 (in particular boundedness of the cost function for decreasing δ) can always be ensured on the associated *infinite-horizon* admissible set. Consequently, Theorem 3.7 can be shown to hold with

$$\mathcal{X}_N = \Omega_\infty = \{x \in \mathcal{X} \mid \exists \boldsymbol{u} = \{u_1, u_2, \ldots\} \text{ s.t. } u_k \in \mathcal{U}, x_k(\boldsymbol{u}, x) \in \mathcal{X}\ \forall k \in \mathbb{N}\}. \tag{3.69}$$

Based on similar arguments as above, it is then possible to show in all three cases again the existence of a sufficiently small relaxation parameter $\bar{\delta}_0$ (the rank condition of Theorem 3.7 is not required). While the exact derivations shall not be discussed here for the sake of brevity, more details can for example be found in (Feller and Ebenbauer, 2017a).

3.5.4 Numerical example

In the following, we briefly discuss and illustrate some of our findings by means of a numerical example. In particular, we consider again the double integrator example that has already been discussed in the context of closed-loop stability in Section 3.4. All problem parameters as well as the design of the underlying relaxed barrier function based MPC schemes are exactly the same as in Section 3.4.4 and are thus not repeated here for the sake of convenience. The following two scenarios are considered.

Scenario 1

In a first step, we compare the different MPC approaches proposed in Section 3.4 with respect to the characteristic sets that were identified and discussed in the above constraint satisfaction analysis. Beside the set $\mathcal{X}_N$ for which any desired level of constraint satisfaction can in principle be achieved by making the relaxation parameter δ arbitrarily small, this includes the sets $\hat{\mathcal{X}}_N(\delta)$ and $\mathcal{X}_0 \subset \mathcal{X}_N$ for which exact or approximate constraint satisfaction can be guaranteed for a *given* value of δ, see Corollary 3.2, respectively Theorem 3.7, and Remark 3.14. For the considered problem setup from Section 3.4.4 and a horizon of $N = 15$, the resulting sets are depicted in Figure 3.4. Therein, the polytopic feasible sets $\mathcal{X}_N^i$ were constructed by making use of the Multi-Parametric Toolbox (Kvasnica et al., 2004), while the sets $\hat{\mathcal{X}}_N^i(\delta)$ and $\mathcal{X}_0^i$ in the plot on the right-hand side were obtained by evaluating both the value function and the presented maximal constraint violation bound from Theorem 3.6 over a grid in the state space and then plotting the respective level sets with MATLAB's `contour` function. Note that whereas the sets with guaranteed *exact* constraint satisfaction are quite small even for $\delta = 10^{-6}$, it can be observed that the sets for which approximate constraint satisfaction can be guaranteed with a tolerance of $\hat{z}_{\text{tol}} = 10^{-3}$ essentially recover the sets $\mathcal{X}_N^i$ that are predicted by the theoretical results presented above. Note that no meaningful constraint satisfaction guarantees can be given for the approach based on a zero terminal cost as the required (conservative) critical horizon is in this case prohibitively large, see also Section 3.4.4. Due to this reason, the characteristic constraint satisfaction sets $\hat{\mathcal{X}}_N^4(\delta)$ and $\mathcal{X}_0^4$ associated to the zero terminal cost approach can not be plotted for the considered example.

Scenario 2

As a second scenario, we perform simulations of the resulting closed-loop behavior for the same initial conditions as in the numerical example presented in Section 3.4.4, however now with a focus on the approach with purely quadratic terminal cost. The only modification to the problem setup is that in the following discussion we consider an enlarged prediction horizon of $N = 30$. The associated simulation results are presented in Figure 3.5. As can be seen from subplots (a) and (b), the enlarged prediction horizon leads for the considered approach with quadratic terminal cost to a significantly improved constraint satisfaction behavior. Furthermore, the level of observed constraint violation

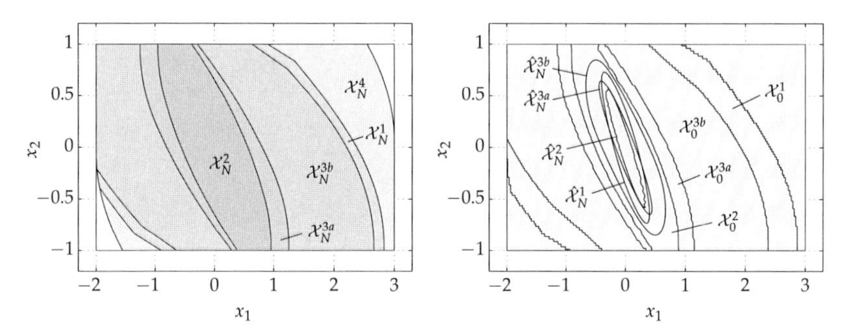

Figure 3.4. *Left:* Sets of initial conditions for which any nonnegative constraint satisfaction tolerance can be satisfied for a sufficiently small choice of the relaxation parameter. The numbering of the sets corresponds to the ordering of the different MPC approaches as listed in Table 3.1 on page 56. That is, $\mathcal{X}_N^1$ relates to the approach based on a nonrelaxed terminal set constraint, $\mathcal{X}_N^2$ to that based on a quadratic terminal cost, $\mathcal{X}_N^3$ to that based on a suitably chosen tail-sequence ("a" for dead-beat and "b" for LQR with zero terminal state constraint), and $\mathcal{X}_N^4$ to the approach based on a zero terminal cost.
Right: The sets $\hat{\mathcal{X}}_N^i = \hat{\mathcal{X}}_N^i(\delta)$ for which *exact* constraint satisfaction is ensured for $\delta = 10^{-6}$, as well as the associated sets $\mathcal{X}_0^i(\delta)$ for which approximate satisfaction of input and state constraints can be guaranteed with a tolerance of $\hat{z}_{\text{tol}} = 10^{-3}$. The same numbering scheme as on the left-hand side is used. The sets $\hat{\mathcal{X}}_N^4$ and $\mathcal{X}_0^4$ for the approach with zero terminal cost cannot be plotted since the required critical horizon is prohibitively large.

depends as predicted heavily on the choice of the relaxation parameter. In particular, subplot (c) reveals that for $x(0) = x_{0,1}$, the "critical" state constraint $x_2 \geq -1$ is only slightly violated for $\delta = 10^{-3}$ and exactly satisfied for $\delta = 10^{-5}$. Also plotted are the associated theoretical upper bounds $\hat{z}_x^4(k)$, which, as stated by Theorem 3.6, can be seen to upper bound the actual constraint violation $z^4(k) = -x_2(k) - 1$ and which decay in fact monotonically over time. Note also that the computed upper bounds are actually not that conservative when the system state is close to the constraint. In addition, subplot (d) shows, for a suitably chosen sequence of sets $\mathcal{X}_0(\kappa) \to \mathcal{X}_N$, the evolution of the parameter $\bar{\delta}_0$ that is according to Theorem 3.7 sufficient for ensuring exact or approximate constraint satisfaction. The depicted values for $\bar{\delta}_0$ were for each set of initial conditions computed based on the algorithm given in Appendix B.8. Although the results are quite conservative, it can be observed that constraint satisfaction with a tolerance of 10^{-3} can be guaranteed for a large range of initial conditions when choosing $\delta = 10^{-5}$.

In total, we are therefore able to observe and reproduce many aspects of the above theoretical analysis also in our numerical simulations, including in particular the direct impact of the relaxation parameter on the resulting constraint satisfaction properties.

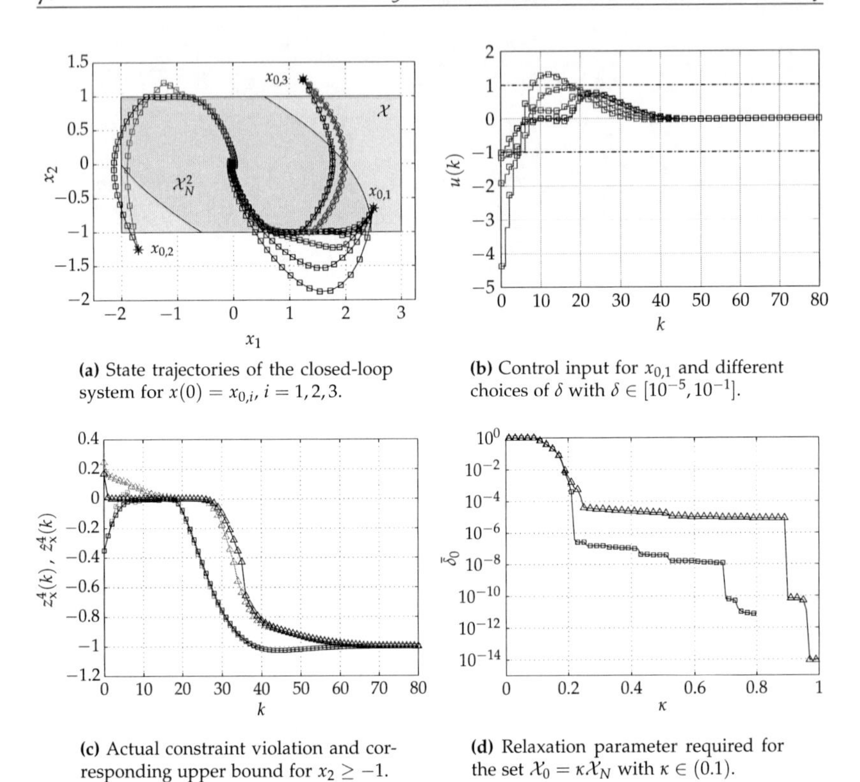

(a) State trajectories of the closed-loop system for $x(0) = x_{0,i}$, $i = 1,2,3$.

(b) Control input for $x_{0,1}$ and different choices of δ with $\delta \in [10^{-5}, 10^{-1}]$.

(c) Actual constraint violation and corresponding upper bound for $x_2 \geq -1$.

(d) Relaxation parameter required for the set $\mathcal{X}_0 = \kappa\mathcal{X}_N$ with $\kappa \in (0.1)$.

Figure 3.5. Subplots (a) and (b): results for the double integrator system when applying the relaxed barrier function based MPC scheme with quadratic terminal cost and horizon $N = 30$. For $x_{0,1}$, the constraint satisfaction properties depend again on the choice of the relaxation parameter. For $x_{0,2}$, the resulting closed-loop behavior is much better than for the case with $N = 15$ (depicted in gray, cf. Section 3.4.4). Also depicted in subplot (a) are for $x_{0,3}$ the closed-loop state trajectories for the approaches based on a nonrelaxed ($\circ$) terminal set constraint, a tail-sequence based terminal cost using a linear dead-beat controller ($\triangledown$), a finite-tail terminal cost with $T = N$ ($\triangle$), and a zero terminal cost ($\diamond$). Subplot (c) compares the actual constraint violation $z_x^4(k)$ ($\square$) for $\delta = 10^{-5}$ with the theoretical upper bound $\hat{z}_x^4(k)$ ($\triangle$) that is provided by Theorem 3.6. Depicted in gray is the resulting behavior for $\delta = 10^{-3}$. Subplot (d) shows the relaxation parameter $\bar{\delta}_0$ that is required to guarantee exact constraint satisfaction ($\square$) and approximate constraint satisfaction with a tolerance of $\hat{z}_{\text{tol}} = 10^{-3}$ ($\triangle$) for initial conditions in $\mathcal{X}_0 = \kappa\mathcal{X}_N$ with $\kappa \in (0,1)$.

3.6 Robustness properties and output feedback

In the previous sections of this chapter, we saw how the concept of relaxed barrier function based MPC leads via various approaches to the characterization of globally defined and asymptotically stabilizing control laws that, in particular, also allow us to ensure important constraint satisfaction properties of the resulting closed-loop system. In the following, we want to turn our attention to yet another central question that we already formulated at the beginning of this chapter. In particular, we want to investigate whether we can say anything about the *robustness* properties of the resulting closed-loop system that goes beyond the fact that a well-defined control input will exist for any realization of the system state. Concerning this point, the first main result of this section is to show that for any of the presented globally stabilizing MPC approaches, the closed-loop system will in fact be *input-to-state stable* with respect to arbitrary additive disturbances. In a second step, we then demonstrate that these favorable robustness properties will be preserved when combining the respective relaxed barrier function based MPC schemes in a certainty equivalence output feedback fashion with suitable state estimation procedures.

The results presented in this section are mainly based on Feller and Ebenbauer (2015b) and Feller et al. (2016). The required stability definitions are given in Appendix A.

3.6.1 Problem setup

In contrast to the nominal stability analysis presented above, we in the following consider disturbance-affected linear discrete-time systems of the form

$$x(k+1) = Ax(k) + Bu(k) + w(k)\,, \tag{3.70}$$

where $x(k) \in \mathbb{R}^n$ and $u(k) \in \mathbb{R}^m$ refer again to the respective state and input vectors, while $w(k) \in \mathbb{R}^n$ denotes an unknown additive disturbance that is acting on the plant. Note that, as it might in principle depend both on the system state and the system input, $w(k)$ allows to model a wide class of internal or external disturbances. Furthermore, we do not assume $w(k)$ to lie in an a priori known set, as it is often done in conventional robust MPC approaches, see also the discussion at the end of this section.

As in Section 3.5, we want to focus in the following analysis on a relaxed barrier function based MPC formulation with a purely quadratic terminal cost. In particular, we consider the barrier function based open-loop optimal control problem

$$\hat{J}_N^*(x) = \min_{U} \sum_{k=0}^{N-1} \hat{\ell}(x_k, u_k) + x_N^\top P x_N \tag{3.71a}$$

$$\text{s.t.}\;\; x_{k+1} = Ax_k + Bu_k,\; x_0 = x\,, \tag{3.71b}$$

where $\hat{\ell} : \mathbb{R}^n \times \mathbb{R}^m \to \mathbb{R}_+$ and $P \in \mathbb{S}^n_{++}$ are chosen according to (3.57) and (3.58). We assume that all the conditions from the corresponding stability theorem Theorem 3.3 are satisfied. The resulting MPC feedback is given by $u(k) = \hat{u}_0^*(x(k))$, which directly leads to the following perturbed closed-loop system dynamics

$$x(k+1) = Ax(k) + B\hat{u}_0^*(x(k)) + w(k)\,. \tag{3.72}$$

Based on the nominal stability results presented in Section 3.4, in particular Theorem 3.3, we know that whenever the disturbance $w(k)$ is identical to zero for all $k \in \mathbb{N}$, the origin of system (3.72) will be globally asymptotically stable and

$$\hat{J}_N^*(x(k+1)) - \hat{J}_N^*(x(k)) \leq -\hat{\ell}(x(k), \hat{u}_0^*(x(k))) \;\; \forall\, k \in \mathbb{N}\,. \tag{3.73}$$

Together with inherent continuity and convexity properties of the value function, these nominal stability results will in the following allow us to derive also important systems theoretic guarantees on the behavior of the perturbed closed-loop system (3.72). While the subsequent analysis focuses on the case of a purely quadratic terminal cost, the required assumptions are in principle satisfied by any of the globally stabilizing MPC approaches presented in Section 3.4 of this thesis. Consequently, the following robustness results apply to these alternative design approaches as well.

3.6.2 Input-to-state stability of the closed-loop system

In the following, we present our main results concerning the robustness properties of the closed-loop system (3.72) with respect to (in principle arbitrary) additive disturbances. The underlying stability analysis relies on the well-known concept of *input-to-state stability* (Sontag, 1989; Jiang and Wang, 2001), for which a rigorous definition as well as a brief introductory summary can be found in Appendix A.2 of this thesis.

Main results

Essentially, the perturbed closed-loop system (3.72) is called input-to-state stable (ISS) with respect to the disturbance sequence $w = \{w(0), w(1), \ldots\}$ if there exist functions $\beta \in \mathcal{KL}$ and $\gamma \in \mathcal{K}$ such that the associated system state satisfies for any $k \in \mathbb{N}_+$

$$\|x(k)\| \leq \beta(\|x(0)\|, k) + \gamma(\|w_{[k-1]}\|)\,. \tag{3.74}$$

Here, $\|w_{[k-1]}\|$ denotes the truncated signal norm of the disturbance sequence, see Appendix A.2. A direct consequence of (3.74) is that there exists an explicit gain between the norm of the respective disturbances and the norm of the resulting closed-loop system state, implying that the system state will converge to the origin for any convergent disturbance sequence. In order to show that this property does indeed hold for system (3.72), we make use of the following auxiliary result.

Lemma 3.6. *Consider the relaxed barrier function based MPC problem (3.71) with the underlying stage and terminal cost function terms chosen according to (3.57) and (3.58). Then, the associated value function $\hat{J}_N^* : \mathbb{R}^n \to \mathbb{R}_+$ is a twice continuously differentiable and convex function. Furthermore, there exist scalars $a \in \mathbb{R}_{++}$ and $b \in \mathbb{R}_{++}$ such that*

$$a\|x\|^2 \leq \hat{J}_N^*(x) \leq b\|x\|^2 \quad \forall\, x \in \mathbb{R}^n\,. \tag{3.75}$$

A proof of this result can be found in Appendix C.13. Note that a direct consequence of Lemma 3.6 is that the origin of the nominal system is in fact globally *exponentially* stable, see Appendix A.1, which is a stricter result that the global asymptotic stability shown in Theorem 3.3. Furthermore, based on Lemma 3.6, we can state the following main result, for which a proof is given in Appendix C.14.

Theorem 3.8. *Consider the relaxed barrier function based MPC problem (3.71) with the underlying stage and terminal cost function terms chosen according to (3.57) and (3.58), and let* $u(k) = \hat{u}_0^*(x(k))$ *be the corresponding feedback law. Then, the resulting perturbed closed-loop system (3.72) is ISS with respect to the disturbance sequence* $\mathbf{w}$*. Furthermore, the associated value function* $\hat{J}_N^* : \mathbb{R}^n \to \mathbb{R}_+$ *is a twice continuously differentiable ISS Lyapunov function.*

Discussion

Theorem 3.8 reveals that the proposed relaxed barrier function based MPC scheme is not only robust against uncertainties and disturbances in the sense that the corresponding open-loop optimal control problem admits a well-defined solution for any possible realization of the system state, but that it also ensures important system theoretic guarantees concerning the robust stability of the closed-loop system. In particular, we were able to prove global exponential stability in the nominal case as well as input-to-state stability of the resulting closed-loop system with respect to arbitrary additive disturbances. Especially the latter result relies to a large part on the applied constraint relaxation and is typically not possible in the context of conventional (robust) MPC approaches.
Compared for example to tube-based robust MPC approaches (Chisci et al., 2001; Mayne and Langson, 2001; Mayne et al., 2005), no a priori known bounds for the disturbance need to be assumed and it is neither necessary to compute robust positive invariant sets nor to perform a possibly conservative tightening of the input and state constraints. Furthermore, compared to works that study the robustness of (linear and nonlinear) MPC approaches based on *regional* ISS concepts (Scokaert et al., 1997; Limón et al., 2002; Magni et al., 2006; Goulart and Kerrigan, 2008; Limón et al., 2009; Lazar et al., 2009; Zeilinger et al., 2010, 2014a), the presented ISS property is shown to hold in the usual sense, i.e., for arbitrary large realizations of the disturbance and for any initial condition. However, it should again be emphasized that no saturation or truncation of input or state variables has been considered, see also Remark 3.16 below.

In the light of the above discussion, the presented ISS results are particularly interesting for scenarios in which nothing is known about the disturbances $w(0), w(1), \dots$ except that they remain bounded and possibly decay to zero over time. This might for example be the case for asymptotically converging observer errors, i.e. $w(k) = B(\hat{u}_0^*(\hat{x}(k)) - \hat{u}_0^*(x(k)))$, an input and state-dependent plant-model mismatch due to linearization, i.e. $w(k) = f(x(k), u(k))$ for some nonlinear $f : \mathbb{R}^n \times \mathbb{R}^n \to \mathbb{R}^n$, or when applying iterative optimization algorithms that result in slightly suboptimal solutions, i.e. $w(k) = B(u(k) - \hat{u}_0^*(x(k)))$. All three scenarios as well as their combination are very common in the context of real-

world applications, which illustrates the practical relevance of the robust stability properties discussed above. We will revisit and discuss all of the aforementioned disturbance classes at several places throughout the remainder of this thesis – in particular in the following section on robust output feedback MPC as well in the Chapters 4 and 5.
Furthermore, the presented results allow us to link the framework of relaxed barrier function based MPC with important ISS-related concepts, for example the small-gain theorem, which may provide powerful tools for the stability analysis of cascaded systems or feedback interconnections. Thus, it can be expected that the presented ISS properties will be of advantage when analyzing interconnections of the proposed MPC schemes with other components that may for themselves be uncertain or subject to internal dynamics. An interesting example, which will also be the topic of the next section, is the combination with suitable state estimation procedures in a certainty equivalence output feedback fashion.

Note that the major difficulty in proving the discussed global ISS properties is given by the fact that the system state is not necessarily guaranteed to remain in a compact robust positively invariant set, e.g., a suitably chosen sublevel set of the associated value function. A consequence of this is that we can in general not assume a bounded Lipschitz gain for the value function, as it has for example been done in the works by Scokaert et al. (1997), Limón et al. (2002), and Zeilinger et al. (2010, 2014a). Instead, the *convexity* of the value function needs to be exploited in order to derive the above global linear growth condition for the gradient. In fact, an interesting side results of the proof of Theorem 3.8 is given by the following more general Lemma.

Lemma 3.7. *Given an arbitrary continuous function $f : \mathbb{R}^n \to \mathbb{R}$ with $f(0) = 0$, consider for $k \in \mathbb{N}$ the associated time-invariant discrete-time dynamical system*

$$x(k+1) = f(x(k)) + w(k)\,, \tag{3.76}$$

where $x(k) \in \mathbb{R}^n$ refers to the system state and $w(k) \in \mathbb{R}^m$ denotes an exogenous input. Assume that there exists a continuously differentiable and convex function $V : \mathbb{R}^n \to \mathbb{R}_+$ such that

$$a\|x\|^2 \leq V(x) \leq b\|x\|^2\,, \tag{3.77a}$$

$$V(f(x)) - V(x) \leq -c\|x\|^2 \tag{3.77b}$$

holds for any $x \in \mathbb{R}^n$ and suitably chosen scalars a, b, $c \in \mathbb{R}_{++}$. Then, system (3.76) is input-to-state stable with respect to the input sequence $w = \{w(0), w(1), \ldots\}$ and $V : \mathbb{R}^n \to \mathbb{R}_+$ is a suitable ISS Lyapunov function.

Thus, global exponential stability with a *convex* and continuously differentiable Lyapunov function always implies ISS with respect to additive disturbances. Although there are some regional ISS results that also exploit convexity properties of the value function, see for example Goulart and Kerrigan (2008), we believe Lemma 3.7 to be an interesting result that, in particular, appears to have not been discussed so far in the literature.

Remark 3.15. As the required properties of the value function also hold for all the other globally stabilizing approaches presented in Section 3.4, Lemma 3.6 and Theorem 3.8 apply to these alternative approaches as well. In particular, note that an explicit quadratic upper bound for the value function may in the case of a nonrelaxed logarithmic terminal set barrier function (see Theorem 3.2) be derived by making use of a suboptimal input sequence that guarantees $x_N(\bar{U}(x), x) = 0$ for any $x \in \mathbb{R}^n$. Under the assumption that $N \geq n$ such a sequence always exist and can be written as a sequence of linear state feedback terms, see Appendix B.6. Together with the quadratic upper bounds provided by Lemma 3.4, it is then possible to derive a similar quadratic upper bound as above.

Remark 3.16. Note that the above robust stability results rely inherently on the assumption that both the control input and the state vector can in principle take arbitrary values, i.e., that there are no saturations or truncations. Although this might not always be feasible in practical applications (as for example in the presence of physical limitations on the control input), we nevertheless believe the presented global ISS results to be strong indicators for possible advantages of the relaxation based approach as well as for the robustness of the corresponding closed-loop system. Furthermore, based on regional ISS concepts and using similar arguments as Limón et al. (2002), it is in principle possible to derive comparable *local* robust stability results that then also allow one to ensure input and state constraint satisfaction for certain classes of (sufficiently small) additive disturbances.

Remark 3.17. Note that the above ISS results do not necessarily imply robustness against arbitrary uncertainties or state-dependent disturbances. This can for example be seen by considering

$$w(k) = (\epsilon I - A)x(k) - B\hat{u}^*(x(k)), \tag{3.78}$$

which obviously destabilizes the closed-loop system (3.74) for any $\epsilon > 1$. However, as in this case also $\|w(k)\| \to \infty$, this is not a contradiction to the ISS property (3.74).

Remark 3.18. Note that it might be possible to prove the above ISS result also based on some form of converse Lyapunov function argument, cf. (Scokaert et al., 1997). However, by choosing the value function itself as Lyapunov function candidate, it is in principle possible to compute the (possibly conservative) ISS gain $\gamma \in \mathcal{K}$ explicitly (Jiang and Wang, 2001). Furthermore, positively invariant sublevel sets of the value function may in this case be used to guarantee the satisfaction of input and state constraints in scenarios with sufficiently small additive disturbance, see Remark 3.16 above.

3.6.3 Robust certainty equivalence output feedback

Up to now, we assumed in all our discussions that the full and exact system state is available via measurement, allowing us to apply at each sampling instant the exact optimal control input. However, in practical applications it is oftentimes only possible to measure a part of the system state directly, while the remaining state information needs to be reconstructed by means of a state estimator. As pointed out by Mayne (2014), one of the main obstacles in designing MPC approaches that guarantee stability properties of the resulting closed-loop system when making use of potentially inexact state estimates, is given by the fact that the controlled system needs to be robust against the typically unknown estimation error. In particular, while the problem of partial state information is in many practical applications handled by employing a certainty equivalence output feedback approach, i.e., by directly replacing the unknown system state with the current state estimate, there exists (in contrast to the linear state feedback case) no universally valid separation principle that allows to ensure stability of the overall closed-loop dynamics by simply combining a stabilizing MPC scheme with a stable state estimator. Naturally, the situation gets even worse if in addition external disturbances are acting on the states and/or outputs of the system.

In this section, we show that in the context of the proposed relaxed barrier function based MPC framework, the presented robustness properties will actually be preserved when combining the respective control schemes in a certainty equivalence output feedback fashion with suitable state estimation procedures. More precisely, we prove that the resulting overall closed-loop system – consisting now of both the state and the state estimation error dynamics – will be robustly stable in the ISS sense under rather mild assumptions on the underlying state estimation, including in particular the important cases of Luenberger observers and Kalman filtering. As the state estimation and controller parts can be designed independently from each other, the main conceptual result is a separation principle for the discussed class of relaxed barrier function based MPC schemes.

Problem setup

We slightly modify our problem setup from above by appending the disturbance affected dynamical system (3.70) with a suitable output equation, leading to

$$x(k+1) = Ax(k) + Bu(k) + w(k)\,, \tag{3.79a}$$
$$y(k) = Cx(k) + v(k)\,. \tag{3.79b}$$

Here, $x(k) \in \mathbb{R}^n$ refers to the system state, $u(k) \in \mathbb{R}^m$ to the system input, and $y(k) \in \mathbb{R}^p$ to the measured system output, all at time instant $k \in \mathbb{N}$. Moreover, $w(k) \in \mathbb{R}^n$ and $v(k) \in \mathbb{R}^p$ model the effect of unknown state and output disturbances. We assume in the following that the pair (A, C) is detectable.

As outlined above, the proposed output feedback MPC scheme is based on combining the concept of relaxed barrier function based MPC in a certainty equivalence feedback fashion with a suitable state estimation procedure. In particular, given at time instant $k \in \mathbb{N}$ a state estimate

$$\hat{x}(k) = x(k) + e(k)\,, \tag{3.80}$$

with $x(k) \in \mathbb{R}^n$ referring to the current exact system state and $e(k) \in \mathbb{R}^n$ denoting the associated state estimation error, the applied feedback is given by

$$u(k) = \hat{u}_0^*(\hat{x}(k)) = \hat{u}_0^*(x(k) + e(k))\,, \tag{3.81}$$

where $\hat{u}_0^*(x)$ denotes the first vector element of the optimal solution to problem (3.71) for given $x \in \mathbb{R}^n$. As before, we focus thereby again on the relaxed barrier function based MPC formulation (3.71) based on a purely quadratic cost. Furthermore, we assume that the unknown estimation error dynamics are governed by an operator $\hat{\Phi} : \mathbb{N}_+ \times \mathbb{R}^n \times \mathbb{R}^n \times \mathbb{R}^p \to \mathbb{R}^n$ according to

$$e(k+1) = \hat{\Phi}(k, e(k), w(k), v(k)) . \tag{3.82}$$

As it is shown below, this formulation allows capturing for example Luenberger-type observers as well as the famous Kalman filter. Based on the above problem setup, the overall closed-loop system – consisting now of both the system state and the estimation error dynamics – can then be described as

$$x(k+1) = Ax(k) + B\hat{u}_0^*(x(k) + e(k)) + w(k) \tag{3.83a}$$

$$e(k+1) = \hat{\Phi}(k, e(k), w(k), v(k)) . \tag{3.83b}$$

In the following section, we will discuss the robust stability properties of system (3.83) by making again use of the concept of input-to-state stability and exploiting the ISS properties shown in Section 3.6.2. Note that it is in this context natural to assume that the above estimation error dynamics are in some sense well-behaved with respect to the corresponding state and output disturbances. Within the ISS framework, this may be captured by the following assumption.

Assumption 3.6. *The state estimation error dynamics are given by a difference equation of the form (3.82) and they are ISS with respect to the disturbance sequences $\boldsymbol{w}$ and $\boldsymbol{v}$.*

Note however that even if Assumption 3.6 holds, it is not immediately clear how the estimation error will affect the closed-loop system state in (3.83a) via the (nonlinear) barrier function based MPC feedback law. While the following discussion of our main results focuses again on the above relaxed barrier function based MPC approach with a purely quadratic terminal cost, the presented results do in principle hold for any of the globally stabilizing approaches proposed in Section 3.4.2.

Main results

Let us begin our analysis by recalling the vectorized formulation of the relaxed barrier function based open-loop optimal control problem (3.71), which is for the considered problem setup given by

$$\hat{J}_N^*(x) = \min_U \hat{J}_N(U, x) \tag{3.84a}$$

$$\hat{J}_N(U, x) = \frac{1}{2} U^\top H U + x^\top F U + \frac{1}{2} x^\top Y x + \varepsilon \hat{B}_{\mathrm{xu}}(U, x) , \tag{3.84b}$$

where $U := [u_0^\top \ \cdots \ u_{N-1}^\top]^\top \in \mathbb{R}^{Nm}$ is the stacked open-loop input vector, while $\hat{B}_{\mathrm{xu}} : \mathbb{R}^{Nm} \times \mathbb{R}^n \to \mathbb{R}_+$ denotes a recentered relaxed logarithmic barrier function for polytopic constraints of the form $GU \leq d + Ex$, see Appendix B.9. The applied control input can consequently be characterized as

$$u(k) = \Pi_0 \, \hat{U}^*(\hat{x}(k)), \qquad \Pi_0 = \begin{bmatrix} I_m & 0 & \cdots & 0 \end{bmatrix}, \tag{3.85}$$

where $\hat{U}^* : \mathbb{R}^n \to \mathbb{R}^{Nm}$ refers to the unique minimizer of (3.84). Concerning the question of how the above estimation error will affect the closed-loop system state, we can then state the following Lemma, which reveals that the optimal solution to problem (3.71), respectively the parametrized minimizer of (3.84), is globally Lipschitz continuous in the system state $x \in \mathbb{R}^n$. A proof of Lemma 3.8 is given in Appendix C.15.

Lemma 3.8. *Consider the relaxed barrier function based MPC problem (3.71) with the underlying stage and terminal cost function terms chosen according to (3.57) and (3.58), and let*

$$\hat{U}^*(x) = \arg\min_{U} \hat{J}_N(U, x) \tag{3.86}$$

be the unique minimizer associated to (3.84). Then, $\hat{U}^* : \mathbb{R}^n \to \mathbb{R}^{Nm}$ is globally Lipschitz continuous with Lipschitz constant $L_U \in \mathbb{R}_{++}$, that is,

$$\|\hat{U}^*(x_1) - \hat{U}^*(x_2)\| \leq L_U \|x_1 - x_2\| \quad \forall\, x_1, x_2 \in \mathbb{R}^n \,. \tag{3.87}$$

The importance of Lemma 3.8 for the stability analysis of the proposed output feedback MPC scheme is emphasized by the well-known fact that the class of Lipschitz continuous nonlinear systems possesses many advantages in the context of observer based feedback design (Pagilla and Zhu, 2004; Roset et al., 2008). In fact, based on Lemma 3.8, we can in combination with the ISS results from the previous section directly state the following theorem, which constitutes our main result concerning the robust stability properties of the extended closed-loop system (3.83).

Theorem 3.9. *Consider the relaxed barrier function based MPC problem (3.71) with the underlying stage and terminal cost function terms chosen according to (3.57) and (3.58). Furthermore, let $u(k) = \hat{u}_0^*(x(k))$ be the associated feedback law and let Assumption 3.6 hold. Then, the resulting overall closed-loop system (3.83) is ISS with respect to the disturbance sequences w and v.*

A proof of this result is presented in Appendix C.16. A direct consequence of Theorem 3.9 is that the overall closed-loop system dynamics (3.83) will be globally asymptotically stable whenever $w(k) = 0$, $v(k) = 0$ for all $k \in \mathbb{N}$. In addition, both the estimation error and the system state are guaranteed to remain bounded and will converge to zero for any convergent realization of bounded state and output disturbances. Furthermore, as long as Assumption 3.6 is satisfied, the state estimator can be designed independently from the underlying control procedure, which reveals that Theorem 3.9 constitutes a separation principle for the discussed class of relaxed barrier function based MPC schemes.

Approaches for ISS state estimation

As shown by the above results, robust stability of the combined system state and estimation error dynamics can be ensured as long as the estimation error is well-behaved (in the ISS sense) with respect to the external state and output disturbances. It is now only natural to ask how restrictive the underlying assumptions on the structure and the robustness of the estimation error dynamics are and which state estimation procedures may actually be used within the proposed robust output feedback MPC scheme. As we will see in the following, Assumption 3.6 is, in fact, satisfied by both Luenberger–type observers and the discrete-time Kalman filter, which constitute probably the most classical state estimation schemes for linear systems and are therefore widely used in practical applications.

1) Luenberger observer

Given a dynamical system of the form (3.79), a classical Luenberger observer computes the state estimate at each sampling instant $k \in \mathbb{N}$ based on the observer equation

$$\hat{x}(k+1) = A\hat{x}(k) + Bu(k) + L(y(k) - \hat{y}(k)), \tag{3.88}$$

where $L \in \mathbb{R}^{n \times p}$ is a suitably chosen estimator gain. It is then easy to show that the state estimation error $e(k) = \hat{x}(k) - x(k)$ evolves in this case according to

$$e(k+1) = (A - LC)e(k) + Lv(k) - w(k), \tag{3.89}$$

which is a difference equation of the form (3.82). Obviously, these error dynamics are asymptotically stable in the nominal case if and only if the gain L is chosen such that $A - LC$ is a Schur matrix, i.e., $|\lambda_i(A - LC)| < 1 \; \forall i = 1, \dots, n$. However, if this is the case, then (3.89) represents an asymptotically stable linear system that is perturbed by the additive disturbance $\tilde{w}(k) = Lv(k) - w(k)$. As $\tilde{w}(k)$ is linear in both $v(k)$ and $w(k)$, the desired ISS property follows directly from the fact that every asymptotically stable linear system is ISS with respect to additive disturbances (Jiang and Wang, 2001). In fact, it can be shown easily that

$$\|e(k)\| \leq c\beta^k \|e(0)\| + \gamma_w \|\boldsymbol{w}_{[k-1]}\| + \gamma_v \|\boldsymbol{v}_{[k-1]}\| \tag{3.90}$$

where $c \in \mathbb{R}_{++}$ and $\beta \in (0, 1)$ are chosen such that $\|A - LC\|^k \leq c\beta^k$ and $\gamma_w, \gamma_v \in \mathbb{R}_{++}$ are defined as $\gamma_w := c/(1 - \beta)$, $\gamma_v := cL/(1 - \beta)$. As noted by Jiang and Wang (2001), suitable constants c and β are guaranteed to exist under the above assumptions on the estimator gain L. By this we can conclude that Assumption 3.6 will actually be satisfied by any nominally stabilizing Luenberger observer of the form (3.88).

2) The Kalman filter

The so-called Kalman filter is a set of recursive equations that allows one to estimate the state of a (possibly perturbed) linear dynamical system based on noisy measurements, see for example the seminal work by Kalman (1960) as well as Chapter 1.4 in the book of Rawlings and Mayne (2009). In particular, the state estimate for a system of the form (3.79) is at each sampling instant $k \in \mathbb{N}$ computed as

$$\hat{x}(k) = \hat{x}^-(k) + K_k\left(y(k) - C\hat{x}^-(k)\right), \tag{3.91}$$

where $y(k) \in \mathbb{R}^p$ is the current measurement, the so-called a priori estimate $\hat{x}^-(k) \in \mathbb{R}^n$ is, for each $k \in \mathbb{N}_+$, given by

$$\hat{x}^-(k) = A\hat{x}(k-1) + Bu(k-1), \tag{3.92}$$

and the time-varying Kalman gain $K_k \in \mathbb{R}^{n \times p}$ is obtained recursively via

$$K_k = P_k^- C^\top \left(CP_k^- C^\top + R_\mathrm{v}\right)^{-1} \tag{3.93a}$$

$$P_k^+ = (I - K_k C)P_k^-, \qquad P_{k+1}^- = AP_k^+ A^\top + Q_\mathrm{w}. \tag{3.93b}$$

Here, $Q_\mathrm{w} \in \mathbb{S}^n_{++}$ and $R_\mathrm{v} \in \mathbb{S}^p_{++}$ denote the covariance matrices of the state and output disturbances, which are assumed to be drawn from a zero mean multivariate normal distribution, i.e., $w(k) \sim \mathcal{N}(0, Q_\mathrm{w})$, $v(k) \sim \mathcal{N}(0, R_\mathrm{v})$ Under the above assumptions, the Kalman filter is an optimal filter in the sense that it minimizes the variance of the estimation error (Kalman, 1960). By defining the a priori and a posteriori estimation errors

$$\zeta(k) := \hat{x}^-(k) - x(k), \qquad e(k) := \hat{x}(k) - x(k) \tag{3.94}$$

and exploiting that $e(k) = (I - K_k C)\zeta(k) + K_k v(k)$ and $\zeta(k+1) = Ae(k) - w(k)$, see (3.91), the state estimation error dynamics of the Kalman filter can be derived as

$$e(k+1) = A(I - K_{k+1}C)e(k) + K_{k+1}v(k+1) - (I - K_{k+1}C)w(k). \tag{3.95}$$

Note that system (3.95) is not exactly of the form (3.82) as it contains a direct feed-through of the output noise $v(k+1)$ on the estimation error. However, we can still apply our results from above by slightly modifying the ISS definition such that the supremum of the output disturbance sequence v is taken over all elements up to time instant k instead of time instant $k-1$. In particular, we can state the following ISS result for the Kalman filter. As we were quite surprisingly not able to find a rigorous proof of this result in the literature, a proof of Lemma 3.9 is presented in Appendix C.17 of this thesis.

Lemma 3.9. *Consider system (3.79) with (A, C) detectable and assume that $\operatorname{rank}(A) = n$. Let the state of system (3.79) be estimated based on (3.91)–(3.93) with $Q_\mathrm{w} \in \mathbb{S}^n_{++}$, $R_\mathrm{v} \in \mathbb{S}^p_{++}$, and $P_0^+ \in \mathbb{S}^n_+$. Then, the Kalman filter estimation error dynamics (3.95) are ISS with respect to the disturbance sequences $\boldsymbol{w}$, $\boldsymbol{v}$ in the sense that there exist $\beta \in (0, 1)$ and $c, \gamma_\mathrm{w}, \gamma_\mathrm{v} \in \mathbb{R}_{++}$ such that for any $e(0) \in \mathbb{R}^n$ and any $k \in \mathbb{N}_+$*

$$\|e(k)\| \leq c\beta^k \|e(0)\| + \gamma_\mathrm{w}\|\boldsymbol{w}_{[k-1]}\| + \gamma_\mathrm{v}\|\boldsymbol{v}_{[k]}\|. \tag{3.96}$$

Remark 3.19. Note that the assumption $\operatorname{rank}(A) = n$ is satisfied whenever system (3.79) is obtained from the discretization of an associated continuous-time system (Kalman, 1960). Moreover, under suitable uniform detectability assumptions, the above ISS result may also be extended to the time-varying case as long as $(A(k), C(k))$ remain bounded.

Discussion

In combination with Theorem 3.9, a direct consequence of the above results is that the overall closed-loop system consisting of both system state and estimation error dynamics will be robustly stable in the ISS sense with respect to external disturbances if the proposed relaxed barrier function based MPC scheme is combined in a certainty equivalence output feedback fashion with a state estimator based on the well-known principles of Luenberger observers or the discrete-time Kalman filter. As a consequence, there exist $\beta \in \mathcal{KL}$ and $\gamma_w, \gamma_v \in \mathcal{K}$ such that for all $z(0) = (x(0), e(0)) \in \mathbb{R}^n \times \mathbb{R}^n$ and all $k \in \mathbb{N}_+$ it holds that

$$\|z(k)\| \leq \beta(\|z(0)\|, k) + \gamma_w(\|\boldsymbol{w}_{[k-1]}\|) + \gamma_v(\|\boldsymbol{v}_{[k-1]}\|), \tag{3.97}$$

where $z(k) = (x(k), e(k)) \in \mathbb{R}^n \times \mathbb{R}^n$ denotes the overall system state. This result implies, in particular, asymptotic stability of the origin of system (3.83) in the disturbance-free case as well as convergence to the origin in the presence of asymptotically vanishing disturbances. Note that no bounds on the disturbances or the initial estimation error need to be known and that the presented results follow directly from the separation principle that is inferred by the robust stability properties of relaxed barrier function based MPC. In these aspects the presented approach differs from many existing output feedback MPC approaches, which often make use of set-membership estimation techniques (Bemporad and Garulli, 2000; Chisci and Zappa, 2002), ensure recursive feasibility and stability based on a priori computable robustly positive invariant error sets (Lee and Kouvaritakis, 2001; Mayne et al., 2006; Goulart and Kerrigan, 2007), or that exploit properties such as inherent robustness or (regional) input-to-state stability of the nominal MPC scheme (Scokaert et al., 1997; Messina et al., 2005; Roset et al., 2008). However, the presented global ISS results rely again heavily on the relaxation of the underlying state and input constraints, and no saturation of the associated variables has been considered in the analysis, see also Remark 3.16 above. Note also that while the presented output feedback results can quite easily be extended to the other globally stabilizing approaches proposed in Section 3.4.2, they do not immediately apply to the globally stabilizing approach based on a nonrelaxed terminal set constraint. The main reason for this is that due to the underlying nonrelaxed logarithmic barrier function it is for this approach far more difficult (if not impossible) to derive, as in Lemma 3.8, a Lipschitz condition on the resulting optimal control input.

3.6.4 Summary

Summarizing, we can conclude that we were, based on the well-known concept of input-to-state stability, able to prove quite remarkable robustness properties of the proposed relaxed barrier function based MPC scheme. In particular, the main results of this section are *i*) that the associated closed-loop system dynamics are in fact globally *exponentially* stable in the nominal case, that *ii*) they are globally input-to-state stable with respect to in principle arbitrary additive disturbance, and that *iii*) these input-to-state stability properties are preserved when combining the MPC approach in a certainty equivalence output feedback fashion with standard state estimation procedures such as Luenberger observers

or Kalman filtering. The key to the latter result was to show that the underlying optimal control input is in the relaxed barrier function case actually globally Lipschitz continuous in the system state. As discussed above, this actually implies a separation principle for the considered class of relaxed barrier function based MPC schemes.
It can be expected that the presented robustness results may also prove useful for more general scenarios that go beyond the considered case of estimation based output feedback and require to analyze interconnections of the proposed MPC schemes with other dynamical or uncertain components. However, as outlined above, the analysis relied on the assumption that both state and input variables can in principle take arbitrary values, which necessitates reconsideration in the presence of physically motivated state or input saturations.

3.6.5 Numerical Example

Let us briefly illustrate some of our findings by means of a numerical example. In order to demonstrate the robustness of the proposed MPC approach with respect to linearization errors, we consider a nonlinear system which is then linearized around the desired set point. In particular, we consider a vertically cascaded three-tank system, in which the goal is to regulate the water level in the bottom tank (subsequently tank three) by controlling the water flow that is entering the topmost tank (subsequently tank one), see Figure 3.6 on the next page. Each tank has a cross-sectional area A_i and an outlet with cross-sectional area a_i. For the considered example, these parameters were taken from a physical three-tank system that is available at the IST, see again Figure 3.6. The system states are given by the respective water heights in the three tanks and are denotes as h_i, $i = 1,2,3$. The input to the system is given by the water flow v that supplied to the topmost tank by means of an electric pump. Based on Torricelli's law, the associated nonlinear continuous-time system dynamics can then be derived as

$$\frac{\mathrm{d}h_1}{\mathrm{d}t} = -\frac{a_1}{A_1}\sqrt{2gh_1} + \frac{1}{A_1}v \tag{3.98a}$$

$$\frac{\mathrm{d}h_2}{\mathrm{d}t} = -\frac{a_2}{A_2}\sqrt{2gh_2} + \frac{a_1}{A_2}\sqrt{2gh_1} \tag{3.98b}$$

$$\frac{\mathrm{d}h_3}{\mathrm{d}t} = -\frac{a_3}{A_3}\sqrt{2gh_3} + \frac{a_2}{A_3}\sqrt{2gh_2}\,. \tag{3.98c}$$

Due to physical constraints on the pump flow and the dimension of the water tanks, the state and input variables are constrained as follows

$$0 \le v \le 60 \text{ ml/s}, \qquad 0 \le h_i \le 28 \text{ cm},\ i = 1,2,3\,. \tag{3.99}$$

Given a desired stationary water level h_3^s for the bottom tank, the required stationary input v^s and the corresponding stationary water heights h_1^s and h_2^s can be computed as

$$v^s = a_3\sqrt{2gh_3^s}, \quad h_1^s = \left(\frac{a_3}{a_1}\right)^2 h_3^s, \quad h_2^s = \left(\frac{a_3}{a_2}\right)^2 h_3^s\,. \tag{3.100}$$

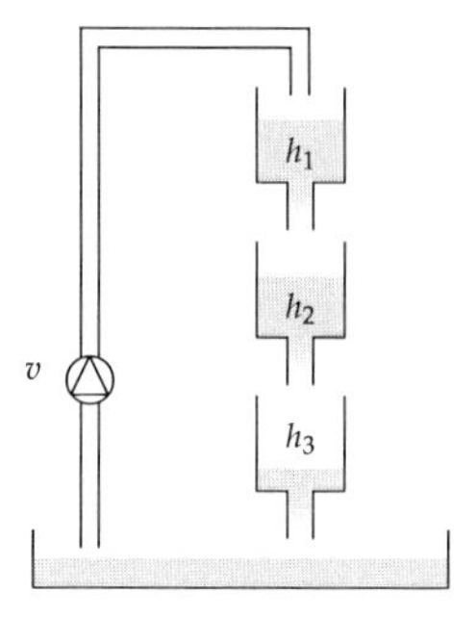

Parameter	Value	Unit
A_1	78.54	[cm²]
A_2	50.27	[cm²]
A_3	50.27	[cm²]
a_1	0.273	[cm²]
a_2	0.284	[cm²]
a_3	0.278	[cm²]
g	981	[cm/s²]

Figure 3.6. Schematic illustration and parameters for the considered three-tank system.

Assuming a desired water level of $h_3^s = 10.80$ cm, this results in $v^s = 40.4675$ ml/s as well as $h_1^s = 11.1992$ cm and $h_2^s = 10.3485$ cm. Linearizing system (3.98) around this desired stationary operating point and discretizing the resulting linear dynamics with a sampling time of $T_s = 5$ s, we arrive at a discrete-time linear system of the form

$$x(k+1) = \begin{bmatrix} 0.8913 & 0 & 0 \\ 0.1540 & 0.8233 & 0 \\ 0.0148 & 0.1608 & 0.8300 \end{bmatrix} x(k) + \begin{bmatrix} 0.0601 \\ 0.0052 \\ 0.0003 \end{bmatrix} u(k), \tag{3.101}$$

where $x = h - h^s \in \mathbb{R}^3$ refers to the system state and $u = v - v^s \in \mathbb{R}$ to the system input. The associated polytopic input and state constraints follow directly from (3.99) and can be characterized by

$$C_u = \begin{bmatrix} 1 \\ -1 \end{bmatrix}, \; d_u = \begin{bmatrix} 19.5325 \\ -40.4675 \end{bmatrix}, \qquad C_x = \begin{bmatrix} 1 & 0 & 0 \\ 0 & 1 & 0 \\ 0 & 0 & 1 \\ -1 & 0 & 0 \\ 0 & -1 & 0 \\ 0 & 0 & -1 \end{bmatrix}, \; d_x = \begin{bmatrix} 16.8008 \\ 17.6515 \\ 17.2000 \\ 11.1992 \\ 10.3485 \\ 10.8000 \end{bmatrix}. \tag{3.102}$$

Note that besides the following numerical simulation results, the relaxed barrier function based MPC approach has for the considered three-tank system also been successfully implemented and tested in an experimental case study (Wenzelburger, 2014).

MPC problem setup

As assumed in the above theoretical results, we make use of the relaxed barrier function based MPC formulation (3.56)–(3.57), where we choose the underlying parameters as

$Q = \text{diag}(1, 1, 10)$, $R = 0.01$, $N = 15$. The barrier function weighting and relaxation parameters are chosen as $\varepsilon = \delta = 10^{-3}$. The weighting vectors for recentering the barrier functions are constructed by making use of the procedure in Appendix B.3, leading to

$$w_\text{u} = \begin{bmatrix} 2.0718 \\ 1.0000 \end{bmatrix}, \qquad w_\text{x} = \begin{bmatrix} 1.0000 & 1.0000 & 1.0000 & 1.5002 & 1.7057 & 1.5926 \end{bmatrix}^\top . \tag{3.103}$$

Furthermore, the quadratic terminal cost matrix $P \in \mathbb{S}^n_{++}$ is computed based on (3.58), which results for the chosen problem setup and the above weighting vectors in

$$P = \begin{bmatrix} 1.0251 & 0.3319 & 0.0971 \\ 0.3319 & 0.5890 & 0.1743 \\ 0.0971 & 0.1743 & 0.4192 \end{bmatrix} \times 10^4 . \tag{3.104}$$

As shown above, this results in a globally defined and strongly convex optimization problem, whose solution characterizes a unique and globally Lipschitz continuous control input. The presented robust stability properties of the resulting closed-loop system are in the following briefly illustrated and discussed by means of two different scenarios. All computations and simulations where performed using MATLAB/Simulink R2015b.

Scenario 1

In order to illustrate the input-to-state stability properties of the perturbed closed-loop system (3.72), we are in the following going to compare the behavior of the nominal, unperturbed system with the resulting closed-loop behavior under the influence of two qualitatively different additive disturbances. First, we simulate the behavior of the closed-loop system when applying the proposed relaxed barrier function based MPC scheme in a sampled-data fashion to the original, nonlinear plant model (3.98). In this case, the disturbances $w(k)$ are therefore given by the plant-model mismatch that is inferred by the performed linearization and discretization. Second, we study the behavior of the closed-loop system when perturbing it with a persistent sequence of normally distributed random disturbances $w(k) \sim \mathcal{N}(0, Q_\text{w})$ with $Q_\text{w} = \text{diag}(0.5, 0.1, 0.1)$. The resulting input and state trajectories for the different disturbance scenarios and initial condition $h(0) = [27.5, 15, 15]^\top$ are presented and compared in Figure 3.7. It can be seen that the plant converges to the desired set point not only for the nominal but also for the nonlinear case. This may be explained by noting that, at least in the considered example, the disturbances (which are in the latter case given by the linearization errors) decay asymptotically to zero as the system state is steered closer and closer to the desired set point. Concerning the case of a persistent random disturbance, it can be observed that the closed-loop system converges only to a neighborhood of the desired set point. However, as shown in the above ISS results, the size of this neighborhood will in general directly depend on the maximal amplitude of the respective disturbances. Note also that both input and state constraints are satisfied despite the underlying relaxation and the quite severe perturbations that are acting on the system.

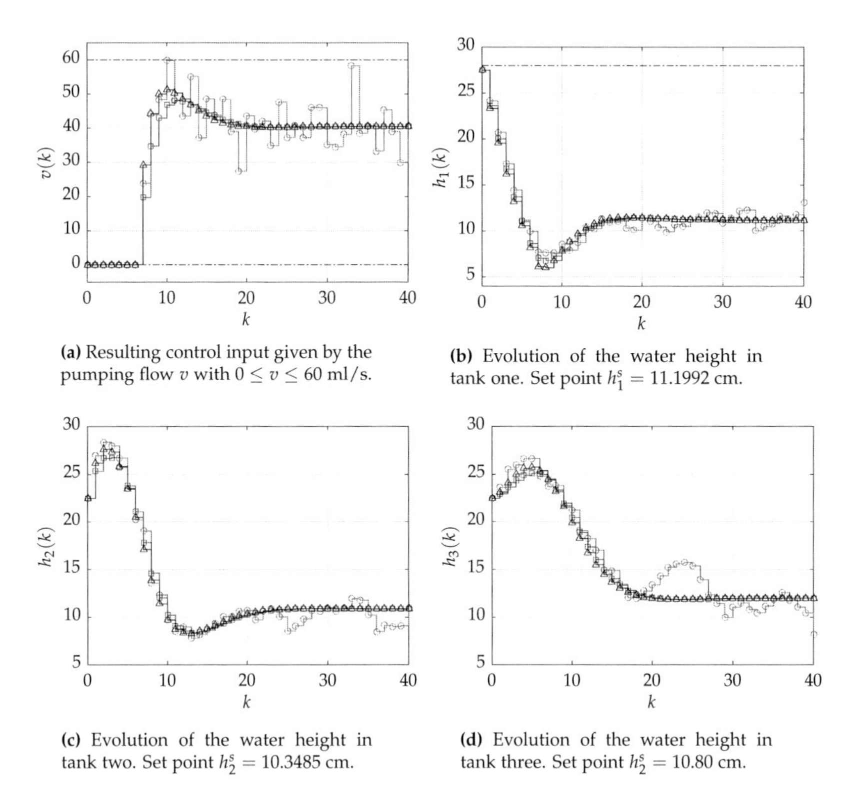

(a) Resulting control input given by the pumping flow v with $0 \leq v \leq 60$ ml/s.

(b) Evolution of the water height in tank one. Set point $h_1^s = 11.1992$ cm.

(c) Evolution of the water height in tank two. Set point $h_2^s = 10.3485$ cm.

(d) Evolution of the water height in tank three. Set point $h_2^s = 10.80$ cm.

Figure 3.7. Simulated closed-loop behavior of the three-tank system for Scenario 1 and initial condition $h(0) = [27.5, 15, 15]^\top$. Note that the results are given in the original (v, h)-coordinates. Depicted are the input and state trajectories associated to the nominal linearized system ($\triangle$), the unperturbed nonlinear plant model ($\square$), and the linear system perturbed by a persistent random disturbance ($\circ$). The dashed horizontal lines in subplots (a) and (b) indicate the underlying input and state constraints. As in the nominal case, convergence to the desired set point is achieved also when controlling the nonlinear plant with the proposed linear MPC scheme. Convergence to a neighborhood of the set point is achieved in the presence of a persistent additive disturbance. As the weight on the control input is small in comparison to the weights on the system states, the pumping flow is quite sensitive to the applied state disturbances. However, note that the respective input and state constraints are satisfied in all cases.

Scenario 2

In order to illustrate the robust output feedback results presented in Section 3.6.3, we consider now a slightly modified example scenario in which only the state of the bottom tank can be measured directly, leading to a problem setup of the form (3.79) with $C = [0, 0, 1]$. The state and output disturbances are chosen as normally distributed random noise with zero mean and covariances $Q_w = \mathrm{diag}(0.5, 0.1, 0.1)$ and $R_v = 10^{-3}$. In the following, we are going to compare the nominal closed-loop behavior based on full state feedback with that related to two different output feedback scenarios. First, we consider the case of output feedback without the additional influence of state and output disturbances. For this scenario, we design an exponentially stable Luenberger observer based on

$$L = \begin{bmatrix} 1.2515 & 1.4320 & 0.8446 \end{bmatrix}^\top , \qquad \lambda(A - LC) = \begin{bmatrix} 0.50 & 0.55 & 0.65 \end{bmatrix}^\top . \tag{3.105}$$

Second, we consider a scenario in which the closed-loop system is affected by the above state and output disturbances and an inherently inexact state estimate is constructed based on the discrete-time Kalman filter. The initial covariance for the Kalman filter is chosen as $P_0^+ = 0$. Furthermore, both state estimation schemes are initialized with $\hat{x}(0) = 0$, which corresponds to $\hat{h}(0) = h^s$. Note that this is quite far away from the true initial state, which is again chosen as $h(0) = [27.5, 15, 15]^\top$. The simulation results for the different scenarios are depicted and compared in Figure 3.8 and Figure 3.9. As predicted by the presented ISS results, both the state and the estimation error converge to zero in the disturbance free case. In the presence of persistent disturbances, it is again possible to achieve convergence to a neighborhood of the desired set point.

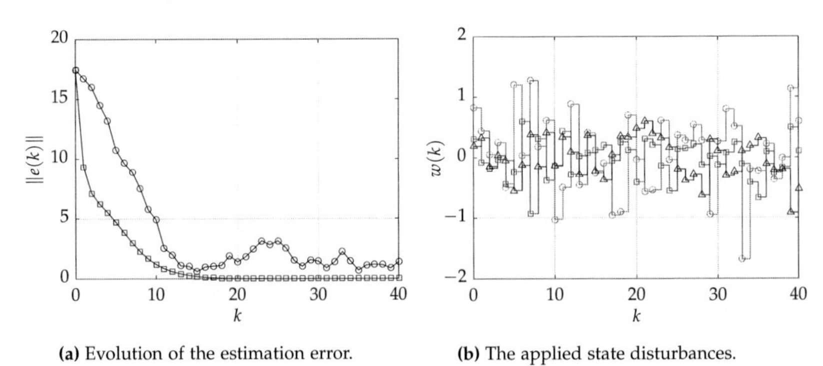

(a) Evolution of the estimation error. **(b)** The applied state disturbances.

Figure 3.8. Evolution of the state estimation error for Luenberger observer (□) and Kalman filter (○) for the above problems setup and $\hat{h}(0) = h^s$. Also depicted are the quite severe state disturbances that are acting on the system in the latter case. The plot marker scheme on the right-hand side is as follows: tank one (○), tank two (□), tank three (△).

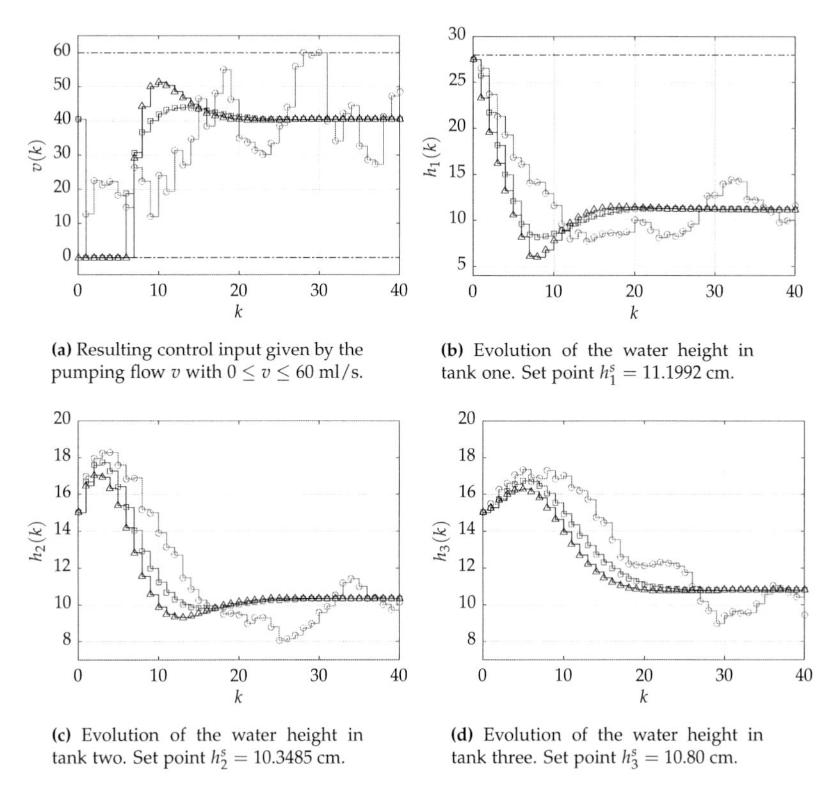

(a) Resulting control input given by the pumping flow v with $0 \leq v \leq 60$ ml/s.

(b) Evolution of the water height in tank one. Set point $h_1^s = 11.1992$ cm.

(c) Evolution of the water height in tank two. Set point $h_2^s = 10.3485$ cm.

(d) Evolution of the water height in tank three. Set point $h_3^s = 10.80$ cm.

Figure 3.9. Simulated closed-loop behavior of the three-tank system for Scenario 2 and initial condition $h(0) = [27.5, 15, 15]^\top$. Depicted are the input and state trajectories associated to the nominal linearized system ($\triangle$), unperturbed output feedback based on a Luenberger observer ($\square$), and output feedback based on the discrete-time Kalman filter in the presence of persistent disturbances given by normally distributed state and measurement noise ($\circ$). The dashed horizontal lines in subplots (a) and (b) again indicate the underlying input and state constraints. When no disturbances are acting on the system, convergence to the desired set-point is achieved also in the estimation-based output feedback case. Convergence to a neighborhood of the set point is achieved when applying output feedback in the presence of a persistent additive disturbance. As in Scenario 1, the underlying input and state constraints are satisfied.

3.7 An outlook on relaxed barrier function based NMPC

All the results presented in this chapter were stated for the case of linear MPC, i.e., under the assumption that the underlying plant can be modeled as a linear discrete-time system. Although a detailed analysis of the proposed relaxed barrier function based MPC framework in the context of nonlinear system dynamics is beyond the scope of this thesis, we want to discuss in this section very briefly whether and how at least some of the above results may be extended to the nonlinear MPC case. In particular, we consider the control of time-invariant discrete-time systems of the form

$$x(k+1) = f(x(k), u(k)), \tag{3.106}$$

where $x(k) \in \mathbb{R}^n$ and $u(k) \in \mathbb{R}^m$ denote the state and input vectors at time $k \in \mathbb{N}$ and $f : \mathbb{R}^n \times \mathbb{R}^m \to \mathbb{R}^n$ is a given, possibly nonlinear function. As it is also common in conventional NMPC approaches, we assume that the system dynamics are continuous and that $f(0,0) = 0$. The goal is to control the state of system (3.106) to the origin while satisfying pointwise input and state constraints of the form $u(k) \in \mathcal{U}$, $x(k) \in \mathcal{X}$ for all $k \in \mathbb{N}$. For the sake of simplicity, we in the following assume that $\mathcal{U}$ and $\mathcal{X}$ are again given as compact polytopic sets of the form discussed above.

The relaxed barrier function based NMPC problem

Based on the above problem setup and following the same idea as in the linear case, the associated relaxed barrier function based open-loop optimal control problem may be formulated as

$$\hat{J}_N^*(x) = \min_{u} \sum_{k=0}^{N-1} \hat{\ell}(x_k, u_k) + \hat{F}(x_N) \tag{3.107a}$$

$$\text{s.t. } x_{k+1} = f(x_k, u_k), \; x_0 = x, \tag{3.107b}$$

where $N \in \mathbb{N}_+$ is again the finite prediction horizon, $x = x(k) \in \mathbb{R}^n$ refers to the current system state, and $\hat{\ell} : \mathbb{R}^n \times \mathbb{R}^m \to \mathbb{R}_+$ and $\hat{F} : \mathbb{R}^n \to \mathbb{R}_+$ denote the barrier function based stage and terminal cost function terms. As before, the stage cost may be formulated as a combination of quadratic terms and suitably chosen relaxed barrier functions for the polytopic constraints, while the terminal cost function needs to be designed in such a way that desired stability properties of the resulting closed-loop system are guaranteed.

Stability of the closed-loop system

Probably the most intuitive way to design a relaxed barrier function based NMPC scheme with guaranteed stability properties is to design a stabilizing *nonrelaxed* barrier function based scheme (using for example the terminal set based approach of Wills and Heath (2004)) and then simply relax in a second step all of the underlying logarithmic barrier functions. In fact, as they do not explicitly rely on the linearity of the underlying system dynamics, the results presented in Section 3.4.1 of this thesis will in principle still

hold in the nonlinear MPC case. This includes in particular the results presented in Theorem 3.1 and Lemma 3.2, which state not only that stability of the closed-loop system can in the full relaxation case be guaranteed with a region of attraction whose size depends on the choice of the relaxation parameter but also that this region will essentially recover the region of attraction of the associated nonrelaxed formulation whenever the relaxation parameter approaches zero. Moreover, under the assumption of suitable controllability properties of system (3.106), stability of the closed-loop system may also be shown by making use of the approach based on a nonrelaxed terminal set constraint (Theorem 3.2). A simple alternative approach for achieving closed-loop stability at least locally is to make us of an additional terminal state constraint of the form $x_N = 0$. Assuming that the resulting modified problem admits a feasible solution upon initialization, it is in this case quite straightforward to show the value function associated to problem (3.107) will decrease monotonically at each time step. As a consequence, the origin of the closed-loop system will be asymptotically stable, with the region of attraction given by the set of initial conditions from where the nonlinear system (3.106) can be steered to the origin in N steps. Following the above idea of a tail-sequence based terminal cost function, the aforementioned zero terminal state constraint may also be relaxed if suitable parametrized input and state sequences are known that steer the state of system (3.106) from any initial condition to the origin in a finite number of steps. In this case, global asymptotic stability of the resulting closed-loop system may be shown based on similar arguments as in the proof of Theorem 3.4. Although of course hard to construct in the general case, it is known that suitable dead-beat control laws do exist for certain classes of nonlinear systems (Grasselli et al., 1980; Nešić, 1996; Nešić and Mareels, 1998).
Note furthermore that the global quadratic upper bound which has been presented for the considered class of quadratically relaxed barrier functions does still hold (Lemma 3.3). Nevertheless, the approach based on a purely quadratic terminal cost can not be applied directly since a suitable upper bound for the infinite-horizon cost-to-go can in the case of nonlinear system dynamics not be computed based on the discrete-time algebraic Riccati equation (3.34). In contrast to this, the approach based on a zero terminal cost may in principle still be used as long as it is possible to derive uniform upper and lower bounds on the value function, cf. Lemma 3.5. Similar to the work of Grüne (2012), this may for example be ensured based on a suitable controllability assumption.

Satisfaction of input and state constraints

As they do not explicitly exploit the linearity of the underlying open-loop system but mainly rely on the guaranteed decay of the value function, it should in principle be possible to extend the presented constraint satisfaction results in a quite straightforward way to the case of nonlinear MPC. In particular, the results that the maximally possible constraint violation will be bounded and monotonically decaying over time (Theorem 3.6) and that exact satisfaction of all input and state constraints may still be achieved at least for a certain set of initial conditions (Theorem 3.7) should still hold.

Robust stability properties and output feedback

In the form as they were given above, the presented ISS and output feedback results can unfortunately not be applied in a straightforward way to the nonlinear MPC case. The main reason for this is that Lemma 3.6 and Theorem 3.8 rely heavily on the convexity of the underlying value function, which will in general be lost in the presence of arbitrary nonlinear system dynamics. However, under the (quite common) assumption that the resulting value function is Lipschitz continuous, it may in principle be possible to derive regional ISS results similar to Scokaert et al. (1997), Limón et al. (2002), and Zeilinger et al. (2010). In addition, the smoothing relaxed barrier function based formulation may be exploited to show continuity of the value function, which automatically implies interesting inherent robustness properties (Grimm et al., 2004; Pannocchia et al., 2011) and is usually not guaranteed in the general nonlinear MPC case.

3.8 Chapter reflection

In this first main part of the thesis, we introduced the novel concept of model predictive control based on relaxed barrier functions and presented a systems theoretical framework for the constructive design and analysis of conceptually different relaxed barrier function based MPC schemes for discrete-time linear systems.

After introducing in the Sections 3.1, 3.2, and 3.3 the considered problem setup, the concept of relaxed logarithmic barrier functions, and the associated relaxed barrier function based MPC formulation, we studied in Section 3.4 the stability properties of the resulting closed-loop system. The first main result of this section was that the desirable stability and constraint satisfaction properties of nonrelaxed barrier function based MPC can always be recovered by choosing the relaxation parameter sufficiently small. The second main result was that the underlying relaxation also enables the formulation of novel, terminal set free MPC schemes which not only allow us to guarantee *global* asymptotic stability of the resulting closed-loop system but also to reformulate the whole MPC problem as unconstrained minimization of a globally defined, twice continuously differentiable, and strongly convex cost function.
In Section 3.5, we analyzed the constraint satisfaction properties of the resulting closed-loop system. The first main result of this section was that, despite the underlying relaxation, the resulting maximal violation of input and state constraint will always be bounded and decaying over time. Furthermore, an explicit upper bound can be computed a priori and there will always exist a compact set of initial conditions for which the constraint will not be violated at all. The second main result was that any desired level of constraint satisfaction may, at least for a certain set of initial conditions, be achieved in a constructive fashion by choosing the relaxation parameter sufficiently small.
In Section 3.6, we studied the robustness properties of the proposed relaxed barrier function based MPC schemes and showed that the resulting closed-loop system is in fact

input-to-state stable with respect to arbitrary additive disturbances. As a second main result of this section, we furthermore showed that these beneficial robustness properties are preserved when combining the proposed MPC approaches in a certainty equivalence output feedback fashion with suitable state estimation procedures – including in particular the important cases of Luenberger observers and Kalman filtering.
In Section 3.7, we gave an outlook on relaxed barrier function based NMPC and briefly discussed whether and how some of the presented results may carry over to the context of nonlinear discrete-time systems.

From a methodological point of view, recurring themes in the presented relaxed barrier function based MPC framework were that both the resulting cost function and the associated optimal control input are always defined, that the underlying relaxed logarithmic barrier functions can be globally upper bounded by a quadratic function, and that as long as there exists a strictly feasible solution to the nonrelaxed problem, there always exists a sufficiently small choice for the relaxation parameter such that the solutions of nonrelaxed and relaxed problem formulation will be identical. In fact, perhaps *the* main key idea behind many of the presented results is that the use of relaxed barrier functions leads to a smoothed and unconstrained formulation of the underlying MPC problem while still allowing us to approximate the original constrained problem arbitrarily close.

Summarizing, we can therefore conclude that the proposed relaxed barrier function based approach is more than an alternative, barrier function based reformulation of conventional MPC schemes or a straightforward extension of already existing nonrelaxed barrier function based MPC concepts – enabling us in particular to formulate conceptually novel and simple MPC schemes that are not necessarily based on an additional terminal set constraint, possess remarkable stability, constraint satisfaction, and inherent robustness properties, and that in addition allow us to characterize the stabilizing control input as the minimizer of a globally defined, twice continuously differentiable and strongly convex cost function. Further beneficial numerical and algorithmic aspects of the relaxed barrier function based formulation will be analyzed in more detail in the following chapter, which forms the second main part of this thesis and in which we present and discuss tailored iteration schemes and optimization algorithms that allow for an efficient and stability preserving algorithmic implementation of the relaxed barrier function based MPC schemes that have been presented above.
Disadvantages of the relaxed barrier function based approach and the presented stability analysis are that both input and state constraints may in principle be violated and that the resulting MPC feedback law will also in the case of a feasible initial condition not necessarily recover the behavior of the associated conventional MPC scheme with hard input and state constraints (as it may for example be achieved when making use of suitably defined (non-smooth) exact penalty functions, cf. (Kerrigan and Maciejowski, 2000; Zeilinger et al., 2014a)).

Chapter 4

Relaxed Barrier Function Based MPC Algorithms with Guaranteed Stability

In this chapter, we complement the more theory-oriented MPC framework that we presented in the first part of the thesis with tailored optimization algorithms and iteration schemes that allow for an efficient and stability-preserving implementation of the respective relaxed barrier function based MPC schemes. In particular, besides identifying and discussing suitable optimization routines for solving the underlying open-loop optimal control problem in an efficient way, our main goal is to derive and analyze stabilizing *MPC algorithms* that explicitly exploit structural properties of the barrier function based formulation and thereby allow to guarantee important systems theoretic properties of the resulting overall closed-loop system. The main result of the chapter is a novel class of *anytime MPC algorithms*, which preserve many of the stability, constraint satisfaction, and robustness properties from the previous chapter independently of the number of performed optimization algorithm iterations – including the case of performing only *one* optimizing update of the control input at each sampling step. Furthermore, the possibility to exploit structure and sparsity in the underlying numerical optimization procedures is also discussed and evaluated by means of a numerical benchmark study.
Besides exploiting many of the theoretical properties of the relaxed barrier function based formulation from Chapter 3, the results presented in this chapter are mainly based on (Feller and Ebenbauer, 2013, 2014a, 2016, 2017b).

4.1 Problem setup and chapter outline

We consider the same problem setup as in Chapter 3, which we briefly recall in the following for the sake of convenience. As above, the system dynamics of the plant are assumed to be described by a discrete-time linear time-invariant system of the form

$$x(k+1) = Ax(k) + Bu(k)\,, \tag{4.1}$$

where $x(k) \in \mathbb{R}^n$ and $u(k) \in \mathbb{R}^m$ denote the vectors of system states and inputs, both at time $k \in \mathbb{N}$, while $A \in \mathbb{R}^{n\times n}$ and $B \in \mathbb{R}^{n\times m}$ denote the given system matrices. The goal is to steer the state of system (4.1) to the origin while, at each $k \in \mathbb{N}$, satisfying polytopic state and input constraints of the form

$$x(k) \in \mathcal{X} := \{x \in \mathbb{R}^n : C_x x \leq d_x\}, \tag{4.2a}$$
$$u(k) \in \mathcal{U} := \{u \in \mathbb{R}^m : C_u u \leq d_u\}, \tag{4.2b}$$

where $C_x \in \mathbb{R}^{q_x \times n}$, $C_u \in \mathbb{R}^{q_u \times m}$ and $d_x \in \mathbb{R}^{q_x}_{++}$, $d_u \in \mathbb{R}^{q_u}_{++}$, with $q_x, q_u \in \mathbb{N}_+$ denoting the number of affine constraints in each case. Throughout this chapter, the aforementioned control problem shall be handled by a relaxed barrier function based MPC approach that makes use of a purely quadratic terminal cost function (cf. Section 3.4.2). In particular, we consider the relaxed barrier function based MPC problem

$$\hat{J}^*_N(x) = \min_U \sum_{k=0}^{N-1} \hat{\ell}(x_k, u_k) + x_N^\top P x_N \tag{4.3a}$$
$$\text{s.t. } x_{k+1} = Ax_k + Bu_k,\ x_0 = x, \tag{4.3b}$$

where

$$\hat{\ell}(x,u) := \|x\|_Q^2 + \|u\|_R^2 + \varepsilon \hat{B}_x(x) + \varepsilon \hat{B}_u(u), \quad \varepsilon \in \mathbb{R}_{++}, \tag{4.4}$$

denotes the globally defined stage cost with $\hat{B}_x : \mathbb{R}^n \to \mathbb{R}_+$ and $\hat{B}_u : \mathbb{R}^m \to \mathbb{R}_+$ being recentered relaxed logarithmic barrier functions for the polytopic sets $\mathcal{X}$ and $\mathcal{U}$ for a given value of the relaxation parameter $\delta \in \mathbb{R}_{++}$. Furthermore, $U = [u_0^\top, \ldots, u_{N-1}^\top]^\top \in \mathbb{R}^{Nm}$ refers to the optimization variable in stacked vector form and $\varepsilon \in \mathbb{R}_{++}$ is the fixed barrier function weighting parameter. Following the design procedure proposed in Section 3.4.2, the terminal cost matrix $P \in \mathbb{S}^n_{++}$ is chosen based on

$$K = -\left(R + B^\top PB + \varepsilon M_u\right)^{-1} B^\top PA \tag{4.5a}$$
$$P = (A+BK)^\top P(A+BK) + K^\top(R + \varepsilon M_u)K + Q + \varepsilon M_x, \tag{4.5b}$$

where the matrices $M_x \in \mathbb{S}^n_{++}$ and $M_u \in \mathbb{S}^m_{++}$ are defined according to Lemma 3.4. We assume in the following that all the underlying problem parameters are given, satisfying in particular the conditions of the associated stability result Theorem 3.3. Note, however, that basically all the results presented in this chapter apply, with some minor modifications, also to the other relaxed barrier function based MPC approaches from Chapter 3.

As already outlined in the previous chapter, problem (4.3) can be rewritten in the following unconstrained and condensed matrix-vector form

$$\hat{J}^*_N(x) = \min_U \hat{J}_N(U,x), \quad \hat{J}_N(U,x) = \frac{1}{2}U^\top HU + x^\top FU + \frac{1}{2}x^\top Yx + \varepsilon \hat{B}_{xu}(U,x), \tag{4.6}$$

where the respective matrices can be computed in a straightforward way from the given problem formulation and $\hat{B}_{xu} : \mathbb{R}^{Nm} \times \mathbb{R}^n \to \mathbb{R}_+$ denotes a suitably defined relaxed logarithmic barrier function for the combined state and input constraints. A detailed derivation is given in Appendix B.9. Assuming that the exact optimal solution is available immediately upon measurement of the current system state, the associated closed-loop system can then be written as

$$x(k+1) = Ax(k) + B\Pi_0 \hat{U}^*(x(k)), \qquad \Pi_0 = \begin{bmatrix} I_m & 0 & \cdots & 0 \end{bmatrix}, \tag{4.7}$$

where $\hat{U}^* : \mathbb{R}^n \to \mathbb{R}^{Nm}$ refers to the unique solution of (4.6) and the multiplication with the matrix $\Pi_0 \in \mathbb{R}^{m \times Nm}$ can be interpreted as a projection that allows one to select only the first input vector. The systems theoretical properties of this closed-loop system have been studied extensively in Chapter 3. However, although quite common in the MPC literature, the assumption that the exact optimal solution can be computed in zero time (or that the required computation time is at least negligible) might not always be satisfied in practical applications. Consequently, the closed-loop dynamics of the actual algorithmic MPC implementation will in general not be of the form (4.7).

Motivated by this gap between MPC theory and MPC practice, the main goal of this chapter is to drop the idealizing assumption that the exact optimal control input is available immediately and to instead analyze the systems theoretical properties of the overall closed-loop system *as it is implemented*. More precisely, we are going to take the step from idealized *MPC schemes*, which essentially treat the required on-line optimization as a static map, to the more general and application-oriented concept of *MPC algorithms*, which explicitly consider and analyze the dynamics of the underlying optimization algorithm as a part of the resulting overall closed-loop system. One natural and interesting question arising in this context is which classes of optimization methods could in principle be applied and whether it is possible to derive explicit numerical complexity estimates for a desired level of optimality. This issue will be addressed in Section 4.2 of this chapter, in which we discuss structural properties of the relaxed barrier function based problem formulation from an algorithmic perspective and identify suitable first and second order methods that can be used for solving the associated optimization problem in an efficient way. In Section 4.3, we then present a suitable *MPC iteration scheme* that formulates the overall closed-loop system (consisting of both the plant and optimization algorithm dynamics) in an integrative fashion, allowing us in particular to capture the effect of performing only a finite number of optimization algorithm iterations at each sampling instant. For this iteration scheme, we then basically discuss the same systems theoretic properties as in Chapter 3, that is, asymptotic stability of the origin, satisfaction of input and state constraints, and robustness with respect to additive disturbances or observer errors. In Section 4.4, we will go even further into the direction of numerical implementation and study in which way inherent structural properties of the proposed class of MPC approaches may be exploited when, instead of (4.6), considering a non-condensed and therefore *sparse* formulation of the associated optimization problem. Returning to the above condensed formulation in Section 4.5, we present an alternative algorithmic approach which is based on asymptotically tracking the associated parametrized optimal solution by means of a suitably defined continuous-time optimization algorithm, leading to the concept of relaxed barrier function based continuous-time MPC algorithms. Finally, the results of the chapter are summarized and discussed in Section 4.6.

4.2 Properties of the underlying optimization problem

In this section, we state and discuss some basic but nevertheless very important structural properties of the barrier function based optimization problem (4.6). While some of these properties were already briefly mentioned and exploited in the first part of this thesis, others are new, and we present all of them here in a self-contained and rigorous form. For the sake of simplicity, we focus in the following on the above condensed problem formulation. However, very similar results can in a straightforward fashion also be derived for the sparse formulation that will be discussed in Section 4.4 of this chapter.

4.2.1 Convexity, regularity, and self-concordance

For any arbitrary but fixed $x \in \mathbb{R}^n$, problem (4.6) represents an unconstrained optimization problem in the variable $U \in \mathbb{R}^{Nm}$. It is well-known that in the context of unconstrained optimization, properties like convexity, smoothness, or self-concordance of the respective objective function are of crucial importance for the design and analysis of efficient optimization methods (Nesterov, 2004). Concerning the proposed relaxed barrier function based problem formulation, we can in this regard state the following results.

Theorem 4.1. *Consider the relaxed barrier function based MPC formulation (4.3)–(4.5) and let* $\hat{J}_N : \mathbb{R}^{Nm} \times \mathbb{R}^n \to \mathbb{R}_+$ *given in (4.6) refer to the associated cost function. Then, for any arbitrary but fixed* $x \in \mathbb{R}^n$, *the cost function* $\hat{J}_N(U, x)$ *is twice continuously differentiable and strongly convex in* U *and there exist* $\mu, L \in \mathbb{R}_{++}$ *such that for any* $U, U' \in \mathbb{R}^{Nm}$

$$\mu I \preceq \nabla^2_U \hat{J}_N(U, x) \preceq LI \tag{4.8a}$$

$$\|\nabla_U \hat{J}_N(U, x) - \nabla_U \hat{J}_N(U', x)\| \leq L\|U - U'\|. \tag{4.8b}$$

Furthermore, there exists $L' \in \mathbb{R}_{++}$ *such that for any* $U, U' \in \mathbb{R}^{Nm}$

$$\|\nabla^2_U \hat{J}_N(U, x) - \nabla^2_U \hat{J}_N(U', x)\| \leq L'\|U - U'\|. \tag{4.9}$$

Finally, for any $x \in \mathbb{R}^n$, *the cost function* $\hat{J}_N(\cdot, x)$ *is self-concordant according to Definition 2.3.*

A proof of this result as well as explicit expressions for the constants μ, L, and L' can be found in Appendix C.18. From the perspective of numerical optimization methods, Theorem 4.1 shows that the resulting cost function is not only globally defined and strongly convex but also that it belongs for any arbitrary but fixed realization of the system state to the class of self-concordant functions with globally Lipschitz continuous gradient and globally Lipschitz continuous Hessian. While the class of smooth and strongly convex functions with Lipschitz continuous gradient has been studied extensively in the context of first order methods (Polyak, 1987; Nesterov, 2004), the self-concordance property as well as the assumption of a Lipschitz continuous Hessian are usually of crucial importance when analyzing the convergence behavior of second order methods like the Newton method (Nesterov and Nemirovskii, 1994; Renegar, 2001; Boyd and Vandenberghe, 2004). In this light, Theorem 4.1 reveals that the relaxed barrier function based MPC formulation inherently leads to a very special and *"nice"* class of optimization problems, for which many powerful and well-understood optimization methods exist. Some of these methods as well as important implications of the above results for the derivation of explicit complexity estimates will be discussed briefly in the following subsection.

4.2.2 Convergence and complexity estimates

A direct consequence of the above results is that the relaxed barrier function based problem (4.6) can be tackled by a multitude of efficient algorithms for smooth and convex optimization. One intuitive and basic example for such an algorithm is the well-known gradient descent method, which can for the given problem at hand be described by

$$U_{k+1} = U_k + s\Delta U_k\,, \qquad \Delta U_k := -\nabla_U \hat{J}_N(U_k, x)\,. \tag{4.10}$$

where $s \in \mathbb{R}_{++}$ refers to a suitably chosen step size. It can be shown that for the class of convex functions with Lipschitz continuous gradient, the choice $s = \frac{1}{L}$ ensures that the iterates of the gradient method converge globally to the optimal solution with convergence rate $\mathcal{O}(\frac{1}{k})$, see for example (Nesterov, 2004). An alternative approach is given by Nesterov's accelerated gradient method, which is a so-called multi-step method that can in its simplest form with constant step sizes be written as

$$U_{k+1} = y_k - \frac{1}{L}\nabla_U \hat{J}_N(y_k, x) \tag{4.11a}$$

$$y_{k+1} = U_{k+1} + \beta(U_{k+1} - U_k)\,, \tag{4.11b}$$

where L is the Lipschitz constant of the gradient and $\beta \in \mathbb{R}_{++}$ a step size parameter. As proven by Nesterov (1983) for suitably chosen β, the accelerated gradient method converges for the class of convex functions with Lipschitz continuous gradient with the improved (and in fact optimal) rate of $\mathcal{O}(\frac{1}{k^2})$, see also (Nesterov, 2004) for a more general and detailed discussion. Furthermore, when considering the class of *strongly* convex function with Lipschitz continuous gradient, oftentimes also denoted as $\mathcal{S}^{1,1}_{\mu,L}$, both the gradient method and the accelerated gradient method can be be shown to converge linearly in the sense that there exist $c \in \mathbb{R}_{++}$ and $\sigma \in (0,1)$ such that

$$\|U_k - \hat{U}^*(x)\| \leq c\sigma^k\,. \tag{4.12}$$

In particular, the gradient method ensures for a step size of $s = 2/(\mu + L)$ a linear decay of the form (4.12) with $c = \|U_0 - \hat{U}^*(x)\|$ and $\sigma = (L-\mu)/(L+\mu)$, see for example Theorem 2.1.14 in (Nesterov, 2004). Similar expressions can also be derived for the accelerated gradient method, see for example Theorem 2.2.3 in the aforementioned reference.

One immediate consequence of the above linear contraction guarantees is that they allow us to derive explicit complexity bounds for a desired level of suboptimality of the solution to problem (4.6). In fact, (4.12) directly implies that the maximal number of algorithm iterations that is required to arrive at an ϵ-suboptimal solution with $\|U_k - \hat{U}^*(x)\| \leq \epsilon$ can be upper bounded by

$$\bar{k} = \log_{\frac{1}{\sigma}} \left(\frac{c}{\epsilon}\right) . \tag{4.13}$$

This shows that for the considered class of relaxed barrier function based MPC problems, the computational complexity of computing an approximate solution with a desired level of optimality based on gradient or accelerated gradient methods can be estimated a priori. This result applies in principle also to other first order methods like the heavy-ball method, the conjugated gradient method, or the Quasi-Newton method, which are not discussed here for the sake of brevity. We want to emphasize at this point again that the required constants μ and L can for the presented MPC approach be computed explicitly and that explicit expressions are given in the proof of Theorem 4.1.

As our above analysis revealed that the cost function in (4.6) enjoys beneficial second order regularity properties, is is only natural to consider also second order optimization methods. In particular, a canonical alternative to the hitherto discussed first order methods is given by the multi-dimensional Newton method for unconstrained optimization, which for the considered optimization problem reads

$$U_{k+1} = U_k + s_k \Delta U_k , \qquad \Delta U_k := -\left(\nabla^2_U \hat{J}_N(U_k, x)\right)^{-1} \nabla_U \hat{J}_N(U_k, x) , \tag{4.14}$$

where $s_k \in (0, 1]$ denotes a suitably chosen, possibly time-varying step size. Making use of an Armijo backtracking line search, the step size may for example be chosen as $s_k = \beta^{j_k}$, where j_k is the first nonnegative integer $j \in \mathbb{N}$ that ensures for suitably chosen line search parameters $\alpha, \beta \in (0, 1)$ satisfaction of the Armijo condition

$$\hat{J}_N(U_k + \beta^j \Delta U_k, x) \leq \hat{J}_N(U_k, x) + \alpha \beta^j \, \nabla_U \hat{J}_N(U_k, x)^\top \Delta U_k . \tag{4.15}$$

More details on suitable line search procedures as well as on the resulting convergence behavior of Newton's method can be found in (Ortega and Rheinboldt, 1970; Nesterov and Nemirovskii, 1994; Nocedal and Wright, 1999; Nesterov, 2004; Boyd and Vandenberghe, 2004) as well as in probably almost any other textbook on unconstrained optimization methods. One main problem in the convergence analysis is that in order to ensure global convergence of the Newton method, one typically needs to distinguish two separate phases – usually referred to as the *damped Newton phase* and the *quadratically convergent phase*. Exploiting strong convexity as well as Lipschitz continuity of both the first and the second derivative (that is, exactly the properties that were for the considered problem setup proven in Theorem 4.1), it is nevertheless possible to derive upper bounds on the number of required algorithm iterations, see for example Nesterov (2004) as well as Section 9.5.3 in the book of Boyd and Vandenberghe (2004). In particular, applying the general bound given in the latter reference to our problem setup, it is straightforward to show that the maximal number of algorithm iterations that is required to arrive at an ϵ-suboptimal solution (meaning now $\hat{J}_N(U_k, x) - \hat{J}_N^*(x) \leq \epsilon$) can be upper bounded by

$$\bar{k} = \frac{L'^2 L^2/\mu^5}{\alpha\beta \min\{1, 9(1-2\alpha)^2\}}\, \epsilon_0 + \log_2\left(\log_2\left(\frac{2\mu^3}{L^2\epsilon}\right)\right), \tag{4.16}$$

where $\epsilon_0 := \hat{J}_N(U_0, x) - \hat{J}_N^*(x)$ denotes the initial error and α, β are the backtracking line search parameter, assuming that $\alpha < \frac{1}{2}$. Note that strong convexity implies $\frac{\mu}{2}\|U_k - \hat{U}^*(x)\|^2 \leq \hat{J}_N(U_k, x) - \hat{J}_N^*(x) \leq \epsilon$, such that (4.16) leads in a straightforward fashion also to an upper bound for the resulting suboptimality with respect to the optimization variable, that is, $\|U_k - \hat{U}^*(x)\| \leq \epsilon' = \sqrt{2\epsilon/\mu}$ for all $k \geq \bar{k}$. Again, all the required constants can be computed explicitly and are provided by Theorem 4.1. Finally, coming back to the last result of Theorem 4.1, we observe that also the whole theory of self-concordant function calculus as well as the important role that self-concordant functions play in the analysis of Newton's method can be applied to our relaxed barrier function based problem setup. In particular, the convexity and regularity properties given in (4.8) and (4.9) can essentially be replaced by the property of self-concordance, which then allows for a refined convergence analysis that does not depend on the constants μ, L, and L', and is therefore invariant under affine coordinate changes (Boyd and Vandenberghe, 2004). An example for such a complexity bound is given in Section 9.6.4 of the aforementioned reference, revealing that the number of Newton iterations that is required to arrive at an ϵ-suboptimal solution can be upper bounded by

$$\bar{k} = \frac{20 - 8\alpha}{\alpha\beta(1-2\alpha)^2}\, \epsilon_0 + \log_2\left(\log_2\left(\frac{1}{\epsilon}\right)\right). \tag{4.17}$$

Note that this complexity bound now depends solely on the desired level of suboptimality with respect to the cost function, the initial error $\epsilon_0 := \hat{J}_N(U_0, x) - \hat{J}_N^*(x)$, as well as on the backtracking line search parameters α and β. Less conservative upper bounds may be derived based on a more sophisticated convergence analysis, see for example (Nesterov and Nemirovskii, 1994; Nesterov, 2004).

4.2.3 A first step towards MPC algorithms with systems theoretic guarantees

The results of the previous two subsections showed that the relaxed barrier function based MPC formulation leads to a quite special and attractive class of optimization problems, which can be handled by means of efficient and well-understood convex optimization methods. In the following, we want to discuss briefly how these findings from the algorithmic domain may be used in taking a first step towards the design and analysis of MPC algorithms that allow not only to estimate the computational complexity of the underlying numerical optimization but also to give guarantees on the stability properties of the resulting closed-loop system. To this end, let us consider the following system

$$x(k+1) = Ax(k) + B\Pi_0\, \hat{U}(k), \qquad \Pi_0 = \begin{bmatrix} I_m & 0 & \cdots & 0 \end{bmatrix}, \tag{4.18}$$

where $\hat{U}(k)$ denotes the (now potentially suboptimal) solution to problem (4.6) at sampling instant $k \in \mathbb{N}$. Following our above discussion, we assume that the underlying optimization is performed by means of a suitable first or second order method, leading to a suboptimality of $\epsilon(k) \in \mathbb{R}_+$, that is,

$$\|\hat{U}(k) - \hat{U}^*(x(k))\| \leq \epsilon(k)\,. \tag{4.19}$$

From a practical point of view, this setup may for example be related to a scenario in which only a fixed number of optimization algorithms is performed at each sampling step, which then causes the optimization error to vary depending on the underlying initialization or warm-start procedure. On the other hand, one might also assume that the number of iterations is always chosen large enough in order to ensure a given desired level of suboptimality ϵ, in which case $\epsilon(k) \leq \epsilon$ for any $k \in \mathbb{N}$. Making use of our theoretical results from the first part of the thesis, we can state the following theorem.

Theorem 4.2. *Consider the relaxed barrier function based MPC formulation (4.3)–(4.5) and assume that the associated optimization problem (4.6) is at each sampling instant $k \in \mathbb{N}$ solved with suboptimality $\epsilon(k)$, resulting in the closed-loop system (4.18)–(4.19). Then, there exist $\beta \in \mathcal{KL}$ and $\gamma \in \mathcal{K}$ such that the closed-loop system state satisfies for any $x(0) \in \mathbb{R}^n$ and any $k \in \mathbb{N}$*

$$\|x(k)\| \leq \beta(\|x(0)\|, k) + \gamma(\|\epsilon\|_{[k-1]})\,. \tag{4.20}$$

In fact, this result follows almost immediately from the ISS properties of the underlying relaxed barrier function based MPC approach, see Section 3.6.2 of the previous chapter. For the sake of completeness, a proof of Theorem 4.2 is presented in Appendix C.19. Note that a rather direct and intuitive consequence of Theorem 4.2 is that asymptotic stability of the origin will be recovered for $\epsilon(k) = 0\ \forall k \in \mathbb{N}$, i.e., whenever problem (4.6) is at each sampling instant solved to exact optimality. Moreover, it provides us with a practical stability result that allows us to characterize the behavior of the closed-loop system in dependence of the optimization error $\epsilon(k)$. Thus, the presented results may be seen as a first step towards algorithmic MPC approaches that allow to drop the idealizing assumption that always the *exact* optimal solution is applied to the plant. However, there are also several important drawbacks related to the presented analysis approach. First, when assuming a fixed upper bound on the number of optimization algorithm iterations, the resulting suboptimality, i.e., the evolution of the sequence ϵ, will typically be unknown and hard to predict. The main reason for this is that the optimal solution, and therefore potentially also the initial optimization error, will change at each sampling instant due to the evolving system state. As the ISS gains in (4.20) may in addition also be quite conservative, it might therefore actually be hard to give in this case more than qualitative guarantees on the behavior of the closed-loop system state. Second, when aiming on the other hand to upper bound or even control the level of suboptimality at each sampling step, it may be hard to derive explicit a priori estimates for the required number of algorithm iterations (the reason being again that the evolution of the initial optimization error is typically not known). Thus, also in this case, the achievable guarantees on the algorithmic properties are more of a qualitative nature in the sense that convergence of the underlying optimization is guaranteed and that the complexity estimates presented in Section 4.2.2 hold. Third, arguing from a more conceptual perspective, it should be noted that the above procedure is no *MPC algorithm* in the sense that we defined at the beginning of this chapter. In particular, the influence of the underlying optimization is solely captured by the resulting optimization error and without taking the respective optimization algorithm *dynamics* explicitly into account. In fact, neglecting the underlying algorithm dynamics and treating the optimization algorithm, respectively the associated optimization error, as an external disturbance to the system makes it difficult to analyze further systems theoretic properties such as closed-loop robustness or the satisfaction of input and state constraints. Novel classes of MPC algorithms which essentially allow us to eliminate all of these drawbacks will be presented and analyzed in Sections 4.3 and 4.4.

4.2.4 Discussion

As one of the main results of this section, Theorem 4.1 revealed that the relaxed barrier function based MPC formulation from Chapter 3 actually results in a twice continuously differentiable, strongly convex, and self-concordant cost function, which satisfies in addition important (and explicitly computable) Lipschitz conditions on the associated gradient and Hessian. As outlined in our rather informal discussion in Section 4.2.2, this enables us to make use of standard convex optimization methods like (accelerated) gradient methods or the Newton method, as well as to directly apply all the theoretical convergence results and complexity estimates that are available for these methods. Thus, as a first main result of this chapter, we can conclude that the relaxed barrier function based MPC approach leads – in addition to the interesting theoretical properties discussed in Chapter 3 – also from the perspective of algorithmic implementation to a highly attractive and numerically beneficial problem formulation. However, as revealed by our discussion in Section 4.2.3, the resulting complexity and stability certificates tend to be rather conservative, which necessitates a more thorough analysis when aiming for the design of MPC algorithms with practically relevant systems theoretic guarantees. The design of such stability preserving MPC algorithms as well as the question whether and how the underlying optimization methods can be tailored further to the relaxed barrier function based formulation will be the topic of the next sections and, in fact, of the whole rest of this chapter.

4.3 Stabilizing iteration schemes and anytime MPC algorithms

As outlined above, the idealizing assumption that the exact optimal control input is available immediately upon measurement of the current system state might not be valid when considering applications in which the optimization algorithm operates on the same time-scale as the system to be controlled – for example due to fast system dynamics and/or the use of low-cost hardware. In this case, special measures have to be taken in order to ensure stability, or even mere recursive feasibility, of the closed-loop system. Two rather common approaches in this context are to speed up the optimization accordingly and then employ again tacitly the aforementioned time-scale separation argument (e.g. Bemporad et al., 2002; Tøndel et al., 2003; Ferreau et al., 2008; Wang and Boyd, 2010; Kögel and Findeisen, 2011a; Richter et al., 2012; Domahidi et al., 2012; Patrinos and Bemporad, 2014; Giselsson, 2014) or to settle for approximate solutions and then take the suboptimality of the applied control input explicitly into account in the respective stability analysis, see for example (Scokaert et al., 1999; McGovern and Feron, 1999; Pannocchia et al., 2011; Zeilinger and Jones, 2011; Zeilinger et al., 2014b; Rubagotti et al., 2014; Bemporad et al., 2015; Alamir, 2015) as well as Section 4.2 of this thesis. However, there are only few works which explicitly consider the *dynamics* of the optimization algorithm and acknowledge the fact that the optimization typically needs to be performed in parallel to the evolving dynamics of the plant. Interesting exceptions are given by the works of Diehl et al. (2005, 2007), in which the authors present, for the case of unconstrained nonlinear systems, a so-called *real-time iteration scheme* that performs only one Newton-based optimizer update per sampling instant. In particular, the computation is split up into a preparation phase (running in parallel with the evolving system dynamics) and a, possibly much shorter, feedback phase (starting as soon as the next state measurement becomes available). The authors rigorously analyze the combined dynamics of plant and optimizer state and provide a proof that the state of the corresponding overall closed-loop system will asymptotically converge to the origin. However, no input and state constraints are considered and the theoretical analysis is, as the authors themselves emphasize, carried out based on some technical assumptions that might be hard to verify for general practical applications. Nevertheless, the results represent a theoretical underpinning of the proposed iteration scheme and nicely illustrate possible benefits that might be gained by combining ideas from the areas of MPC stability theory and optimization algorithms.

In this section, we follow a similar approach and show that the relaxed barrier function based framework presented in this thesis leads in a quite natural way to the design of stabilizing MPC iteration schemes with desirable systems theoretic guarantees. In contrast to the aforementioned works by Diehl et al., we focus in the following on the above linear MPC problem setup with polytopic input and state constraints and analyze, based on easily verifiable standard linear MPC assumptions, not only the (robust) stability properties of the combined state-optimizer dynamics but also the satisfaction of input and state constraints – both when performing at each sampling instant only a limited number of optimization algorithm iterations. The overall main result is a novel class of *anytime MPC algorithms* that allow one to guarantee important systems theoretic properties of the closed-loop system after each internal iteration of the underlying optimization algorithm and exhibit therefore inherently various desirable real-time capabilities. The section is structured as follows. An MPC iteration scheme for the condensed relaxed barrier function based problem formulation is introduced and discussed in Section 4.3.1. The stability, constraint satisfaction, and robustness properties of the associated overall closed-loop system are discussed in the Sections 4.3.2, 4.3.3, and 4.3.4. Possible choices for the underlying optimization algorithm are discussed in Section 4.3.5, while a summarizing discussion and a numerical example are presented in the Sections 4.3.6 and 4.3.7.

4.3.1 An MPC iteration scheme based on relaxed barrier functions

In the following, we present a rather general and abstract MPC iteration scheme that exploits many of the characteristic properties of the relaxed barrier function based formulation and allows us to analyze the dynamics of both the controlled system and the underlying optimization algorithm in an integrative fashion. In particular, in order to design an MPC algorithm that is able to return a meaningful control input directly when the next state measurement becomes available, the rather intuitive approach is to predict at each sampling instant the next nominal system state, start the optimization procedure for this predicted state by making use of a suitably shifted warm-start solution, and then, at the end of the available computation time, simply apply the first element of the current, possibly suboptimal, optimizer state. Aiming for a more systems theoretic interpretation and analysis, we write the corresponding closed-loop system dynamics as

$$x(k+1) = A\,x(k) + B\Pi_0\,U(k)\,, \qquad x(0) = x_0\,, \tag{4.21a}$$

$$U(k+1) = \Phi^{i_{\mathrm{T}}(k)}\left(U(k), x(k)\right)\,, \qquad U(0) = U_0\,, \tag{4.21b}$$

where (4.21a) describes the dynamics of the controlled linear plant with the matrix $\Pi_0 \in \mathbb{R}^{m \times Nm}$ being defined as in (4.7). Note that the system state dynamics are directly coupled with the dynamics of the optimization algorithm, which are described by (4.21b). Here, $\Phi^{i_{\mathrm{T}}(k)} : \mathbb{R}^{Nm} \times \mathbb{R}^{n} \to \mathbb{R}^{Nm}$ denotes a suitable *optimization algorithm operator* that describes how the next optimizer state is computed at each sampling instant $k \in \mathbb{N}$ based on the current optimizer state $U(k)$ and the measured system state $x(k)$ by performing $i_{\mathrm{T}}(k) \in \mathbb{N}_+$ optimization algorithm iterations. To be more precise, for given *termination index* $i_{\mathrm{T}}(k) \in \mathbb{N}_+$, we define the operator $\Phi^{i_{\mathrm{T}}(k)}$ recursively as

$$\Phi^{0}(U,x) = \Psi_{\mathrm{s}}(U,x) \tag{4.22a}$$

$$\Phi^{i}(U,x) = \Psi_{\mathrm{o}}\Big(\Phi^{i-1}(U,x), Ax + B\Pi_0 U\Big)\,. \tag{4.22b}$$

Here, (4.22a) characterizes how a suitable warm-start solution for the predicted next system state $x^+ = Ax + B\Pi_0 U$ is obtained in the first step via the *shift operator*

$$\Psi_{\mathrm{s}}(U,x) = \begin{bmatrix} u_1^\top & u_2^\top & \cdots & u_{N-1}^\top & k_f^\top(U,x) \end{bmatrix}^\top\,, \tag{4.23}$$

with $k_f : \mathbb{R}^{Nm} \times \mathbb{R}^{n} \to \mathbb{R}^{m}$, while all subsequent iterates are computed via a suitably chosen *optimizer update operator* $\Psi_{\mathrm{o}} : \mathbb{R}^{Nm} \times \mathbb{R}^{n} \to \mathbb{R}^{Nm}$. A graphical illustration of the main idea is given in Figure 4.1 on the next page, highlighting in particular also the different roles of the shift and optimizer update operators. Obviously, the choice of the function $k_f(U,x)$ as well as of the optimizer update operator $\Psi_{\mathrm{o}}(U,x)$ is crucial for analyzing and influencing the behavior of the closed-loop system (4.21). The main idea behind the following results is to ensure that both the applied warm-start procedure and the subsequent iterative optimization lead to a guaranteed monotonic decay in the cost function (4.6), despite the fact that the system state will change between two consecutive sampling instants.

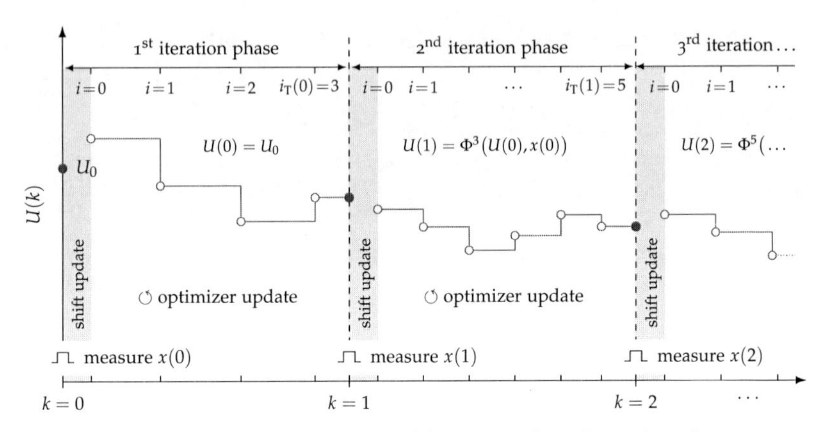

Figure 4.1. A graphical illustration of the proposed MPC iteration scheme.

In particular, recalling the stability results presented in Chapter 3, we propose to choose the warm-start solution simply as the shifted, suboptimal candidate solution from the respective MPC stability proof, that is

$$k_f(U,x) = Kx_N(U,x) = K_U U + K_x x \tag{4.24}$$

where $K_U := K[A^{N-1}B, \cdots, B]$, $K_x = KA^N$, and the auxiliary control gain $K \in \mathbb{R}^{m \times n}$ is chosen according to (4.5). Furthermore, we make the following assumption concerning the applied optimizer update operator.

Assumption 4.1. *Given the cost function* $\hat{J}_N : \mathbb{R}^{Nm} \times \mathbb{R}^n \to \mathbb{R}_+$ *in (4.6), the optimizer update operator* $\Psi_o : \mathbb{R}^{Nm} \times \mathbb{R}^n \to \mathbb{R}^{Nm}$ *satisfies for any* $(U,x) \in \mathbb{R}^{Nm} \times \mathbb{R}^n$ *the descent condition*

$$\hat{J}_N(\Psi_o(U,x),x) - \hat{J}_N(U,x) \leq -\gamma(U,x)\,, \tag{4.25}$$

where $\gamma : \mathbb{R}^{Nm} \times \mathbb{R}^n \to \mathbb{R}_+$ *is a suitable function with* $\gamma(U,x) \geq 0$ *for all* $(U,x) \in \mathbb{R}^{Nm} \times \mathbb{R}^n$ *and* $\gamma(U,x) = 0$ *if an only if* $\nabla_U \hat{J}_N(U,x) = 0$.

While we do for the moment not specify the optimizer update operator explicitly, different suitable realizations will be discussed later in Section 4.3.5, where we will also relate the function $\gamma(\cdot,\cdot)$ directly to the step size and the search direction of several well-known line-search methods. For the time being, we may simply think of the resulting operator $\Phi^{i_T(k)}$ as an iterative optimization procedure that improves the current solution by applying a fixed number of suitable optimizer updates to the shifted warm-start solution (4.23). We now begin our systems theoretic analysis of the overall closed-loop system by showing that, under the above assumptions, the origin of system (4.21) will be asymptotically stable *independently* of the number $i_T(k)$ of performed optimizer updates.

4.3.2 Stability properties of the closed-loop system

Based on the properties of the relaxed barrier function based MPC formulation and the above assumptions, we can state the following main result concerning the stability properties of the proposed MPC iteration scheme. A proof is given in Appendix C.20.

Theorem 4.3. *Consider the relaxed barrier function based MPC formulation (4.3)–(4.5) and let $\hat{J}_N : \mathbb{R}^{Nm} \times \mathbb{R}^n \to \mathbb{R}_+$ given in (4.6) refer to the associated cost function. Moreover, consider the closed-loop system (4.21) with the operator $\Phi^{i_T(k)}$ being defined according to (4.22)–(4.24), and let Assumption 4.1 hold. Then, for any sequence $\boldsymbol{i}_T = \{i_T(0), i_T(1), \ldots\}$, the origin $(U, x) = (0, 0)$ of system (4.21) is globally asymptotically stable.*

On the one hand, a direct consequence of Theorem 4.3 is that a fixed limit on the number of optimization algorithm iterations may be chosen by the control engineer without jeopardizing the stability properties of the overall closed-loop system – including the case of performing only *one* optimizing step at each sampling instant. On the other hand, $i_T(k)$ may also be defined implicitly as the (possibly time-varying) number of iterations after which the optimization algorithm is forced to return a solution. In this case, the number of performed optimizer updates can be seen as an uncertain and a priori unknown variable that will in general depend on the available computation time as well as on the available hardware resources, which may in practical applications both be varying over time. In the light of the above stability result, the proposed iteration scheme can therefore be seen as a stabilizing *anytime MPC algorithm* (Fontanelli et al., 2008; Bemporad et al., 2015) in the sense that it returns a valid and asymptotically stabilizing control input after each of its internal iterations and successively produces solutions with decreasing levels of suboptimality. Depending on the sampling time and the available computation power, the resulting control inputs will range from potentially highly suboptimal solutions ($i_T(k) = 1$) to the exact optimal – or at least only slightly suboptimal – solutions related to $i_T(k) \gg 1$. This in particular shows that the closed-loop behavior and performance of conventional MPC schemes that are based on applying the *exact* optimal solution can be recovered whenever the oftentimes inherently assumed time-scale separation argument holds.

As revealed by the proof of Theorem 4.3, the crucial ingredients to the obtained stability results and the associated anytime properties are that the relaxed barrier function based formulation inherently leads to a continuous, positive definite, and radially unbounded cost function, which is guaranteed to decrease under the warm-starting shift update given in (4.23)–(4.24). Moreover, as will be discussed in more detail in Section 4.3.5, Assumption 4.1 can for the underlying strongly convex and unconstrained problem formulation be satisfied based on numerically efficient and off-the-shelf line-search methods. The presented stability results follow therefore almost directly from the properties of the relaxed barrier function based MPC approach, and no additional assumptions or stabilizing modifications are needed. In this, the proposed iteration scheme differs from many existing approaches and results from the literature. In particular, it should be noted that the proposed iteration scheme differs from early termination of a conventional interior point method in the sense that the barrier functions are relaxed and recentered, that the associated weighting parameter ε is kept fixed instead of being gradually decreased, that the barrier functions are explicitly taken into account in the design of the underlying MPC formulation, and that the warm-starting is systematically used for ensuring stability of the overall closed-loop system. Furthermore, it differs from other suboptimal MPC approaches, as for example presented by Scokaert et al. (1999), Zeilinger et al. (2014b), and Bemporad et al. (2015), in the sense that the decay of the cost function is not enforced explicitly by adding a stabilizing contraction constraint to the MPC formulation.

It should be emphasized that procedures like computing the control input with a one-step-ahead prediction or performing only a limited number of optimization algorithm iterations are of course well-known in the literature, see for example (Zavala and Biegler, 2009; Scokaert et al., 1999; Wang and Boyd, 2010; Pannocchia et al., 2011) and, in particular, (Diehl et al., 2005, 2007). However, no iteration scheme of the form (4.21) with stability properties as in Theorem 4.3 has been presented so far. In fact, it appears that the underlying recentered barrier function based formulation with a fixed choice for the barrier function weighting parameter is actually very beneficial for deriving the presented stability results. Only the recentering enabled us to use the resulting positive definite cost function as Lyapunov function for the closed-loop system, while only the choice of a fixed barrier function parameter allowed to establish the necessary decay of the cost function under the proposed warm-starting procedure. Further beneficial properties of the proposed iteration scheme such as constraint satisfaction and robustness to external disturbances and observer errors will be discussed in the following two sections.

Remark 4.1. Note that Theorem 4.3 implies that if $\|U_0\| \leq \alpha_0(\|x_0\|)$ for some $\alpha_0 \in \mathcal{K}_\infty$, then the controlled system state dynamics (4.21a) will be asymptotically stable in the sense that there exists $\beta_\mathrm{x} \in \mathcal{KL}$ such that for any $x(0) = x_0 \in \mathbb{R}^n$ and any $k \in \mathbb{N}$

$$\|x(k)\| \leq \beta_\mathrm{x}(\|x_0\|, k)\,. \tag{4.26}$$

In particular, if $\beta \in \mathcal{KL}$ is the $\mathcal{KL}$ function related to the stable overall system dynamics (4.21), then $\beta_\mathrm{x}(r,s) = \beta((\alpha_0 + \mathrm{id})(r), s)$. Suitable initializations with the above property are for example given by $U_0 = \hat{U}^*(x_0)$, that is, the optimal input vector for the initial condition x_0, by $U_0 = \bar{K}x_0$ with $\bar{K} \in \mathbb{R}^{Nm \times n}$, or simply by $U_0 = 0$. Note however that (4.21a) is not an autonomous system, such that we can strictly speaking not apply the standard definition for asymptotic stability of an equilibrium point.

Remark 4.2. As usual in MPC, a conceptually very simple "algorithm" leading to similar stability results can be obtained by applying at each sampling instant simply the shifted open-loop candidate solution without performing any optimization, i.e., $i_\mathrm{T}(k) \equiv 0$. However, this will typically result in a significant degradation of closed-loop performance (especially for bad initializations U_0) and is also not robust in the presence of plant-model mismatch or external disturbances. Thus, in practice one typically performs at least one optimizing iteration in order to provide feedback of the current system state. The above results ensure that asymptotic stability of the overall closed-loop system can in this case be guaranteed for any number of optimization algorithm iterations.

4.3.3 Constraint satisfaction in closed-loop operation

The importance of upper bounds on the maximal level of constraint violations as well as some interesting results for the *exact* barrier function based state feedback case have already been discussed in Chapter 3. In the following, we present similar results for the above MPC iteration scheme. Interestingly, the resulting constraint satisfaction guarantees now not only depend on the choice of the relaxation parameter but also on the initialization U_0 of the optimizer iteration. We begin our studies with the following Theorem, which provides an upper bound on the possible maximal constraint violation that is in addition shown to be monotonically decreasing over time, cf. Theorem 3.6 in Section 3.5.

Theorem 4.4. *Consider the relaxed barrier function based MPC formulation (4.3)–(4.5) and let $\hat{J}_N : \mathbb{R}^{Nm} \times \mathbb{R}^n \to \mathbb{R}_+$ given in (4.6) refer to the associated cost function. Moreover, consider the closed-loop system (4.21) with the operator $\Phi^{i_\mathrm{T}(k)}$ being defined according to (4.22)–(4.25), and let $\{x(0), x(1), \ldots\}$ and $\{u(0), u(1), \ldots\}$ with $u(k) = \Pi_0 U(k)$ denote the resulting closed-loop state and input trajectories. Then, for any initialization $(U_0, x_0) \in \mathbb{R}^{Nm} \times \mathbb{R}^n$, any sequence $i_\mathrm{T} = \{i_\mathrm{T}(0), i_\mathrm{T}(1), \ldots\}$, and any $k \in \mathbb{N}$, it holds that*

$$C_\mathrm{x} x(k) \leq d_\mathrm{x} + \hat{z}_\mathrm{x}(k), \quad C_\mathrm{u} u(k) \leq d_\mathrm{u} + \hat{z}_\mathrm{u}(k), \tag{4.27}$$

where the elements of the maximal constraint violation vectors $\hat{z}_\mathrm{x}(k) \in \mathbb{R}^{q_\mathrm{x}}$ and $\hat{z}_\mathrm{u}(k) \in \mathbb{R}^{q_\mathrm{u}}$ decrease strictly monotonically over time. Furthermore, for each $k \in \mathbb{N}$, the respective elements can be computed explicitly by solving a convex optimization problem, and there exists a finite $k_0 \in \mathbb{N}$ such that $\hat{z}_\mathrm{x}(k) \leq 0$, $\hat{z}_\mathrm{u}(k) \leq 0$ for any $k \geq k_0$.

In fact, as it relies again mainly on the monotonic decay of the cost function, the proof of this result is very similar to that of Theorem 3.6. The full proof as well as explicit characterizations of the convex optimization problems that may be used for computing the upper bounds are given in Appendix C.21. Note that Theorem 4.4 implies that the maximally possible violations of the state and input constraints will for all time instants $k \in \mathbb{N}$ be upper bounded by $\hat{z}_\mathrm{x}(0)$ and $\hat{z}_\mathrm{u}(0)$, which can for any given initial state $x(0) \in \mathbb{R}^n$ be computed by solving the convex optimization problems stated in (C.122) for $\hat{\alpha}(0) := \hat{J}_N^*(x(0)) - x(0)^\top P_\mathrm{uc}^* x(0)$. Here, $P_\mathrm{uc}^* \in \mathbb{S}_{++}^n$ denotes again the solution to the associated infinite-horizon LQR problem. Thus, for any initialization (U_0, x_0) and any choice of the relaxation parameter δ, an upper bound for the maximal possible constraint violation can be computed a priori.

In addition, being effectively the counterpart to Theorem 3.7, the following result shows that we can in many cases again control and guarantee a desired level of approximate (or even exact) constraint satisfaction by choosing the relaxation parameter as well as the initialization of the optimizer dynamics in a suitable way. In particular, the core message of the following theorem is that, for a suitable set of feasible initial conditions, the input and state constraints can be satisfied with any desired tolerance if we choose δ small enough and initialize the iteration scheme good enough, that is, close enough to the optimal solution. A proof of Theorem 4.5 can be found in Appendix C.22.

Theorem 4.5. *Consider the relaxed barrier function based MPC formulation (4.3)–(4.5) and let $\hat{J}_N : \mathbb{R}^{Nm} \times \mathbb{R}^n \to \mathbb{R}_+$ given in (4.6) refer to the associated cost function. Moreover, consider the closed-loop system (4.21) with the operator $\Phi^{i_T(k)}$ being defined according to (4.22)–(4.25), and let $\{x(0), x(1), \ldots\}$ and $\{u(0), u(1), \ldots\}$ with $u(k) = \Pi_0 U(k)$ denote the resulting closed-loop state and input trajectories. Assume that $\begin{bmatrix} A^{N-1}B & \cdots & AB & B \end{bmatrix}$ has full row rank n and let the set $\mathcal{X}_N \subseteq \mathcal{X}$ be defined as*

$$\mathcal{X}_N := \left\{ x \in \mathcal{X} \mid \exists U \in \mathbb{R}^{Nm} \text{ s.t. } u_k \in \mathcal{U}, x_k(U,x) \in \mathcal{X}, x_N(U,x) = 0 \right\} . \tag{4.28}$$

Then, for any compact set $\mathcal{X}_0 \subseteq \mathcal{X}_N^\circ$ and any $\hat{z}_{\mathrm{x,tol}} \in \mathbb{R}_+^{q_x}$, $\hat{z}_{\mathrm{u,tol}} \in \mathbb{R}_+^{q_u}$, there exists $\bar{\delta}_0 \in \mathbb{R}_{++}$ and $\gamma \in \mathcal{K}_\infty$ such that for any relaxation parameter $0 < \delta \leq \bar{\delta}_0$, any $x_0 \in \mathcal{X}_0$, any initialization $U_0 \in \mathcal{B}_{\gamma(\delta)}^{Nm}(\hat{U}^*(x_0))$, any $\boldsymbol{i}_{\mathrm{T}} = \{i_{\mathrm{T}}(0), i_{\mathrm{T}}(1), \ldots\}$, and any $k \in \mathbb{N}$, it holds that

$$C_{\mathrm{x}}\, x(k) \leq d_{\mathrm{x}} + \hat{z}_{\mathrm{x,tol}} , \qquad C_{\mathrm{u}}\, u(k) \leq d_{\mathrm{u}} + \hat{z}_{\mathrm{u,tol}} . \tag{4.29}$$

Note that, similar to Section 3.5, the above constraint satisfaction results, particularly the restriction of x_0 to the set $\mathcal{X}_N^\circ$ and of U_0 to a ball around $\hat{U}^*(x_0)$, are more of a conceptual nature and tend to be rather conservative when compared to the actual closed-loop behavior. In fact, as also illustrated by the numerical examples presented in Section 4.3.7, a satisfactory behavior of the closed-loop system may also be achieved for suboptimal initializations of the form $U_0 = K x_0$ as long as the relaxation parameter δ is chosen sufficiently small. Nevertheless, the fact that the maximal possible constraint violation is monotonically decaying over time while any desired level of constraint satisfaction can in principle be achieved by a suitable choice of the underlying parameters represents an important and non-trivial conceptual result that, in particular, also provides a theoretical justification for constraint satisfaction behavior that can typically be observed in both numerical simulations and practical applications.

Remark 4.3. Note that the underlying barrier function relaxation is not really crucial for the obtained stability results, which reveals that the discussed iteration scheme may also be applied to the nonrelaxed barrier function based MPC approaches in (Wills, 2003; Wills and Heath, 2004) and (Feller and Ebenbauer, 2013, 2014b, 2015a), leading to local versions of the above stability results under strict satisfaction of input and state constraints. However, in this case the optimizer iteration needs to be initialized with a strictly feasible U_0, which may be hard to compute a priori. Moreover, all the robustness properties discussed in the following section as well as some of the numerical advantages discussed in Section 4.3.5 will be lost when making use of nonrelaxed logarithmic barrier functions.

4.3.4 Robustness properties

In the following, we want to study the robust stability properties of the proposed, nominally stabilizing MPC iteration scheme introduced in Section 4.3. Similar to Section 3.5, we make use of suitable input-to-state stability concepts and study robustness with respect to both external and internal disturbances (e.g. estimation errors). To this end, let us consider the following disturbance-affected version of (4.21)

$$x(k+1) = A\,x(k) + B\Pi_0\,U(k) + w(k)\,, \tag{4.30a}$$

$$U(k+1) = \Phi^{i_{\mathrm{T}}(k)}\left(U(k), x(k) + e(k)\right)\,, \tag{4.30b}$$

in which $w(k) \in \mathbb{R}^n$ may be seen as external input or state disturbance, while $e(k) \in \mathbb{R}^n$ models the effect that the underlying optimization is not performed for the exact system state. In particular, $e(k)$ may be given by the state measurement or estimation error $e(k) = \hat{x}(k) - x(k)$, where $\hat{x}(k)$ denotes the current state measurement or estimate. Especially in the latter case, $e(k)$ thus rather represents an internal than an external disturbance. Concerning the stability properties of system (4.30), we have the following result.

Theorem 4.6. *Consider the relaxed barrier function based MPC formulation (4.3)–(4.5) and let $\hat{J}_N : \mathbb{R}^{Nm} \times \mathbb{R}^n \to \mathbb{R}_+$ given in (4.6) refer to the associated cost function. Moreover, consider the closed-loop system (4.30) with the operator $\Phi^{i_{\mathrm{T}}(k)}$ defined according to (4.22)–(4.25). Assume that $\Phi^{i_{\mathrm{T}}(k)} : \mathbb{R}^{Nm} \times \mathbb{R}^n \to \mathbb{R}^{Nm}$ is continuous. Then, for any sequence $\boldsymbol{i}_{\mathrm{T}} = \{i_{\mathrm{T}}(0), i_{\mathrm{T}}(1), \ldots\}$, system (4.30) is integral input-to-state stable with respect to the disturbance sequences $\boldsymbol{w}$ and $\boldsymbol{e}$.*

The concept of integral input-to-state stability is recalled in Appendix A.2, while a proof of this result can be found in Appendix C.23. A direct consequence of Theorem 4.6 is that there exist $\beta \in \mathcal{KL}$ and $\sigma \in \mathcal{K}_\infty$ such that for any $k \in \mathbb{N}$

$$\|z(k)\| \leq \beta(\|\zeta\|, k) + \sum_{i=0}^{k-1} \sigma(\|(w(i), e(i))\|) \tag{4.31}$$

for all bounded sequences $\boldsymbol{w}$, $\boldsymbol{e}$, where $z = (x, U)$ and $\zeta = (x(0), U(0)) \in \mathbb{R}^n \times \mathbb{R}^{Nm}$. A direct implication of this result is the the so-called *bounded energy converging state property* (Sontag, 1998; Angeli, 1999), which states that

$$\sum_{i=0}^{\infty} \sigma(\|(w(i), e(i))\|) < \infty \Rightarrow \lim_{k \to \infty} \|z(k)\| = 0\,. \tag{4.32}$$

In order to illustrate the possible implications of this result, let us assume that the disturbances are exponentially decaying, i.e., there exist $c \in \mathbb{R}_{++}, \rho \in [0, 1)$ such that for all $k \in \mathbb{N}$ and all $(w(0), e(0)) \in \mathbb{R}^n \times \mathbb{R}^n$

$$\|(w(k), e(k))\| \leq c\rho^k \|(w(0), e(0))\|\,. \tag{4.33}$$

Assume furthermore that the $\mathcal{K}_\infty$-function σ in (4.31) is polynomially upper bounded in the sense that there exist some $c_r, \kappa \in \mathbb{R}_{++}$ such that $\sigma(r) \leq c_r \max\{r, r^\kappa\}\ \forall r \in \mathbb{R}_+$. Under these assumptions, it is then straightforward to show that the infinite sum on the left-hand side of (4.32) is in fact bounded, which leads to the following corollary.

Corollary 4.1. *Let all assumptions of Theorem 4.6 hold true. Moreover, let (4.31) hold with $\sigma \in \mathcal{K}_\infty$ being polynomially upper bounded and let the disturbance sequences $\boldsymbol{w}$ and $\boldsymbol{e}$ be exponentially decaying according to (4.33). Then, for any sequence $\boldsymbol{i}_{\mathrm{T}} = \{i_{\mathrm{T}}(0), i_{\mathrm{T}}(1), \ldots\}$ and any initialization $(x(0), U(0)) \in \mathbb{R}^n \times \mathbb{R}^{Nm}$, the state of system (4.30) remains bounded and satisfies $\lim_{k \to \infty} \|(x(k), U(k))\| = 0$.*

A particular interesting consequence of Corollary 4.1 is that the overall system state will converge to the origin if there are no (or exponentially decaying) external disturbances and the proposed MPC iteration scheme is combined in a certainty equivalence output feedback fashion with an exponentially stable state estimator. This reveals to some extent a *weak separation principle* for the considered class of MPC iteration schemes in the sense that state estimator and control algorithm can be designed independently from each other, ensuring convergence of the closed-loop system as long as the state estimation error converges exponentially. Note that the latter is for example ensured by the well-known examples of Luenberger observers and Kalman filtering, see Section 3.6.3.

Remark 4.4. Further consequences of the above IISS result are that system (4.30) is *robustly globally asymptotically stable* in the sense that there exist $\beta \in \mathcal{KL}$ and a continuous positive definite function ρ such that $\|z(k))\| \leq \beta(\|\zeta\|, k)$ for all $k \in \mathbb{N}$ and all $\zeta \in \mathbb{R}^n \times \mathbb{R}^{Nm}$ if $\|(w(k), e(k))\| \leq \rho(\|z(k))\|)$, as well as *semi-globally practically asymptotically stable in the worst case size of inner and outer perturbations* in the sense that there exists $\beta \in \mathcal{KL}$ and for any pair $(\Delta_0, \Delta) \in \mathbb{R}^2_{++}$ there exists $\epsilon \in \mathbb{R}_{++}$ such that if $\|(w(k), e(k))\| \leq \epsilon$ and $\|\zeta\| \leq \Delta_0$, then $\|z(k))\| \leq \max\{\beta(\|\zeta\|, k), \Delta\}$ for all $k \in \mathbb{N}$. See (Messina et al., 2005) for more details.

Remark 4.5. Note that $\Phi^{i_T(k)}$ and therefore also the right-hand side of (4.30b) may be discontinuous if $i_T(k)$ or the underlying optimizer step size are chosen based on $(U(k), x(k))$, e.g., if the optimization operator contains some form of gradient-based stopping criterion. However, as long as $\Phi^{i_T(k)}$ is guaranteed to be bounded, arguments like in (Messina et al., 2005) may be used to ensure a similar IISS result also in this case.

Nevertheless, the obtained integral ISS results are obviously weaker than the robust stability results that we presented in Section 3.6 for the exact optimization case. The following discussion can be seen as a first step towards deriving stronger ISS results also for the case when taking suboptimality of the applied control input into account. In particular, we show that by introducing the optimization error that is made at each step as an additional disturbance, the closed-loop system can be shown to satisfy again a conventional ISS property with respect to the previously considered class of disturbances and this very optimization error. As the optimization error can be made arbitrarily small by increasing the number of optimization algorithm iterations, we thus recover in the limit the ISS result that we presented in Section 3.6 for the case of relaxed barrier function based MPC schemes that apply the exact optimal solution.

Theorem 4.7. *Consider the relaxed barrier function based MPC formulation (4.3)–(4.5) and let $\hat{J}_N : \mathbb{R}^{Nm} \times \mathbb{R}^n \to \mathbb{R}_+$ given in (4.6) refer to the associated cost function. Moreover, consider the closed-loop system (4.30) with the operator $\Phi^{i_T(k)}$ defined according to (4.22)–(4.25), and let*

$$\epsilon(k) = \Phi^{i_T(k)}\left(U(k), \hat{x}(k)\right) - \hat{U}^*\left(A\hat{x}(k) + \Pi_0 U(k)\right) \tag{4.34}$$

denote the optimization error that is made at sampling instant $k \in \mathbb{N}$. Then, for any sequence $\boldsymbol{i}_T = \{i_T(0), i_T(1), \ldots\}$, the resulting closed-loop system (4.30) is input-to-state stable with respect to the disturbance sequences $\boldsymbol{w}$, $\boldsymbol{e}$ and the optimization error sequence $\boldsymbol{\epsilon} := \{\epsilon(-1), \epsilon(0), \ldots\}$, where the initial optimization error is defined as $\epsilon(-1) = U(0) - \hat{U}^*(x(0))$.

A proof of this result is given in Appendix C.24. Let us shortly discuss some implications of Theorem 4.7 as well as its relation to Theorem 4.6. Instead of *integral* ISS with respect to w, e, the above arguments allowed us to show ISS with respect to w, e, and ϵ by introducing the latter, representing the sequence of optimization errors, as an additional disturbance. As shown in the proof, this implies that there exist $\beta \in \mathcal{KL}$ and $\gamma_w, \gamma_v, \gamma_\epsilon \in \mathcal{K}$ such that for any $k \in \mathbb{N}_+$

$$\|z(k)\| \leq \beta(\|\zeta\|, k) + \gamma_w(\|w_{[k-1]}\|) + \gamma_e(\|e_{[k-1]}\|) + \gamma_\epsilon(\|\epsilon_{[-1,k-1]}\|) \tag{4.35}$$

for all bounded sequences w, e, where $z = (x, U)$ and $\zeta = (x(0), U(0)) \in \mathbb{R}^n \times \mathbb{R}^{Nm}$. While this seems on the first glance to be a much stronger result, we would like to point out that the disturbances $w(k)$ and $e(k)$ will typically affect the underlying warm start solutions and, hence, for fixed $i_T(k)$, also the resulting optimization error $\epsilon(k)$. In particular, it is not immediately clear that the optimization error will not grow arbitrarily large for bounded disturbances if only a limited number of optimization algorithm iterations is performed at each step. On the other hand, for any $x(k)$, $U(k)$ and any $\bar{\epsilon} \in \mathbb{R}_{++}$ there exists a bounded $i_T(k)$ such that $\epsilon(k) \leq \bar{\epsilon}$. Thus, the above arguments reveal a direct link between the optimization error that is inferred by the iteration scheme at each sampling step (depending on $i_T(k)$) and the robust stability properties of the resulting closed-loop system. While this result for itself may be rather intuitive, the above derivations make the dependency explicit and allow, in particular, to conclude that by choosing $i_T(k)$ arbitrarily large (i.e., making $\epsilon(k)$ arbitrarily small), the proposed MPC iteration scheme is able to recover the ISS properties of conventional relaxed barrier function based MPC, cf. Section 3.6.2. Note however that (4.35) depends also on the initial optimization error $\epsilon(-1) = U(0) - \hat{U}^*(x(0))$, which can not be influenced by the number of optimization algorithm iterations. Thus, in order to fully recover the exact optimal behavior, the iteration scheme would also need to be to initialized in an optimal way.

4.3.5 Choosing the optimizer update operator

Apart from Assumption 4.1, up to now we provided no explicit characterization of the optimizer update operator $\Psi_o : \mathbb{R}^{Nm} \times \mathbb{R}^n \to \mathbb{R}^{Nm}$. In the following, we show that $\Psi_o(\cdot, \cdot)$ may actually be chosen based on a quite general class of optimization methods that guarantee in each step a decay in the respective cost function. In particular, we propose to perform the optimizer update based on a line-search method of the form

$$\Psi_o(U, x) = U + s\, p(U, x), \tag{4.36}$$

where $s \in \mathbb{R}_{++}$ is a step size parameter and $p : \mathbb{R}^{Nm} \times \mathbb{R}^n \to \mathbb{R}^{Nm}$ is chosen according to a a suitable search direction rule. Note that the step size s is typically not fixed but needs to be chosen based on the current search direction at each iteration step. In order to ensure a cost function decrease at each optimizer update, we are primarily interested in search direction and step size selection procedures which ensure that p and s satisfy the so-called strong Wolfe conditions (Nocedal and Wright, 1999), which are for the considered problem setup given by

$$\hat{J}_N(U+sp,x) \leq \hat{J}_N(U,x) + c_1 s \, \nabla_U \hat{J}_N(U,x)^\top p \tag{4.37a}$$

$$|\nabla_U \hat{J}_N(U+sp,x)^\top p| \leq c_2 |\nabla_U \hat{J}_N(U,x)^\top p| \tag{4.37b}$$

where $c_1, c_2 \in \mathbb{R}_{++}$ with $0 < c_1 < c_2 < 1$ are some positive constants. In fact, the Armijo condition (4.37a) ensures that (4.25) will hold with $\gamma(U,x) = -c_1 s \, \nabla_U \hat{J}_N(U,x)^\top p(U,x)$ if $p = p(U,x)$ is chosen as a *descent direction*, i.e., if $\nabla_U \hat{J}_N(U,x)^\top p(U,x) < 0$ whenever $\nabla_U \hat{J}_N(U,x) \neq 0$. The additional curvature condition (4.37b) may be needed in order to ensure that the selected step size s is not too small (Nocedal and Wright, 1999). Suitable choices for $p = p(U,x)$ are for example provided by the gradient method (G), the conjugated gradient method (CG), the Newton method (N), or the Quasi-Newton method (QN), which are for the given problem setup characterized via

$$G: \quad p(U,x) = -\nabla_U \hat{J}_N(U,x), \tag{4.38a}$$

$$CG: \quad p(U,x) = -\nabla_U \hat{J}_N(U,x) + \beta_{CG} \, p_{-1}, \tag{4.38b}$$

$$N: \quad p(U,x) = -\left(\nabla_U^2 \hat{J}_N(U,x)\right)^{-1} \nabla_U \hat{J}_N(U,x), \tag{4.38c}$$

$$QN: \quad p(U,x) = -\hat{B} \, \nabla_U \hat{J}_N(U,x). \tag{4.38d}$$

Here, $\hat{B} \in \mathbb{S}_{++}^{Nm}$ is a suitably updated approximation of the inverse Hessian matrix, for example computed based on the well-known BFGS formula, while $\beta_{CG} \in \mathbb{R}$ in (4.38b) is a scalar that ensures that $p(U,x)$ is conjugate to the previously used search direction p_{-1}, see (Nocedal and Wright, 1999) for more details. As the cost function $\hat{J}_N(\cdot,x)$ is continuously differentiable and bounded from below and $p = p(U,x)$ is assumed to be a descent direction, there always exists a nonempty interval of step lengths such that the strong Wolfe condition (4.37) will be satisfied, see Lemma 3.1 in Nocedal and Wright (1999). For more details and a discussion of suitable step size selection procedures we refer to Chapter 3.4 of the aforementioned reference. Note that, when applying Quasi-Newton or conjugate gradient methods, satisfaction of the curvature condition (4.37b) is essential in order to ensure that the respective p always represents a descent direction. Thus, in combination with a suitable step size selection procedure, all the aforementioned approaches ensure that the optimizer update operator (4.36) satisfies Assumption 4.1 and can therefore be used in the proposed MPC iteration scheme.

A Newton-based iteration scheme with bounded complexity

In order to give an illustrative example and to shed some more light on the underlying step size selection procedure, we consider in the following an approach in which the Newton-based search direction from above is combined with a suitable backtracking line search. In particular, the optimizer update operator is chosen as

$$\Psi_o(U,x) = U - s\left(\nabla_U^2 \hat{J}_N(U,x)\right)^{-1} \nabla_U \hat{J}_N(U,x), \tag{4.39}$$

in which a suitable step size is found by increasing an integer parameter j starting from $j = 0$ until $s = \rho^j$, $\rho \in (0,1)$, leads to satisfaction of the Armijo condition (4.37a). As discussed by Nocedal and Wright (1999), the additional condition (4.37a) can in this case be neglected as the backtracking approach inherently ensures that the selected step size is short enough but "not too short". It is now interesting to note that we can for the considered class of relaxed barrier function based MPC problems derive an explicit bound on the maximal number of backtracking line search iterations. In particular, based on the already discussed convexity and regularity properties of the cost function, it can be shown that (4.37a) will be satisfied for all step lengths

$$s \leq \bar{s} := 2\mu(1 - c_1)/L, \tag{4.40}$$

where $c_1 \in (0,1)$ refers to the design parameter in (4.37a) and $\mu, L \in \mathbb{R}_{++}$ are the convexity and Lipschitz constants given in Theorem 4.1, see Appendix B.10 for a more detailed derivation. As a consequence, the above backtracking line search will always terminate with a step size not smaller than $2\rho\mu(1 - c_1)/L$, from which it follows immediately that the maximal number of line search iterations is upper bounded by

$$j \leq \bar{j} := \lceil 1 + \log_\rho (2\mu(1 - c_1)/L) \rceil . \tag{4.41}$$

Thus, in addition to systems theoretic properties presented in the previous sections, we can in this case also derive an explicit estimate on the maximal numerical complexity of the resulting MPC algorithm. In particular, the overall numerical complexity that is inferred by the proposed Newton-based iteration at each sampling step is bounded by

$$C_{\text{iter}}(k) \leq C_{\text{shift}} + i_{\text{T}}(k) \left(j_{\max} C_{\text{bt}} + C_{\text{nd}} \right) , \tag{4.42}$$

in which C_{shift} and C_{bt} refer to the complexities of performing one shift update and one backtracking line search iteration, respectively, while C_{nd} denotes the complexity of computing the Newton direction. In general, it can be expected that the complexity of the Newton step will usually dominate the other terms. In particular, inverting the Hessian matrix at each internal iteration may become rather costly for large prediction horizons. This problem will be discussed in Section 4.4, where we also present a sparse formulation of the above iteration scheme that allows one to speed up the Newton direction computation by exploiting the underlying problem structure. We summarize our findings by the following corollary.

Corollary 4.2. *Consider the relaxed barrier function based MPC formulation (4.3)–(4.5) and let $\hat{J}_N : \mathbb{R}^{Nm} \times \mathbb{R}^n \to \mathbb{R}_+$ given in (4.6) refer to the associated cost function. Moreover, consider the closed-loop system (4.21) with the operator $\Phi^{i_{\text{T}}(k)}$ defined according to (4.22)–(4.24) and (4.39), where the step size s is at each iteration chosen based on a Armijo backtracking line search with parameters $c_1, \rho \in (0,1)$. Then, for any sequence $\boldsymbol{i}_{\text{T}} = \{i_{\text{T}}(0), i_{\text{T}}(1), \ldots\}$, the following assertions hold true: i) the origin $(U, x) = (0,0)$ of system (4.21) is globally asymptotically stable, ii) the constraint satisfaction and robustness results of Section 4.3.3 and 4.3.4 apply, and iii) the overall numerical complexity at each sampling step is bounded by (4.42).*

Remark 4.6. Note that the above complexity results may be quite conservative due to conservativeness of the underlying constants μ and L. Nevertheless, in principle they allow one to compute a rough estimate of the number of optimization algorithm iterations that can be executed for a given sampling time, or, vice versa, to choose a suitable sampling time for a desired number of optimizer updates (assuming a fixed and known level of computing power).

4.3.6 Summary

Based on the relaxed barrier function based MPC framework introduced in Chapter 3, we presented and analyzed in this section a novel class of MPC iteration schemes which perform only a limited (possibly small) number of optimization algorithm iteration between two consecutive sampling instants and which allow us to analyze in an integrative fashion the dynamics of both the controlled plant and the underlying optimization algorithm. As a first main result, we discussed the stability properties of the associated state-optimizer dynamics and showed that by combining a suitable warm-start procedure with gradient or Newton based line-search methods, asymptotic stability of the origin can be guaranteed *independently* of the number of performed optimization algorithm iterations. In addition, we studied the corresponding constraint satisfaction properties and showed that, assuming an appropriate initialization of the optimization algorithm, both approximate and exact constraint satisfaction may in many cases again be recovered by a suitable design of the underlying relaxed barrier functions, i.e., by choosing the relaxation parameter sufficiently small. As a third main result, we analyzed the robustness of the presented stability results with respect to both internal and external disturbances and showed that the proposed iteration scheme may (under rather mild assumptions) still be combined in a certainty equivalence output feedback fashion with an exponentially stable state estimation procedure. Moreover, although the presented robustness properties are weaker than those presented in the previous chapter (i.e., going from ISS down to integral ISS), we were able to show that all the beneficial robustness and output feedback properties of "exact" relaxed barrier function based MPC schemes will be recovered when allowing the underlying optimization to converge to the true optimal solution. As our fourth main result, we also discussed suitable choices for the underlying *optimizer update operator*, showing that all the presented theoretical results will hold when making use of conventional line-search methods that ensure at every iteration a monotonic decay in the underlying cost function. In particular, a brief analysis of the case of a Newton method with backtracking line search illustrated nicely that we can characterize the behavior of the proposed MPC algorithm (that is, the combination of the proposed iteration scheme with a specific optimization method) down to the level of the applied line search – and this both from a systems theoretic and algorithmic perspective, see Corollary 4.2.

As outlined above, the discussed iteration scheme can be seen as an interesting prototype for a novel class of stabilizing *anytime MPC algorithms* in the sense that it always returns a (robustly) stabilizing control input and improves the quality of the returned control, and thus the expected performance of the closed-loop system, at each of its internal iterations. This shows the potential relevance of the presented approach for practical applications in which the time and hardware resources that are available for the on-line optimization may be limited or even varying over time. Another important scenario are applications in which it is, for example due to safety reasons, indispensable to analyze the behavior of the applied control algorithm both from a systems theoretic and an algorithmic perspective and to be able to give strict guarantees on the behavior of the actually implemented overall closed-loop system.

4.3.7 Numerical examples

We briefly discuss and illustrate some of our findings from above by revisiting two of the numerical examples that we already considered in Chapter 3. We focus deliberately on the above systems theoretic properties as well as on the overall behavior of the closed-loop system, leaving a more detailed analysis of the numerical performance to the next section.

Double integrator

In the following, we consider again the double integrator example that has already been discussed in Section 3.4 and 3.5 of the previous chapter and which be briefly restate here for the sake of convenience. The linear discrete-time system dynamics are given as

$$x(k+1) = \begin{bmatrix} 1 & T_s \\ 0 & 1 \end{bmatrix} x(k) + \begin{bmatrix} T_s^2 \\ T_s \end{bmatrix} u(k), \qquad T_s = 0.1, \tag{4.43}$$

while the associated polytopic input and state constraints are assumed to be given by box constraints of the form

$$C_\mathrm{u} = \begin{bmatrix} 1 \\ -1 \end{bmatrix}, \; d_\mathrm{u} = \begin{bmatrix} 1 \\ 1 \end{bmatrix}, \qquad C_\mathrm{x} = \begin{bmatrix} 1 & 0 \\ 0 & 1 \\ -1 & 0 \\ 0 & -1 \end{bmatrix}, \; d_\mathrm{x} = \begin{bmatrix} 3 \\ 1 \\ 2 \\ 1 \end{bmatrix}. \tag{4.44}$$

The basic MPC problem parameters, i.e., the prediction horizon and the quadratic stage cost weighting matrices, are chosen as $N = 30$, $Q = \mathrm{diag}(1,\ 0.1)$, and $R = 0.1$, while the barrier function weighting and relaxation parameters are (if not stated otherwise) chosen as $\varepsilon = \delta = 10^{-3}$. The relaxed barrier function based MPC problem formulation is constructed according to (4.3)–(4.6), with the underlying recentering vectors are constructed by making use of the procedure in Appendix B.3, leading to w_x and w_u stated in (3.49). Based on the different line search approaches from Section 4.3.5, the proposed MPC iteration scheme is implemented and tested for the same scenarios that were already considered in the numerical examples of Chapter 3. Here, a backtracking line search with $\rho = 0.5$ is used for the gradient and Newton method, while the more involved step size selection procedure of Moré and Thuente (1994) is used for both the conjugated gradient and the Quasi-Newton method. The Wolfe condition parameters for the line search are chosen as $c_1 = 10^{-3}$, $c_2 = 0.9$. The inverse Hessian approximation $\hat{B}$ in the Quasi-Newton method is initialized with H^{-1}, updated by means of the BFGS formula (Nocedal and Wright, 1999, Chapter 8), and then handed over between consecutive sampling instants.

Scenario 1

We consider an initial condition of $x(0) = x_{0,1} = [2.5, \; -0.65]^\top$ and analyze the behavior of the resulting closed-loop system when making use of the Newton-based iteration scheme discussed in Section 4.3.5, performing only *one* optimizer update at each sampling instant, i.e., $i_\mathrm{T}(k) = 1 \; \forall k \in \mathbb{N}$. The underlying optimizer iteration is initialized with a suboptimal (and in many cases infeasible) input vector $U_0 = \bar{K}x_0$. In particular, we made use of

$$\bar{K} = [K^\top \;\; (A+BK)^\top K^\top \; \ldots \; (A+BK)^{N-1^\top} K^\top]^\top \tag{4.45}$$

which corresponds to an initialization with the open-loop input sequence resulting from the controller gain $K \in \mathbb{R}^{m \times n}$ in (4.5a). As before, we vary the value for the relaxation parameter δ and observe how this affects the resulting constraint satisfaction properties. For the presented simulation results, we chose five different values according to $\delta \in \{10^{-1}, 2.5 \times 10^{-2}, 10^{-2}, 10^{-3}, 10^{-5}\}$. As can be seen from the subplots (a) and (b) of Figure 4.2, there is a quite severe violation of state and input constraints if δ is chosen too large. On the other hand, the resulting closed-loop trajectories seem to satisfy the constraints as soon as the relaxation parameter is chosen small enough (here as soon as $\delta \leq 10^{-3}$). Thus, although the initialization is far from optimal and only one optimization algorithm iteration is performed at each sampling step, we can basically observe the same closed-loop behavior as for example in Section 3.5.4. In particular, a quite satisfactory constraint satisfaction can be achieved whenever $\delta \leq 10^{-3}$. Note on the other hand that strict satisfaction of the input constraints in subplot (b) can only be achieved when making use of an optimal initialization, cf. Theorem 4.5. Furthermore, it can be seen that the system state, as ensured by Theorem 4.3, converges to the origin in all cases.

Scenario 2

As a second scenario, we consider again the *infeasible* initial condition $x_{0,2} = [-1.7, \; -1.25]^\top$ and compare the resulting closed-loop behavior when making use of different line-search methods in the underlying optimization. In particular, we employ in addition to the already introduced Newton-based iteration approach also the other approaches from above, that is, gradient method, conjugated gradient method, and Quasi-Newton method (all performing only one optimizing iteration per sampling instant). As can be seen from Figure 4.2 (and as guaranteed by the respective stability results), asymptotic convergence to the origin is achieved independently of the underlying line-search method. However, as can be observed from both subplot (a) and subplot (c), the best performance is achieved when making use of the proposed Newton-based iteration scheme. In particular, the respective input and state trajectories "activate" some of the constraints and are very similar to those of the exact optimal case, cf. Figure 3.3 and 3.5. The behavior for the gradient and the conjugated gradient method are almost identical for the considered scenario.

Scenario 3

We consider again the infeasible initial condition $x_{0,3} = [1.25, 1.25]^\top$, for which we now want to compare the resulting closed-loop behavior when making use of different initializations U_0 of the optimizer iteration. In particular, beside the proposed suboptimal initialization $U_0 = \bar{K}x_{0,3}$, we now also consider $U_0 = 0$ as well as optimal initialization with $U_0 = \hat{U}^*(x_{0,3})$. The associated simulation results can be found in subplots (a) and (d) of Figure 4.2. While the observed closed-loop trajectories are very similar from a qualitative point of view, they also indicate that the resulting control performance may heavily depend on the initialization U_0. However, to our experience, a suboptimal initialization of the form $U_0 = \bar{K}x_0$ often works quite well in practice, such that knowledge of the optimal initialization is not necessarily required.

In addition, the two subplots of Figure 4.3 show, averaged over 200 randomly chosen initial conditions $x_{0,i} \in \mathcal{X}$, the strictly monotonic decay of the cost function over time when applying the different line-search methods discussed above, based on the initializations $U_0 = \bar{K}x_0$ (left) and $U_0 = \hat{U}^*(x_0)$ (right). Again, only one optimizer update is performed per sampling step. It can be seen that particularly the Newton method based optimization does result in a good closed-loop performance, especially in the case of optimal initialization. Also the Quasi-Newton and the conjugated gradient method lead to acceptable results, although it should be noted that for these methods the behavior is much better if we allow for a few more, say five to ten, iterations. This nicely illustrates the fact that these approaches may need some internal iterations in order to reconstruct the second order information that is in the Newton method directly provided by the Hessian. Note that the observed behavior is in principle also reflected on the level of maximal constraint violations (not shown here due to space limitations). In particular, while one Newton iteration is often enough to achieve almost the same constraint satisfaction properties as the fully optimized solution, the Quasi-Newton and conjugated gradient based methods tend to require a few more, say five to ten, iterations. On the other hand, the gradient method performs rather poorly, which, in principle, had to be expected as the underlying optimization problem may be quite ill-conditioned for the considered choice of the relaxation parameter. One possibility to alleviate this problem of ill-conditioning might be to consider instead a steepest-decent method associated to the quadratic norm $\|\cdot\|_H$, see for example (Boyd and Vandenberghe, 2004, Chapter 9.4). However, in total the presented results illustrate that already a small number of optimization algorithm iterations may well be sufficient for a satisfactory performance of the resulting closed-loop system – an observation that has in fact also been made in the context of MPC based on conventional interior point methods (Wang and Boyd, 2010).

Thus, although the resulting closed-loop behavior may in general depend on the initialization, the number of optimizer iterations, and the line-search method that is used within the underlying optimizer update, we are, in summary, able to reproduce and confirm many of the systems theoretic properties discussed above. In particular, asymptotic stability of the overall closed-loop system is observed for all initial conditions and independently of both the applied optimization procedure and the number of optimizer iterations. Our simulations furthermore suggest that the Newton-based line search often converges already after only five to eight iterations, which nicely relates to the observations reported in (Wang and Boyd, 2010). On the other hand, the above results also show a promising behavior of the optimizer update approaches based on a Quasi-Newton or conjugated gradient line search, which suggests that knowledge and inversion of the full Hessian matrix might not always be necessary.

Three-tank system

In order to demonstrate also the inherent robustness properties of the proposed MPC iteration scheme, let us revisit the three-tank example from Section 3.6.5. The goal is again to regulate the water levels of a vertically cascaded tank system using the input flow v entering the topmost tank. As before, the nonlinear plant model is linearized around the desired set point $h_s = [11.19, 10.34, 10.80]^\top$ cm, $v_s = 40.47\,\text{ml/s}$, and discretized with a sampling time of $T_s = 5\,\text{s}$, leading to a system of the form (4.1) with $x = h - h_s \in \mathbb{R}^3$ and $u = v - v_s \in \mathbb{R}$. The physical constraints on the system are $0 \leq h_i \leq 28\,\text{cm}$, $i = 1,2,3$, and $0 \leq v \leq 60\,\text{ml/s}$, which results in box constraints on the states and inputs. Also all remaining problem parameters are chosen exactly as in Section 3.6.5, and the relaxed barrier function based MPC problem formulation is constructed according to (4.3)–(4.6). The proposed iteration scheme is then implemented based on a Newton-based line-search method, performing only one Newton step between two consecutive sampling instants, that is, $i_{\mathrm{T}}(k) = 1$ if not stated otherwise. The initialization of the underlying optimizer iteration is chosen as $U(0) = \bar{K}x_0$ with $\bar{K} \in \mathbb{R}^{m \times n}$ defined according to (4.45).

Inspired by Section 3.6.5, the following three scenarios are considered: *i*) control of the nominal (i.e., unperturbed) linear system based on full state measurement, *ii*) certainty equivalence output feedback based on $C = [0, 0, 1]$ and an exponentially stable Luenberger observer initialized with $\hat{x}(0) = 0$, and *iii*) the same output feedback scheme under the influence of normally distributed random state and output disturbances, see Section 3.6.5 for more detail on the considered scenarios. Some selected simulation results can be found in Figure 4.4 and 4.5, respectively. For the latter scenario with both estimation errors and external disturbances, we investigate here also the influence of the number of Newton iterations on the behavior and performance of the resulting closed-loop system. In particular, note that Figure 4.4 reveals that the maximal transient excitation of the overall system state may actually *grow* when increasing the number of performed optimization algorithm iterations. However, this may be explained by observing that the estimation error is actually quite large at the beginning, leading to a wrong and unnecessarily large control input. The results therefore suggest that performing less optimizer iterations and applying suboptimal rather than fully optimized control inputs might even be beneficial in the presence of uncertain state information. Apart from this new insight, the obtained simulation results also illustrate and confirm nicely many of the theoretical robustness and output feedback results presented above.

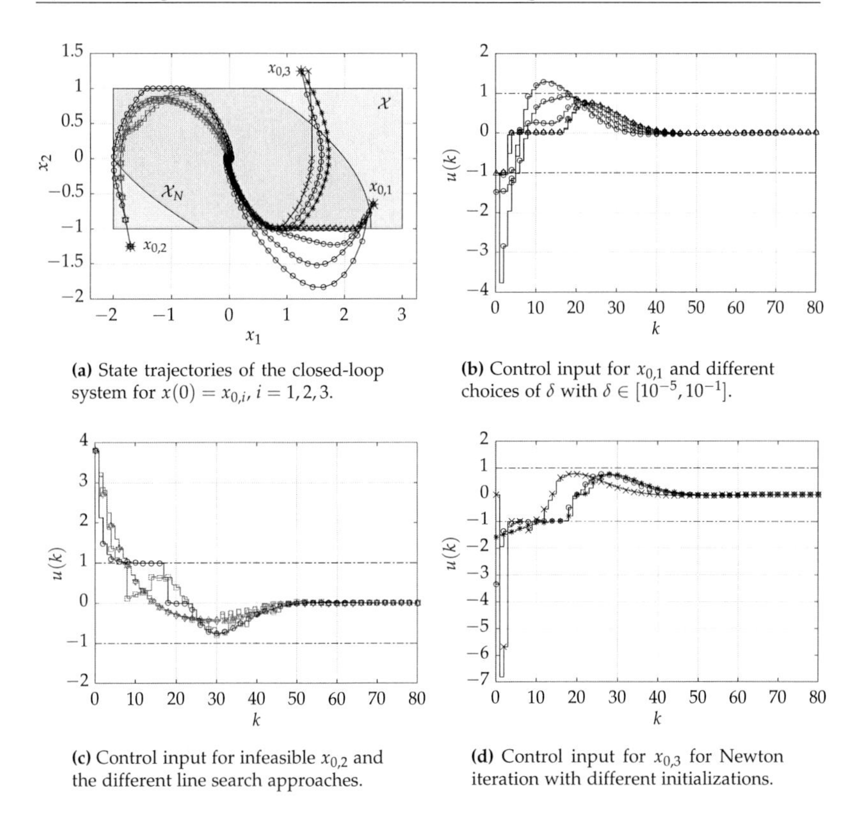

(a) State trajectories of the closed-loop system for $x(0) = x_{0,i}$, $i = 1, 2, 3$.

(b) Control input for $x_{0,1}$ and different choices of δ with $\delta \in [10^{-5}, 10^{-1}]$.

(c) Control input for infeasible $x_{0,2}$ and the different line search approaches.

(d) Control input for $x_{0,3}$ for Newton iteration with different initializations.

Figure 4.2. Results for the double integrator system when applying the proposed MPC iteration scheme with $i_T(k) = 1$. Depicted in (a) are the resulting closed-loop state trajectories ($\circ$) when applying only one Newton iteration per sampling step and making use of a suboptimal initialization of the form $U_0 = \bar{K} x_{0,i}$. For $x_{0,1} = [2.5, -0.65]$, we vary the relaxation parameter $\delta \in \{0.1, 2.5 \times 10^{-2}, 10^{-2}, 10^{-3}\}$ and plot for comparison also the trajectory ($\triangle$) resulting from optimal initialization via $U_0 = \hat{U}^*(x_{0,1})$, which results for this example in strict satisfaction of both state and input constraints. For $x_{0,2} = [-1.7, -1.25]$, we use $\delta = 10^{-3}$ and plot for comparison also the trajectories for the alternative line search approaches, that is, gradient method ($\triangledown$), conjugated gradient method ($\triangle$), and Quasi-Newton method ($\square$). For $x_{0,3} = [1.25, 1.25]$, we compare the behavior of the Newton-based iteration scheme for the initializations $U_0 = \bar{K} x_{0,3}$ ($\circ$), $U_0 = \hat{U}^*(x_{0,3})$ ($*$), and $U_0 = 0$ ($\times$). The associated input trajectories can be found in the subplots (b) – (d).

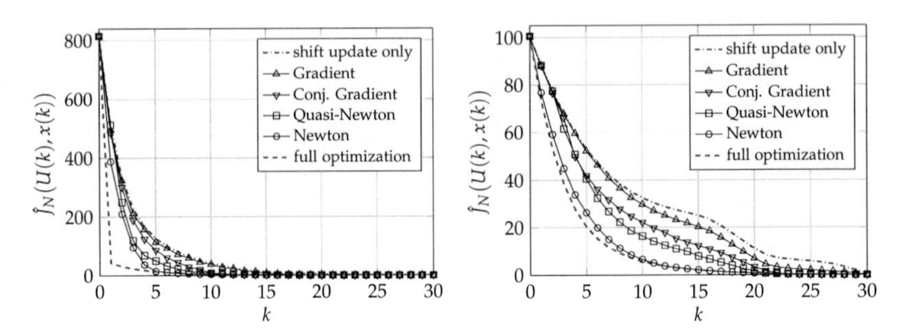

Figure 4.3. Decay of the cost function for the double integrator example when performing at each sampling instant only one optimizing iteration of the respective line-search method. *Left:* Averaged evolution of the cost function for 200 random initial conditions $x_{0,i}$ when using a suboptimal initialization $U_0 = \bar{K}x_{0,i}$. Newton-based optimization achieves the fastest decay, whereas the behavior for the gradient method is almost similar to that when applying only the shift operator, i.e., $i_T(k) = 0\ \forall k \in \mathbb{N}$. *Right:* Averaged evolution of the cost function for the same initial conditions and line-search methods, but now using $U_0 = \hat{U}^*(x_{0,i})$. Performing a single Newton step at each sampling instant achieves almost the same convergence behavior as when applying the exact optimal solution.

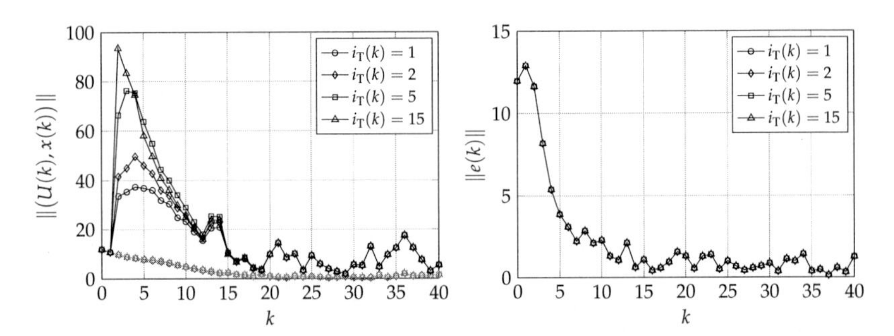

Figure 4.4. Behavior of the overall closed-loop system state and estimation error for the three-tank example under persistent disturbances and estimation-based output feedback. *Left:* Averaged evolution of the overall system state for 200 random initial conditions $x_{0,i}$ when varying the number of optimization algorithm iterations (the gray lines correspond to the norm of the system state $x(k) = h(k) - h_s$). *Right:* Averaged evolution of the corresponding estimation error. Performing less optimizer iterations and applying suboptimal control laws might actually be beneficial in the presence of uncertain state information.

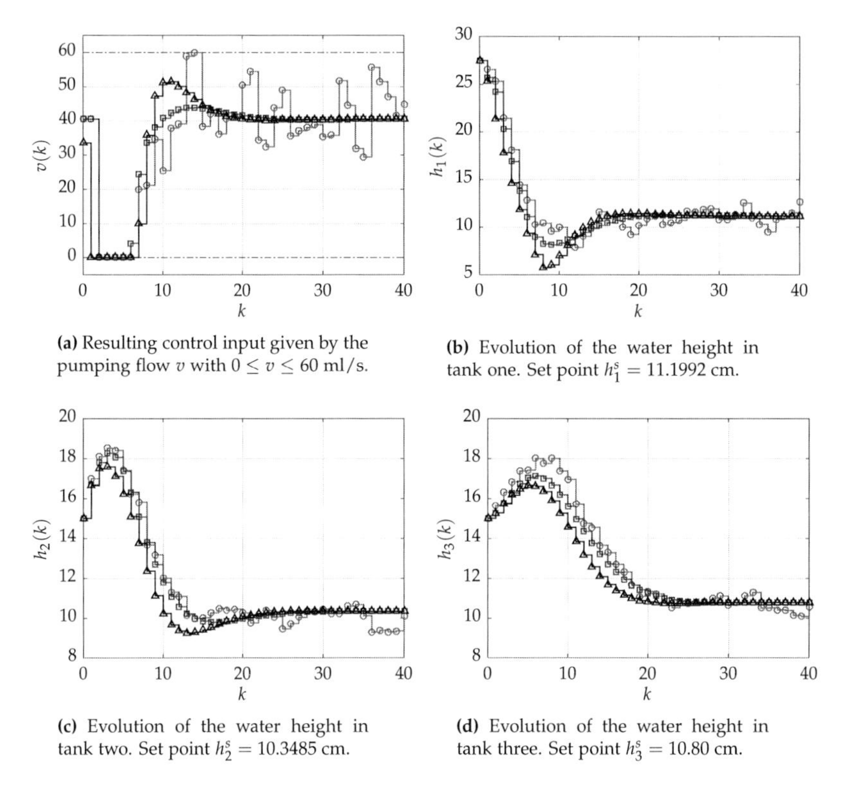

(a) Resulting control input given by the pumping flow v with $0 \leq v \leq 60$ ml/s.

(b) Evolution of the water height in tank one. Set point $h_1^s = 11.1992$ cm.

(c) Evolution of the water height in tank two. Set point $h_2^s = 10.3485$ cm.

(d) Evolution of the water height in tank three. Set point $h_3^s = 10.80$ cm.

Figure 4.5. Simulated closed-loop behavior of the three-tank system for the three considered scenarios and initial condition $h(0) = [27.5, 15, 15]^\top$ cm. Only one optimizing Newton step is performed at each sampling instant and the optimizer iteration is initialized with $U_0 = \bar{K}x_0$. Depicted are the input and state trajectories associated to the nominal linearized system ($\triangle$), unperturbed output feedback based on an exponentially stable Luenberger observer ($\square$), and the same output feedback scheme under persistent state and measurement disturbances ($\circ$). When no disturbances are acting on the system, convergence to the desired set point is achieved also in the estimation-based output feedback case. Furthermore, convergence to a neighborhood of the set point is achieved in the persistently perturbed output feedback case. Despite estimation errors, disturbances, and the application of merely suboptimal solutions, the underlying input and state constraints are almost exactly satisfied in all cases.

4.4 Anytime MPC algorithms exploiting sparsity

In the previous section, we showed how the proposed relaxed barrier function based MPC iteration scheme for the the *condensed*, unconstrained formulation (4.6) in a quite natural way leads to a novel class of anytime MPC algorithms with important systems theoretic and algorithmic guarantees. While the presented theoretical results were formulated for a quite general class of line-search methods, the numerical examples revealed that the best closed-loop behavior can typically be achieved when making use of second order methods like the Newton method. However, it is a well-known fact that computing the required Newton search directions may become computationally demanding when considering higher-dimensional systems and/or long prediction horizons. In such a case, the computational burden can often be reduced significantly by turning instead to the *non-condensed*, equality constrained formulation and explicitly exploiting the special structure of the resulting sparse optimization problem (Wright, 1997a; Rao et al., 1998; Wang and Boyd, 2010; Domahidi et al., 2012).

In this section, we extend the presented iteration scheme approach to the equality constrained and sparse formulation of the underlying relaxed barrier function based MPC problem and show that this leads to a very similar class of anytime MPC algorithms with significantly improved numerical performance.

4.4.1 Sparse problem formulation

As discussed above, the linearity of the underlying system dynamics in principle allows us to eliminate the dynamic constraints (4.3b) from the respective open-loop optimal control problem, resulting in the unconstrained optimization problem (4.6) in the optimization variable $U = [u_0^\top, \ldots, u_{N-1}^\top]^\top \in \mathbb{R}^{Nm}$. However, as advocated by the aforementioned references, we may as well also deliberately keep the predicted system states as optimization variables. To this end, let us define the new vector of optimization variables as

$$w = \begin{bmatrix} u_0^\top & x_1^\top & u_1^\top & x_2^\top & \cdots & u_{N-1}^\top & x_N^\top \end{bmatrix}^\top \in \mathbb{R}^{n_w}, \quad n_w := N(m+n), \tag{4.46}$$

which allows rewriting the relaxed barrier function based open-loop optimal control problem (4.3) as

$$\hat{J}_N^*(x) = \min_w \hat{J}_N(w,x), \quad \hat{J}_N(w,x) = \frac{1}{2} w^\top H w + x^\top F w + \frac{1}{2} x^\top Y x + \varepsilon \hat{B}_{\mathrm{xw}}(w,x) \tag{4.47a}$$

$$\text{s.t. } Cw = b(x). \tag{4.47b}$$

Explicit expressions for the respective matrices and vectors as well as for the overall relaxed barrier function $\hat{B}_{\mathrm{xw}} : \mathbb{R}^{n_w} \times \mathbb{R}^n \to \mathbb{R}_+$ can be found in Appendix B.9 and will also be discussed in more detail in Section 4.4.3. From a numerical optimization perspective, the main advantage of formulation (4.47) over the condensed formulation (4.6) is that it exhibits a lot of structure and that many of the underlying matrices are actually sparse. In particular, $H \in \mathbb{S}_{++}^{n_w}$ is a block-diagonal matrix and $C \in \mathbb{R}^{Nn \times n_w}$ is block-banded, while the underlying inequality constraints also have a block-diagonal structure, see Appendix B.9. Furthermore, the associated cost function given in (4.47a) still possesses the same beneficial convexity and regularity properties as in the condensed formulation. This is made more precise by the following corollary.

Corollary 4.3. *Consider the relaxed barrier function based MPC formulation (4.3)–(4.5) and let* $\hat{J}_N : \mathbb{R}^{n_w} \times \mathbb{R}^n \to \mathbb{R}_+$ *given in (4.47a) refer to the cost function of the associated equality constrained formulation. Then, for any arbitrary but fixed* $x \in \mathbb{R}^n$*, the cost function* $\hat{J}_N(w,x)$ *is twice continuously differentiable and strongly convex in* w *and there exist* $\mu, L \in \mathbb{R}_{++}$ *such that for any* $w, w' \in \mathbb{R}^{n_w}$

$$\mu I \preceq \nabla^2_w \hat{J}_N(w,x) \preceq LI \tag{4.48a}$$

$$\|\nabla_w \hat{J}_N(w,x) - \nabla_w \hat{J}_N(w',x)\| \leq L\|w - w'\|\,. \tag{4.48b}$$

Furthermore, there exists $L' \in \mathbb{R}_{++}$ *such that for any* $w, w' \in \mathbb{R}^{n_w}$

$$\|\nabla^2_w \hat{J}_N(w,x) - \nabla^2_w \hat{J}_N(w',x)\| \leq L'\|w - w'\|\,. \tag{4.49}$$

Finally, for any $x \in \mathbb{R}^n$*, the cost function* $\hat{J}_N(\cdot,x)$ *is self-concordant according to Definition 2.3.*

Corollary 4.3 as well as explicit expressions for the constants μ, L, and L' follow directly from the proof of Theorem 4.1 for the condensed case, which is given in Appendix C.18. While the additional equality constraints prevent a direct application of unconstrained optimization procedures such as the gradient method, the above results reveal that problem (4.47) exhibits many beneficial properties for the application of Newton's method. In particular, the residuals of the associated KKT conditions are given by

$$r_d = \nabla_w \hat{J}_N(w,x) + C^\top \nu \tag{4.50a}$$

$$r_p = Cw - b(x)\,, \tag{4.50b}$$

where $r_d \in \mathbb{R}^{n_w}$ and $r_p \in \mathbb{R}^{Nn}$ denote the dual and primal residual, respectively, and $\nu \in \mathbb{R}^{Nn}$ refers to the Lagrange multiplier of the equality constraints (4.47b). Due to the above convexity and regularity results, the KKT conditions are in this case necessary and sufficient conditions for optimality and characterize the optimal solution $\hat{w}^*(x)$ uniquely. Aiming for an algorithmic and numerically efficient solution, we now apply Newton's method to (4.50), which, for fixed x and given (w,ν), results in the so-called *KKT system* for the search directions $\Delta w \in \mathbb{R}^{n_w}$ and $\Delta\nu \in \mathbb{R}^{Nn}$:

$$\begin{bmatrix} \nabla^2_w \hat{J}_N(w,x) & C^\top \\ C & 0 \end{bmatrix} \begin{bmatrix} \Delta w \\ \Delta \nu \end{bmatrix} = - \begin{bmatrix} r_d \\ r_p \end{bmatrix}. \tag{4.51}$$

In a conventional Newton-iteration, the primal and dual variables are then successively updated according to

$$w^+ = w + s\Delta w\,, \qquad \nu^+ = \nu + s\Delta\nu\,, \tag{4.52}$$

where $s \in (0,1)$ is a suitably chosen step size, which may for instance be determined by a backtracking line search that ensures a certain decay in the norm of the primal and dual residuals (Boyd and Vandenberghe, 2004; Wang and Boyd, 2010). After each update of the form (4.52), the KKT system (4.51) is used to compute new search directions and the whole procedure is repeated until some stopping criterion is satisfied. For the above problem setup, the maximal number of iterations that is required to arrive at an ϵ-suboptimal solution can be upper bounded based on the characteristic constants provided in Corollary 4.3 or by exploiting the property of self-concordance, cf. Section 4.2.2. In fact, when assuming a feasible initialization, exactly the same complexity estimates as in the unconstrained case apply (Boyd and Vandenberghe, 2004, Section 10.2). Furthermore, it can be shown that primal feasibility, that is, satisfaction of the affine equality constraints (4.47b), will always be achieved after a finite number of Newton iterations. In this context, note also that primal feasibility is preserved by the proposed Newton search direction in the sense that if $Cw = b(x)$, then it always holds that $Cw^+ = C(w + s\Delta w) = b(x)$.

In the following two sections, we first present a Newton-based MPC iteration scheme that guarantees basically the same systems theoretic and algorithmic properties as in the condensed case, and then, in a second step, discuss how the sparsity of the KKT system (4.51) can be used for speeding up the underlying search direction computation.

4.4.2 A stabilizing iteration scheme for the sparse formulation

The main problem when aiming for a stabilizing MPC iteration scheme in the spirit of Section 4.3 is that the primal-dual search direction computed via (4.51) does not necessarily lead to a reduction in the cost function given in (4.47a). In particular, if the primal residual r_p is not equal to zero, i.e., if the underlying dynamic constraints are not satisfied, the above iteration procedure is nothing else than an *infeasible start Newton method*, for which it is typically only possible to guarantee a decay in the respective residual vectors (Boyd and Vandenberghe, 2004). Existence of a sufficiently small step size that allows for a guaranteed monotonic decay in the cost function can therefore only be guaranteed if $r_p = Cw - b(x) = 0$. This observation will also play a central role in the warm-starting strategy of the subsequently presented MPC iteration scheme.

In accordance with Section 4.3, we again formulate our iteration scheme in form of an overall closed-loop system that allows us to analyze the interplay of plant and optimizer states in an integrative fashion. In particular, we consider the following combined state-optimizer dynamics

$$x(k+1) = A\,x(k) + B\Pi_0\,w(k)\,, \qquad x(0) = x_0\,, \tag{4.53a}$$
$$w(k+1) = \Phi^{i_T(k)}\left(w(k), x(k)\right)\,, \qquad w(0) = w_0\,, \tag{4.53b}$$

where (4.53a) describes the dynamics of the controlled linear plant, (4.53b) describes the dynamics of the underlying optimization algorithm, and $\Pi_0 := \begin{bmatrix} I_m & 0 & \cdots & 0 \end{bmatrix} \in \mathbb{R}^{m \times n_w}$. Furthermore, $\Phi^{i_T(k)} : \mathbb{R}^{n_w} \times \mathbb{R}^n \to \mathbb{R}^{n_w}$ denotes a suitable *optimization algorithm operator* that describes how the next optimizer state is at each sampling instant $k \in \mathbb{N}$ computed based on the current optimizer state $w(k)$ and the measured system state $x(k)$ by performing $i_T(k) \in \mathbb{N}_+$ optimization algorithm iterations. As before, we define the operator $\Phi^{i_T(k)}$ for a given termination index $i_T(k) \in \mathbb{N}_+$ recursively as

$$\Phi^0(w,x) = \Psi_s(w,x) \tag{4.54a}$$

$$\Phi^i(w,x) = \Psi_o\left(\Phi^{i-1}(w,x), Ax + B\Pi_0\, w\right) . \tag{4.54b}$$

Here, (4.54a) characterizes how a suitable warm-start solution for the predicted next system state $x^+ = Ax + B\Pi_0\, w$ is obtained in the first step via a suitably defined *shift operator* $\Psi_s : \mathbb{R}^{n_w} \times \mathbb{R}^n \to \mathbb{R}^{n_w}$. In order to ensure that the optimizer update iteration (4.54b) is initialized with a warm-start solution that satisfies the equality constraints (4.47b), we propose to choose the shift update as

$$\Psi_s(w,x) = \begin{bmatrix} u_1 \\ Ax^+ + Bu_1 \\ u_2 \\ A^2x^+ + ABu_1 + Bu_2 \\ \vdots \\ u_{N-1} \\ A^{N-1}x^+ + A^{N-2}Bu_1 + \ldots + Bu_{N-1} \\ K(A^{N-1}x^+ + A^{N-2}Bu_1 + \ldots + Bu_{N-1}) \\ (A+BK)(A^{N-1}x^+ + A^{N-2}Bu_1 + \ldots + Bu_{N-1}) \end{bmatrix} = \Gamma_x x + \Gamma_w w\, , \tag{4.55}$$

where $K \in \mathbb{R}^{m \times n}$ is chosen according to (4.5). Explicit expressions for $\Gamma_x \in \mathbb{R}^{n_w \times n}$ and $\Gamma_w \in \mathbb{R}^{n_w \times n_w}$ can be derived in a straightforward way by inserting $x^+ = Ax + Bu_0$ into the stacked vector on the left-hand side and recalling the definition of w given in (4.46). Note that what the above shift update operator actually does is shifting the input variables u_i in the same way as in the condensed formulation (appending the auxiliary linear control law at the end) and then computing the associated predicted plant states based on the underlying linear system dynamics. In this way, it is ensured that the shifted warm-start solution $w^+ = \Psi_s(w,x)$ for the predicted next system state x^+ will always satisfy the equality constraints of problem (4.47), i.e., $Cw^+ = b(x^+)$. In all subsequent optimization algorithm iterations, we then propose to employ the following Newton-based optimizer update operator

$$\Psi_o(w,x) = w + s\Delta w(w,x)\, , \tag{4.56}$$

where $\Delta w(w,x)$ is the primal Newton search direction obtained from (4.51) and $s \in (0,1)$ is a suitable step size that is chosen such that the associated next optimizer state leads to a decrease in the underlying cost function. Note that as the applied warm-starting via (4.55) ensures that the equality constraints are satisfied upon initialization ($r_p = 0$), the resulting search direction Δw always represents a descent direction, which implies that such a step size always exists. In particular, we can state the following auxiliary result, for which a proof is given in Appendix C.25.

Lemma 4.1. *Consider the relaxed barrier function based MPC formulation (4.3)–(4.5) and let* $\hat{J}_N : \mathbb{R}^{n_w} \times \mathbb{R}^n \to \mathbb{R}_+$ *refer to the cost function of the associated equality constrained optimization problem (4.47). Let* $(w, x) \in \mathbb{R}^{n_w} \times \mathbb{R}^n$ *be arbitrary but fixed and assume that (4.47b) holds. Furthermore, let* $\Delta w \in \mathbb{R}^{n_w}$ *be the Newton search direction according to (4.50)–(4.52). Then, for any* $c_1 \in (0, 1)$*, there always exists a nonempty interval of step lengths such that*

$$\hat{J}_N(w + s\Delta w, x) \leq \hat{J}_N(w, x) - c_1 s \|\Delta w\|^2_{\nabla^2_w \hat{J}_N(w,x)} . \tag{4.57}$$

In particular, the Armijo-type condition (4.57) will be satisfied for any $s \leq \bar{s} := 2\mu(1 - c_1)/L$*, where* $\mu, L \in \mathbb{R}_{++}$ *are defined according to Corollary 4.3.*

Note that if $Cw \neq b(x)$, then the shift update operator (4.55) will in general not lead to a decay in the underlying cost function, see also the proof of Theorem 4.8 below. However, at least in the nominal case, this only concerns the transition from a possibly infeasible initial condition (w_0, x_0) to the first shifted optimizer initialization (w_0^+, x_0^+). In all subsequent sampling instants and iterations, the shift update will then lead both to a decrease in the cost function and to a feasible initialization of the associated subsequent Newton iteration (in fact, it will result in exactly the same shift update strategy as in the condensed formulation). As therefore both the shift update operator and all subsequent iterations of the optimizer update operator lead to a guaranteed monotonic decay in the cost function, we can basically derive the same systems theoretical guarantees as in Section 4.3. In particular, as a first main result we can state that the origin of the overall closed-loop system (4.53) will be globally asymptotically stable – again independently of the number of Newton iterations that are performed at each sampling step.

Theorem 4.8. *Consider the relaxed barrier function based MPC formulation (4.3)–(4.5) and let* $\hat{J}_N : \mathbb{R}^{n_w} \times \mathbb{R}^n \to \mathbb{R}_+$ *refer to the cost function of the associated equality constrained optimization problem (4.47). Moreover, consider the closed-loop system (4.53) with the operator* $\Phi^{i_T(k)}$ *being defined according to (4.54)–(4.56), and let the step size be chosen such that (4.57) holds at each iteration of the underlying Newton method. Then, for any sequence* $\boldsymbol{i}_T = \{i_T(0), i_T(1), \ldots\}$*, the origin* $(w, x) = (0, 0)$ *of system (4.53) is globally asymptotically stable.*

A proof of Theorem 4.8 can be found in Appendix C.26. Following similar arguments as in Section 4.3.3, we can furthermore prove that the maximally possible constraint violation will be both bounded and strictly monotonically decaying and that any desired level of approximate (or even exact) constraint satisfaction may in many cases be guaranteed by choosing the relaxation parameter as well as the initialization of the optimizer dynamics in a suitable way, compare Theorems 4.4 and 4.5.

Theorem 4.9. *Consider the relaxed barrier function based MPC formulation (4.3)–(4.5) and let* $\hat{J}_N : \mathbb{R}^{n_w} \times \mathbb{R}^n \to \mathbb{R}_+$ *refer to the cost function of the associated equality constrained optimization problem (4.47). Moreover, consider the closed-loop system (4.53) with the operator* $\Phi^{i_T(k)}$ *being defined according to (4.54)–(4.56), and let the step size be chosen such that (4.57) holds at each iteration of the underlying Newton method. Let* $\{x(0), x(1), \ldots\}$ *and* $\{u(0), u(1), \ldots\}$ *with* $u(k) = \Pi_0 w(k)$ *denote the resulting closed-loop state and input trajectories. Then, for any* *initialization* $(w_0, x_0) \in \mathbb{R}^{n_w} \times \mathbb{R}^n$, *any sequence* $\boldsymbol{i}_\mathrm{T} = \{i_\mathrm{T}(0), i_\mathrm{T}(1), \ldots\}$, *and any* $k \in \mathbb{N}$, *it holds that*

$$C_\mathrm{x} x(k) \leq d_\mathrm{x} + \hat{z}_\mathrm{x}(k), \quad C_\mathrm{u} u(k) \leq d_\mathrm{u} + \hat{z}_\mathrm{u}(k), \tag{4.58}$$

where the elements of the maximal constraint violation vectors $\hat{z}_\mathrm{x}(k) \in \mathbb{R}^{q_\mathrm{x}}$ *and* $\hat{z}_\mathrm{u}(k) \in \mathbb{R}^{q_\mathrm{u}}$ *decrease strictly monotonically over time. Furthermore, for each* $k \in \mathbb{N}$, *the respective elements can be computed explicitly by solving a convex optimization problem, and there exists a finite* $k_0 \in \mathbb{N}$ *such that* $\hat{z}_\mathrm{x}(k) \leq 0$, $\hat{z}_\mathrm{u}(k) \leq 0$ *for any* $k \geq k_0$.

Theorem 4.10. *Consider the relaxed barrier function based MPC formulation (4.3)–(4.5) and let* $\hat{J}_N : \mathbb{R}^{n_w} \times \mathbb{R}^n \to \mathbb{R}_+$ *refer to the cost function of the associated equality constrained optimization problem (4.47). Moreover, consider the closed-loop system (4.53) with the operator* $\Phi^{i_\mathrm{T}(k)}$ *being defined according to (4.54)–(4.56), and let the step size be chosen such that (4.57) holds at each iteration of the underlying Newton method. Let* $\{x(0), x(1), \ldots\}$ *and* $\{u(0), u(1), \ldots\}$ *with* $u(k) = \Pi_0 w(k)$ *denote the resulting closed-loop state and input trajectories. Assume that* $\begin{bmatrix} A^{N-1}B & \cdots & AB & B \end{bmatrix}$ *has full row rank n and define the set* $\mathcal{X}_N \subseteq X$ *as*

$$\mathcal{X}_N := \{x \in \mathcal{X} \mid \exists w \in \mathbb{R}^{n_w} \text{ s.t. } u_k \in \mathcal{U}, x_k(w,x) \in \mathcal{X}, x_N(w,x) = 0\} \,. \tag{4.59}$$

Then, for any compact set $\mathcal{X}_0 \subseteq \mathcal{X}_N^\circ$ *and any* $\hat{z}_\mathrm{x,tol} \in \mathbb{R}_+^{q_x}$, $\hat{z}_\mathrm{u,tol} \in \mathbb{R}_+^{q_u}$, *there exists* $\bar{\delta}_0 \in \mathbb{R}_{++}$ *and* $\gamma \in \mathcal{K}_\infty$ *such that for any* $0 < \delta \leq \bar{\delta}_0$, *any* $x_0 \in \mathcal{X}_0$, *any optimizer initialization* $w_0 \in \mathcal{B}^{n_w}_{\gamma(\delta)}(\hat{w}^*(x_0))$ *with* $Cw_0 = b(x_0)$, *any* $\boldsymbol{i}_\mathrm{T} = \{i_\mathrm{T}(0), i_\mathrm{T}(1), \ldots\}$, *and any* $k \in \mathbb{N}$, *it holds that*

$$C_\mathrm{x}\, x(k) \leq d_\mathrm{x} + \hat{z}_\mathrm{x,tol}\,, \qquad C_\mathrm{u}\, u(k) \leq d_\mathrm{u} + \hat{z}_\mathrm{u,tol}\,. \tag{4.60}$$

The proofs of Theorem 4.9 and Theorem 4.10 are very similar to those of Theorem 4.4 and Theorem 4.5 for the condensed formulation. For the sake of completeness, they are nevertheless provided in detail in the Appendices C.27 and C.28, including in particular also explicit expressions for the convex optimization problems that can be used to compute upper bounds on the maximal constraint violation a priori.

Finally, as in the condensed case, we can also state interesting results concerning the inherent robustness properties of the proposed MPC iteration scheme. Analogously to Section 4.3.4, we consider the following perturbed version of system (4.53)

$$x(k+1) = A\, x(k) + B\Pi_0\, w(k) + d(k)\,, \tag{4.61a}$$

$$w(k+1) = \Phi^{i_\mathrm{T}(k)}\big(w(k), x(k) + e(k)\big)\,, \tag{4.61b}$$

in which $d(k) \in \mathbb{R}^n$ may be seen as external input or state disturbance, while $e(k) \in \mathbb{R}^n$ models the effect that the underlying optimization is not performed for the exact system state, e.g., due to an asymptotically decaying but unknown estimation error. As we know from Theorem 4.8 that the unperturbed closed-loop system is globally asymptotically stable (0-GAS), we can again immediately conclude that (4.61) exhibits inherent robustness properties in the integral ISS sense, cf. Theorem 4.6. Furthermore, it is again possible to show that the ISS properties of the idealizing MPC scheme applying the exact solution can always be recovered by performing in the limit an infinite number of optimizer update iterations, cf. Theorem 4.7. The proofs are given in Appendix C.29 and Appendix C.30.

Theorem 4.11. *Consider the relaxed barrier function based MPC formulation (4.3)–(4.5) and let* $\hat{J}_N : \mathbb{R}^{n_w} \times \mathbb{R}^n \to \mathbb{R}_+$ *refer to the cost function of the associated equality constrained optimization problem (4.47). Moreover, consider the closed-loop system (4.53) with the operator* $\Phi^{i_T(k)}$ *being defined according to (4.54)–(4.56), and let the step size be chosen such that (4.57) holds at each iteration of the underlying Newton method. Assume that* $\Phi^{i_T(k)} : \mathbb{R}^{n_w} \times \mathbb{R}^n \to \mathbb{R}^{n_w}$ *is continuous. Then, for any sequence* $\boldsymbol{i}_T = \{i_T(0), i_T(1), \ldots\}$*, system (4.61) is integral input-to-state stable with respect to the disturbance sequences* $\boldsymbol{d}$ *and* $\boldsymbol{e}$*.*

Theorem 4.12. *Consider the relaxed barrier function based MPC formulation (4.3)–(4.5) and let* $\hat{J}_N : \mathbb{R}^{n_w} \times \mathbb{R}^n \to \mathbb{R}_+$ *refer to the cost function of the associated equality constrained optimization problem (4.47). Moreover, consider the closed-loop system (4.53) with the operator* $\Phi^{i_T(k)}$ *being defined according to (4.54)–(4.56), and let the step size be chosen such that (4.57) holds at each iteration of the underlying Newton method. Denote* $\hat{x}(k) = x(k) + e(k)$ *and let*

$$\epsilon(k) = \Phi^{i_T(k)}(w(k), \hat{x}(k)) - \hat{w}^*(A\hat{x}(k) + \Pi_0 w(k)) \tag{4.62}$$

denote the optimization error that is made at sampling instant $k \in \mathbb{N}$*. Then, for any sequence* $\boldsymbol{i}_T = \{i_T(0), i_T(1), \ldots\}$*, the resulting closed-loop system (4.61) is input-to-state stable with respect to the disturbance sequences* $\boldsymbol{d}$*,* $\boldsymbol{e}$ *and the optimization error sequence* $\boldsymbol{\epsilon} := \{\epsilon(-1), \epsilon(0), \ldots\}$*, where the initial optimization error is defined as* $\epsilon(-1) = w(0) - \hat{w}^*(x(0))$*.*

Thus, summarizing, we can conclude that the proposed Newton-based MPC iteration scheme for the sparse and equality constrained problem formulation allows us to state exactly the same system theoretic guarantees as in the condensed, unconstrained case. In particular, the above results again constitute a novel class of anytime MPC algorithms in the sense that important (robust) stability and constraint satisfaction properties can be ensured independently of the number of optimizing Newton iterations that are actually performed at each sampling step. The main conceptual difference to Section 4.3 is given by the slightly modified shift update operator, which ensures that the underlying equality constraints will also be satisfied in the case of an infeasible initialization or in the presence of disturbances or plant model mismatch. Furthermore, the resulting KKT system (4.51) is due to the sparse problem formulation now highly structured, which can be exploited for computing the required Newton search directions in a tailored and highly efficient manner. How exactly this sparsity based complexity reduction may be achieved will be the topic of the following section.

Remark 4.7. Note that some of the ideas underlying the proposed iteration scheme (e.g., making use of a warm-start solution, keeping the barrier parameter ε fixed, and performing only a limited number of optimizing Newton iterations) are in fact similar to the fast MPC approach presented by Wang and Boyd (2010). However, while the aforementioned work argues on a rather heuristic level, admitting in particular that the authors "do not fully understand why this control works so well", the results presented in this section reveal that the proposed relaxed (and recentered) barrier function based approach allows for a rigorous and integrated analysis of the resulting overall closed-loop system. As a consequence, the proposed Newton-based MPC iteration scheme allows one to make use of the same techniques for exploiting sparsity as in a conventional interior-point method, while at the same time ensuring all the systems theoretic properties that have been discussed above.

Remark 4.8. Note that an alternative to the presented relaxed primal-dual approach is to handle the equality constraints (4.47b) by means of a quadratic penalty function, e.g., by considering a modified cost function of the form

$$\hat{J}'_N(w,x) = \min_{w} \frac{1}{2} w^\top H w + x^\top F w + \varepsilon \hat{B}_{\mathrm{w}}(w) + \frac{1}{2} x^\top Y x + \frac{1}{\varepsilon} \|Cw - b(x)\|^2 . \tag{4.63}$$

This penalty function based formulation again reduces the MPC problem to an unconstrained optimization problem, which then simplifies the design and analysis of a suitable associated MPC iteration scheme that is very similar to the condensed case and possesses the same theoretical properties as discussed above. Moreover, it can be shown in a rather straightforward way that the Hessian of the above cost function will be banded with bandwidth $2n + m - 1$, which can be exploited for speeding up the required Newton search direction, see for example Appendix C3 in (Boyd and Vandenberghe, 2004). However, as the equality constraints are in the above approach only handled in a rather indirect way and parts of the underlying problem structure are destroyed by the penalty function, in this thesis we focus on the more common primal-dual approach that has been discussed above as well as in (Rao et al., 1998; Wang and Boyd, 2010; Domahidi et al., 2012).

4.4.3 Fast search direction computation by exploiting sparsity

In this section, we elaborate in greater detail on the structure of the KKT system (4.51) and show how this structure can be used in order to speed up the computation of the Newton search direction $\Delta w(w,x)$ that is used in the proposed Newton-based optimizer update operator (4.56). In particular, the main result will be to show that the overall complexity of the Newton step is of order $N(n+m)^3$ and does, therefore, grow only linearly with the prediction horizon N. The main idea of exploiting problem structure and sparsity in the context of interior-point methods for MPC has already been reported by several researchers, see for example Rao et al. (1998), Wang and Boyd (2010), and Domahidi et al. (2012), and essentially all of the elimination and factorization techniques that will be discussed in the following can be found in the aforementioned works. However, besides contributing to a self-contained presentation, the subsequent discussion illustrates that all the structural properties and benefits of conventional primal-dual interior point methods are in fact preserved by the proposed relaxed barrier function based formulation.

Remark 4.9. Note that the effort of computing the residuals and the KKT matrix itself will typically also grow with increasing problem dimension. However, the numerical complexity of this step will usually be dominated by the complexity of actually solving the resulting KKT system. All the complexity considerations in the following will therefore solely focus on the task of solving (4.51) for a given KKT matrix and residual vector.

Let us now take a closer look at the structural properties of the matrices contained in the KKT system (4.51). Obviously, the Hessian matrix of the cost function with respect to the optimization variable w is given by

$$\nabla^2_w \hat{J}_N(w,x) = H + \varepsilon G^\top \operatorname{diag}\left(w_r^1 \hat{D}(z_1(w,x)), \dots, w_r^q \hat{D}(z_q(w,x))\right) G\,, \tag{4.64}$$

where the matrices $H \in \mathbb{S}^{n_w}_{++}$ and $G \in \mathbb{R}^{q \times n_w}$ are provided in Appendix B.9, $w_r \in \mathbb{R}^q_{++}$ refers to the underlying recentering vector, and $\hat{D} : \mathbb{R} \to \mathbb{R}_+$ denotes the Hessian of the primordial relaxed barrier function, that is,

$$\hat{D}(z) = \begin{cases} \frac{1}{z^2} & z > \delta \\ \frac{1}{\delta^2} & z \le \delta \end{cases}. \tag{4.65}$$

As before, the respective arguments associated to the underlying linear inequality constraints are defined as $z_i(w,x) = -G^i w + E^i x + d^i$, $i = 1, \dots, q$, see Appendix B.9. For the sake of convenience, we recall that

$$H = 2\begin{bmatrix} R & \cdots & & & \cdots & 0 \\ \vdots & Q & & & & \vdots \\ & & \ddots & & & \\ & & & Q & & \\ \vdots & & & & R & \vdots \\ 0 & \cdots & & & \cdots & P \end{bmatrix}, \qquad G = \begin{bmatrix} 0 & \cdots & & & \cdots & 0 \\ C_\mathrm{u} & \cdots & & & \cdots & 0 \\ \vdots & C_\mathrm{x} & & & & \vdots \\ & & \ddots & & & \\ \vdots & & & & C_\mathrm{u} & \vdots \\ 0 & \cdots & & & \cdots & C_\mathrm{x} \end{bmatrix}, \tag{4.66}$$

which reveals that the overall Hessian matrix given in (4.64) is a positive definite and block-diagonal matrix with alternating blocks of size $m \times m$ and $n \times n$, respectively. Thus, from a structural point of view, the only differences to a conventional interior-point approach using nonrelaxed logarithmic barrier functions are given by the inclusion of the recentering vector w_r and the fact that the second derivative of the natural logarithm is replaced by its relaxed counterpart (4.65). Furthermore, recall that the matrix $C \in \mathbb{R}^{Nn \times n_w}$ describing the linear equality constraints is given by

$$C = \begin{bmatrix} -B & I & 0 & 0 & 0 & 0 & \cdots & 0 & 0 & 0 \\ 0 & -A & -B & I & 0 & 0 & \cdots & 0 & 0 & 0 \\ 0 & 0 & 0 & -A & -B & I & \cdots & 0 & 0 & 0 \\ \vdots & \vdots & \vdots & \vdots & \vdots & \vdots & \ddots & \vdots & \vdots & \vdots \\ 0 & 0 & 0 & 0 & 0 & 0 & \cdots & I & 0 & 0 \\ 0 & 0 & 0 & 0 & 0 & 0 & \cdots & -A & -B & I \end{bmatrix}, \tag{4.67}$$

which as well exhibits a considerable amount of structure. Following the approaches presented by Wang and Boyd (2010) and Domahidi et al. (2012), we now consider the following more compact representation of the KKT system (4.51) introduced above:

$$\begin{bmatrix} \Phi & C^\top \\ C & 0 \end{bmatrix} \begin{bmatrix} \Delta w \\ \Delta \nu \end{bmatrix} = - \begin{bmatrix} r_d \\ r_p \end{bmatrix} . \tag{4.68}$$

Here, $\Phi := \nabla_w^2 \hat{J}_N(w, x)$ is nothing else than a short-hand notation for the positive definite and block-diagonal Hessian matrix discussed above. The goal is now to compute the search directions $\Delta w \in \mathbb{R}^{n_w}$ and $\Delta \nu \in \mathbb{R}^{Nn}$ in an efficient way by exploiting the structure of the KKT matrix on the left-hand side. Note that if we do not exploit any structure and solve the above system directly with a dense LDL$^\top$ factorization, the resulting numerical cost is of order $N^3(2n+m)^3$ (Boyd and Vandenberghe, 2004, Appendix C3).

The procedure that we are going to present in the following makes use of block elimination in combination with a suitable factorization of the underlying submatrices and is very similar to the one presented by Wang and Boyd (2010). As a first step, recall that Φ is a positive definite, and therefore nonsingular, matrix, which allows us to solve (4.68) via block elimination. In particular, we obtain that

$$\Delta w = \Phi^{-1}(-r_d - C^\top \Delta \nu) , \tag{4.69}$$

while the dual search direction $\Delta \nu$ can be computed from

$$Y \Delta \nu = \beta , \qquad Y := C \Phi^{-1} C^\top , \tag{4.70a}$$

$$\beta := r_p - C \Phi^{-1} r_d . \tag{4.70b}$$

It is well-known that the block elimination method typically leads to a significant reduction in numerical complexity under the condition that the submatrix Φ can be easily factored, see for example (Wright, 1997b; Boyd and Vandenberghe, 2004). The overall solution procedure then consists of the following three steps:

i) Construct the Schur complement $Y = C\Phi^{-1}C^\top$ as well as $\beta = r_p - C\Phi^{-1} r_d$,

ii) Compute the dual search direction $\Delta \nu$ by solving $Y \Delta \nu = \beta$,

iii) Compute the primal search direction Δw by solving $\Phi \Delta w = -r_d - C^\top \Delta \nu$.

In the following, we show that by exploiting the underlying problem structure, both the matrix Φ and the Schur complement Y can be factored in a systematic and numerically efficient way. To this end, note that Φ is a block-diagonal matrix of the form

$$\Phi = \begin{bmatrix} \tilde{R}_0 & 0 & 0 & \cdots & 0 & 0 & 0 \\ 0 & \tilde{Q}_1 & 0 & \cdots & 0 & 0 & 0 \\ 0 & 0 & \tilde{R}_1 & \cdots & 0 & 0 & 0 \\ \vdots & \vdots & \vdots & \ddots & \vdots & \vdots & \vdots \\ 0 & 0 & 0 & \cdots & \tilde{Q}_{N-1} & 0 & 0 \\ 0 & 0 & 0 & \cdots & 0 & \tilde{R}_{N-1} & 0 \\ 0 & 0 & 0 & \cdots & 0 & 0 & \tilde{Q}_N \end{bmatrix} , \tag{4.71}$$

where the respective diagonal blocks $\tilde{R}_i \in \mathbb{S}^m_{++}$ and $\tilde{Q}_i \in \mathbb{S}^n_{++}$ follow directly from the Hessian matrix defined in (4.64)–(4.66). The associated inverse is also block-diagonal and thus can be written as

$$\Phi^{-1} = \begin{bmatrix} \tilde{R}_0^{-1} & 0 & 0 & \cdots & 0 & 0 & 0 \\ 0 & \tilde{Q}_1^{-1} & 0 & \cdots & 0 & 0 & 0 \\ 0 & 0 & \tilde{R}_1^{-1} & \cdots & 0 & 0 & 0 \\ \vdots & \vdots & \vdots & \ddots & \vdots & \vdots & \vdots \\ 0 & 0 & 0 & \cdots & \tilde{Q}_{N-1}^{-1} & 0 & 0 \\ 0 & 0 & 0 & \cdots & 0 & \tilde{R}_{N-1}^{-1} & 0 \\ 0 & 0 & 0 & \cdots & 0 & 0 & \tilde{Q}_N^{-1} \end{bmatrix}. \tag{4.72}$$

In combination with the structure of the matrix C given in (4.67), this implies that the Schur complement $Y = C\Phi^{-1}C^\top$ is a block-tridiagonal matrix. In particular,

$$Y = \begin{bmatrix} Y_{11} & Y_{12} & 0 & \cdots & 0 & 0 \\ Y_{21} & Y_{22} & Y_{23} & \cdots & 0 & 0 \\ 0 & Y_{32} & Y_{33} & \cdots & 0 & 0 \\ \vdots & \vdots & \vdots & \ddots & \vdots & \vdots \\ 0 & 0 & 0 & \cdots & Y_{N-1,N-1} & Y_{N-1,N} \\ 0 & 0 & 0 & \cdots & Y_{N,N-1} & Y_{N,N} \end{bmatrix}, \tag{4.73}$$

where

$$Y_{11} = B\tilde{R}_0^{-1}B^\top + \tilde{Q}_1^{-1}, \tag{4.74a}$$

$$Y_{ii} = A\tilde{Q}_{i-1}^{-1}A^\top + B\tilde{R}_{i-1}^{-1}B^\top + \tilde{Q}_i^{-1}, \qquad i = 2, \ldots, N, \tag{4.74b}$$

$$Y_{i,i+1} = Y_{i+1,i}^\top = -\tilde{Q}_i^{-1}A^\top, \qquad i = 1, \ldots, N-1. \tag{4.74c}$$

Thus, the required submatrices can for instance be obtained by computing the Cholesky factorization of each of the blocks of the matrix Φ and then solving

$$\tilde{Q}_i X_1 = I \quad \tilde{Q}_i X_2 = A^\top, \qquad \tilde{R}_i X_3 = B^\top \tag{4.75}$$

for $X_1 = \tilde{Q}_i^{-1}$, $X_2 = \tilde{Q}_i^{-1}A^\top$, and $X_3 = \tilde{R}_i^{-1}B^\top$ by means of forward and backward substitution. Finally, the latter two expressions can be multiplied from the left by A and B in order to form $A\tilde{Q}_i^{-1}A^\top$ as well as $B\tilde{R}_i^{-1}B^\top$. As all of these operations need to be performed N times and the complexity of each Cholesky factorization is cubic in the respective dimension, the overall numerical complexity associated to step i) is $N(\frac{1}{3}(n^3 + m^3) + 4n^2 + 2m^2 + n^3 + n^2m)$, which is of order $N(n+m)^3$.
Let us now turn our attention to step ii). Here, the dual search direction $\Delta\nu$ is obtained by computing the Cholesky factorization of the Schur complement Y, followed by forward and backward substitution. As mentioned above, Y is block-tridiagonal with N block-rows of size $n \times n$. The following standard procedure for factorizing $Y = LL^\top$ in a

structured and efficient way has already been applied by both Wang and Boyd (2010) and Domahidi et al. (2012) in the context of nonrelaxed primal-dual interior-point methods. Due to the block-tridiagonal structure of Y, the associated Cholesky factor L has the following lower bidiagonal block-structure

$$L = \begin{bmatrix} L_{11} & 0 & 0 & \cdots & 0 & 0 \\ L_{21} & L_{22} & 0 & \cdots & 0 & 0 \\ 0 & L_{32} & L_{33} & \cdots & 0 & 0 \\ \vdots & \vdots & \vdots & \ddots & \vdots & \vdots \\ 0 & 0 & 0 & \cdots & L_{N-1,N-1} & 0 \\ 0 & 0 & 0 & \cdots & L_{N,N-1} & L_{N,N} \end{bmatrix}, \tag{4.76}$$

where the submatrices $L_{ii} \in \mathbb{R}^{n \times n}$ on the diagonal are lower triangular with positive diagonal entries. It then follows directly from $Y = LL^\top$ that the blocks of L may be recursively computed as

$$L_{11}L_{11}^\top = Y_{11}, \tag{4.77a}$$

$$L_{ii}L_{i+1,i}^\top = Y_{i,i+1}, \qquad i = 1, \ldots, N-1, \tag{4.77b}$$

$$L_{ii}L_{ii}^\top = Y_{ii} - L_{i,i-1}L_{i,i-1}^\top, \qquad i = 2, \ldots, N. \tag{4.77c}$$

In particular, L_{11} can be determined by Cholesky factorization of Y_{11}. Then, L_{21} can be found by solving $L_{11}L_{21}^\top = Y_{12}$ by means of forward substitution, which then in turn allows obtaining L_{22} via Cholesky factorization of $Y_{22} - L_{21}L_{21}^\top$. This procedure can be carried out until all block-elements of L are determined. As each of the required factorizations has a complexity of $\frac{1}{3}n^3$ (which dominates the complexity of the additional forward substitution steps), the overall complexity of step *ii*) is of order Nn^3. Note that the same complexity result can be obtained by treating Y as a banded matrix with bandwidth $3n$. As the blockwise Cholesky factorization of Φ has already been constructed in step *i*), the remaining third step of computing the primal search direction Δw can be performed by a simple forward and backward substitution scheme, resulting in a (negligible) numerical effort of $N(2n^2 + 2m^2)$ floating point operations for step *iii*).

Thus, we can finally conclude that the overall numerical complexity of computing the required primal-dual search direction $(\Delta w, \Delta \nu)$ based on the above elimination and factorization scheme is of order $N(n+m)^3$. Compared to the complexity of solving (4.68) directly with a dense LDL$^\top$ factorization, this reveals that the numerical effort of each Newton step now scales only linearly with the prediction horizon. Note that the overall complexity reduces even further when considering diagonal weight matrices Q and R in combination with simple box constraints on the states and inputs. In this case, also the matrix Φ will be diagonal, which implies that all the operations related to its factorization can be neglected in the above complexity analysis. As a result, the overall numerical complexity will in this case reduce to order $N(n^3 + n^2 m)$, which grows linearly in both the prediction horizon and the input dimension (Wang and Boyd, 2010). Finally, an even further complexity reduction might be achieved by making use of partial condensing techniques that allow exploiting different levels of sparsity (Axehill, 2015).

4.4.4 Implementation details

A simple version of the proposed sparsity exploiting MPC iteration scheme has been implemented in C, using the freely available LAPACK library (Anderson et al., 1990, 1999) as well as the included version of the Basic Linear Algebraic Subprograms (BLAS) for carrying out the underlying numerical linear algebra operations. The code is based on the stabilizing anytime iteration scheme proposed in Section 4.4.2 and uses the elimination and factorization procedures from Section 4.4.3 for computing the required Newton search directions. In particular, the latter part is closely related to the function `fmpc_step` by Wang and Boyd (2010), which in our implementation has been adapted to the relaxed barrier function based framework. A simple Armijo backtracking line search is used for identifying a suitable step length at each Newton step. Apart from the discussed approaches for exploiting the underlying problem structure, standard code with nested loops is used and no further tuning of the code has been carried out. For the sake of simplicity, the current version of the code only supports a simplified problem setup with box constraints on the states and inputs.

In order to illustrate the effectiveness of the proposed elimination and factorization procedures, a similar C implementation has been written for the condensed MPC iteration scheme from Section 4.3. This will in the following allow us to compare the numerical performance of condensed and sparse formulation for varying problem dimensions.

4.4.5 Numerical example and benchmarking

In order to illustrate the behavior of the proposed sparsity exploiting MPC iteration scheme and to compare its overall numerical and closed-loop performance to the corresponding dense formulation as well as to existing tailored (fast) MPC approaches, we consider a canonical numerical benchmark problem which has also been investigated by both Domahidi et al. (2012) and Wang and Boyd (2010). In particular, the considered problem setup is to control a chain of M identical unit masses that are connected by springs, assuming that there are $M-1$ actuators which allow to exert forces between each pair of adjacent masses, see Figure 4.6 below. We assume that all springs exhibit the same force characteristics and we let $k_s \in \mathbb{R}_{++}$ denote the corresponding spring constant. Furthermore, let $x_i(t)$ refer for any $i \in \{1, \ldots, M\}$ to the displacement of mass i from its tension-free reference position. The dynamical behavior of the overall oscillating masses

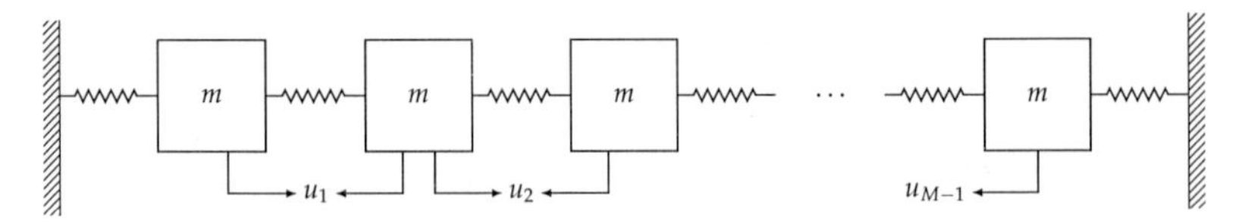

Figure 4.6. Schematic illustration of the oscillating masses example with M masses.

system is then described by a linear continuous-time system of the form

$$\dot{x}(t) = \begin{bmatrix} 0_M & I_M \\ A_{21} & 0_M \end{bmatrix} x(t) + \begin{bmatrix} 0_{M\times M-1} \\ B_2 \end{bmatrix} u(t) \tag{4.78}$$

where $x(t) = [x_1(t), \ldots x_M(t), \dot{x}_1(t), \ldots, \dot{x}_M(t)]^\top \in \mathbb{R}^{2M}$ represents the overall system state and $u(t) = [u_1(t), \ldots, u_{M-1}(t)]^\top \in \mathbb{R}^{M-1}$ denotes the vector of control inputs. The sub-matrices $A_{21} \in \mathbb{R}^{M\times M}$ and $B_2 \in \mathbb{R}^{M\times M-1}$ can be derived in a straightforward manner from Figure 4.6 and are given by

$$A_{21} = k_s \begin{bmatrix} -2 & 1 & & & \cdots & 0 \\ 1 & -2 & 1 & & & \vdots \\ & 1 & -2 & 1 & & \\ & & \ddots & \ddots & \ddots & \\ \vdots & & & 1 & -2 & 1 \\ 0 & \cdots & & & 1 & -2 \end{bmatrix}, \quad B_2 = \begin{bmatrix} 1 & & & \cdots & 0 \\ -1 & 1 & & & \vdots \\ & -1 & 1 & & \\ & & \ddots & \ddots & \\ \vdots & & & -1 & 1 \\ 0 & \cdots & & & -1 \end{bmatrix}. \tag{4.79}$$

For all the results discussed in the following, we assume that there is no damping and that $k_s = 1$. The above continuous-time system is discretized with a sampling time of $T_s = 0.5$ s, which leads to a linear discrete-time system of the form (4.1) with dense matrices A and B. Furthermore, in accordance with the above references, we assume that the associated state and input vectors are subjected to box constraints of the form

$$-4 \cdot \mathbb{1}_n \leq x(k) \leq 4 \cdot \mathbb{1}_n, \qquad -0.5 \cdot \mathbb{1}_m \leq u(k) \leq 0.5 \cdot \mathbb{1}_m \tag{4.80}$$

with $n = 2M$ and $m = M - 1$. The benefit of the considered example is that it allows one to scale the state and input dimensions of the resulting overall dynamical system by varying the number of masses M. Related to this problem setup, we will in the following compare the behavior of the proposed sparsity exploiting relaxed barrier function based MPC algorithm to both the condensation-based approach from Section 4.3 and to two existing and often-used optimization routines that are specifically tailored to the linear MPC context – namely *qpOASES* (Ferreau et al., 2014), which is an efficient open-source implementation of the online active set strategy proposed by Ferreau et al. (2008), and *FORCES Professional* (Domahidi and Jerez, 2014), which is a by now semi-commercial implementation of the sparsity exploiting primal-dual interior point method presented by Domahidi et al. (2012). As the overall behavior of the resulting closed-loop system is for the sparse formulation actually identical to that of the condensed formulation discussed above, the following two scenarios mainly focus on the numerical performance, i.e., on the required computation time, as well as on the closed-loop performance in dependence on the number of iterations. In particular, in order to avoid unnecessary repetitions, the constraint satisfaction and robustness properties of the sparsity exploiting anytime MPC algorithm are not considered in detail (see Section 4.3.7 for the condensed case).

Scenario 1

The main goal of the first benchmark scenario is to investigate whether and to which extent the proposed sparsity exploiting search direction computation really leads to a reduced numerical complexity when increasing more and more the size of the underlying MPC problem. In addition, we want to analyze whether the proposed relaxed barrier function based approach is actually competitive when comparing it in terms of the required computation time to the existing fast MPC solvers mentioned above. To this end, we formulate the MPC problem for the oscillating masses example for different values of M and different choices of the prediction horizon N. Given the respective problem dimensions, the cost function weight matrices are in each scenario chosen as $Q = I_n$ and $R = I_m$. The associated relaxed barrier function based MPC algorithms are set up according to the MPC design presented in Section 4.1, combined with the condensed and sparsity exploiting Newton-based optimization routines described in Sections 4.3 and 4.4, respectively. In both approaches, we set $\varepsilon = \delta = 10^{-3}$ and make use of a backtracking line search with parameters $\rho = 0.9$, $c_1 = 0.01$. The underlying Newton iterations are stopped and the respective solution is declared as optimal as soon as the gradient, respectively the primal residual, lies below a threshold of $\epsilon_{\text{tol}} = 10^{-3}$. In the following, we use the abbreviations $rbMPC_d$ and $rbMPC_s$ for referring to the dense and sparse formulation, respectively. Both approaches were implemented in C (see Section 4.4.4) and then compiled into executable MEX functions by using the GCC compiler with the options "`-v -largeArrayDims`".

As a benchmark solver that does also rely on the primal-dual interior point approach and explicitly exploits the structure and sparsity of the non-condensed MPC problem formulation we make use of the multistage interior-point method underlying the semi-commercial FORCES Pro code generator (Domahidi and Jerez, 2014). The resulting tailored interior-point QP algorithm is based on the results presented by Domahidi et al. (2012) and Wang and Boyd (2010) and the applied complexity reduction techniques are therefore very similar to the ones discussed in Section 4.4.3 of this thesis. For the considered benchmark study, the required multistage problem structure is formulated according to the standard procedure documented in the FORCES Pro user manual and the respective MEX code is generated by calling the server-based FORCES Pro code generation tool. All tolerances are set to 10^{-3} and the code is generated based on a temporary FORCES Pro Prototyping license (compiler optimization level 3). As an additional benchmark solver that does, in particular, not rely on the interior-point or barrier function paradigm, the active set QP solver qpOASES (version 3.2) is used (Ferreau et al., 2007–2015). The underlying active set strategy relies on the condensed MPC formulation and therefore solves a parametric QP of the form (2.11) at each call or sampling instant. Standard option settings are used and the solver is called from within MATLAB by making use of the provided MEX interface.

In order to allow for a fair comparison and to prevent the relaxed barrier function based MPC approach from influencing the behavior of the two comparison solvers (e.g., via the choice of the terminal cost matrix P), the MPC problems for FORCES and qpOASES are set up based on a conventional and often-used linear MPC design procedure. In particular, no stabilizing terminal set constraint is imposed and the terminal cost matrix P is chosen based on the associated infinite-horizon LQR problem, that is, as the positive definite solution to the DARE (4.5) when setting $M_x = 0$, $M_u = 0$. Note however that this setup does not allow to state any stability guarantees for the closed-loop system.

Related to this overall problem setup, the main approach that is underlying our first benchmarking scenario is to assess and compare the numerical performance of the different solvers by measuring the required execution times for different dimensions of the considered oscillating masses example. Therein, our benchmarking study follows exactly the same approach as the numerical performance tests presented in (Domahidi et al., 2012). All the numerical calculations were carried out using MATLAB R2015b on a Desktop PC with 3.4 GHz Intel Core i7 3770 processor and 8 GB RAM, running Debian Linux 7.9 (64 bit). In order to achieve the fastest possible computation times, all outputs to the MATLAB command window were disabled. Detailed numerical results are summarized in Tables 4.1 and 4.2 as well as in the associated Figure 4.7.

Following the approach of Domahidi et al. (2012), Table 4.1 lists the average runtimes that are required for executing ten iterations of the respective optimization algorithms – that is, ten active set, interior-point, or relaxed barrier function based Newton iterations. While the recorded runtimes tend to be very similar if both M and N are small, the sparsity exploiting algorithms FORCES Pro and rbMPC_s quickly outperform the condensation-based solvers as soon as the problem size increases. This becomes particularly apparent when considering long prediction horizons, which illustrates the complexity analysis results from Section 4.4.3. In particular, performing ten iterations of the proposed sparsity exploiting anytime MPC algorithm is for the case $N = M = 30$ more then 30 times faster than performing ten iterations of the corresponding iteration scheme for the condensed formulation. In order to not only compare the pure computation time per iteration, Table 4.2 in addition lists the average runtimes that are required by each of the discussed optimization algorithms to fully converge to the optimal solution (zero initialization, no warm-starting). In addition, a graphical representation of the obtained performance results for varying M is given in Figure 4.7. The performance results from Table 4.1 are essentially recovered, which confirms the observed complexity reduction also for the full convergence case. At first glance, it may appear surprising that our self-programmed algorithm rbMPC_s is in fact even faster than the efficient and mature FORCES Pro solver. However, at this point we want to emphasize again that, strictly speaking, the two algorithms are not solving exactly the same optimization problem. By keeping the barrier parameter ε fixed and making use of relaxed recentered barrier functions, the proposed relaxed barrier function based approach rather computes an approximate solution to the original MPC problem (however, with all the stability and constraint satisfaction guarantees discussed above). In addition, FORCES Pro is by now a semi-commercial solver that has a much broader range of applicability and is, in contrast to our solver, not specifically tailored to the considered problem class with simple box constraints.

We close our observations with the remark that the numerical performance of the different solvers may change when making use of suitable warm-starting techniques. This is in particular true for the online active set strategy underlying qpOASES as well as for the relaxed barrier function based iteration schemes proposed in this thesis, which are both specifically tailored for solving *sequences* of closely related optimization problems. Nevertheless, the reported computation times allow us to analyze and illustrate the overall behavior and complexity of the different solvers with respect to the "worst case" of cold-starting with general initial conditions. Furthermore, they also demonstrate that by exploiting the underlying sparsity structure, the numerical complexity *per iteration* can be significantly reduced for higher-dimensional problems.

Scenario 2

In contrast to the first scenario, we now fix the dimension of the considered MPC problem and analyze how for the proposed sparsity exploiting anytime MPC algorithm both the numerical and the overall control performance of the closed-loop system depend on the maximal number of performed Newton iterations. In particular, we choose $N = 10$, $M = 15$ and then simulate the behavior of the resulting closed-loop system for different values of $i_{\mathrm{T,max}}$, where $i_{\mathrm{T,max}} \in \mathbb{N}_+$ denotes an upper bound on the maximal number of performed Newton iterations in the sense that $i_{\mathrm{T}}(k) \leq i_{\mathrm{T,max}}$ for all sampling instants $k \in \mathbb{N}$. Following our approach from Section 4.3.7, the underlying MPC iteration scheme (4.53) is initialized using a suboptimal initialization of the form $w_0 = \bar{K}_w x_0$, where we choose the gain $K_w \in \mathbb{R}^{n_w \times n}$ as

$$\bar{K}_w = \left[K^\top \ (A+BK)^\top \ (A+BK)^\top K^\top \ \cdots \ (A+BK)^{N-1^\top} K^\top \ (A+BK)^{N^\top}\right]^\top . \tag{4.81}$$

Related to this problem setup, Figure 4.8 shows the averaged evolution of the cost function for different values of $i_{\mathrm{T,max}}$ as well as the average number of Newton iterations that is at each sampling step required to arrive at the optimal solution. It can be observed that a quite small number of optimizer iterations is sufficient to achieve a close-to-optimal closed-loop behavior. Furthermore, the number of optimizer iterations that is required for optimal performance is decreasing over time and converges to a "steady state" of only one performed Newton iteration per sampling instant. In addition, Figure 4.9 shows the average computation time per sampling instant as well as the averaged cumulative stage cost when varying the maximal number of Newton iterations $i_{\mathrm{T,max}}$ in both cases. Here, the cumulative stage cost is for any given initial condition $x(0)$ defined as

$$\ell_\Sigma = \frac{1}{T+1} \sum_{k=0}^{T} \hat{\ell}(x(k), \Pi_0 w(k)) , \tag{4.82}$$

where the summation horizon was in this example chosen uniformly as $T = 20$. As expected, the computation time per sampling instant increases monotonically with $i_{\mathrm{T,max}}$ and saturates as soon as the maximally required number of Newton iterations is reached.

Furthermore, we observe that the averaged cumulative stage cost decreases rapidly also if only a small number of optimizing iterations is performed at each sampling step. In fact, close-to-optimal closed-loop performance can be achieved by performing only one Newton step per sampling instant, allowing for an average computation time of below one millisecond. However, it should be noted that the behavior of the cumulative stage cost may in the general case not necessarily be monotonic in $i_{\mathrm{T,max}}$ and may also heavily depend on the considered problem setup (e.g. the prediction horizon).

In order to get an impression of the closed-loop behavior in terms of the applied control inputs, we moreover simulated the proposed iteration scheme with $i_{\mathrm{T}}(k) = 1$ for two specific initial conditions and compared the resulting control trajectories to that of a conventional MPC scheme that applies at each sampling step a fully optimized control input computed via FORCES Pro. First, we consider a scenario in which only the first and the last mass of the considered chain of masses deviate at initialization from their reference position. In particular, $x_1(0) = x_M(0) = 2$ while $x_i(0) = 0$ for $i = 2, \ldots, M-1$ and $\dot{x}_i(0) = 0$ for $i = 1, \ldots, M$. As an alternative to the initialization based on (4.81), we also investigate the case where the iteration scheme is initialized with the optimal FORCES Pro solution for the considered initial state x_0, that is, $w_0 = w_F^*(x_0)$. Representative for the overall closed-loop behavior, the resulting control input trajectories $u_1(k)$ related to the first pair of masses are depicted in Figure 4.10. The same scenario was then also repeated for a random initial condition $x_0 = x(0) \in \mathbb{R}^n$, for which the corresponding control input trajectories are shown in Figure 4.11. In all cases, the control input of the proposed anytime MPC algorithm (performing now only one Newton iteration per sampling step) is very similar to that of the associated conventional MPC scheme based on FORCES Pro.

4.4.6 Summary

As the main result of this section we can conclude that by keeping the predicted system states as optimization variables and exploiting the sparsity structure of the then resulting equality constrained optimization problem, we can derive a class of relaxed barrier function based anytime MPC algorithms that allows for a potentially much more efficient computation of the underlying optimizer update operator, while still guaranteeing the same systems theoretical properties as the condensation-based approach presented in Section 4.3. In particular, we transferred the principal idea of the proposed MPC iteration scheme to the non-condensed sparse formulation and showed that by making use of a suitably defined shift operator and a Newton-based optimizer update, it is still possible to guarantee a monotonic decay of the associated cost function at each iteration. In this way, we were able to essentially recover all the stability, constraint satisfaction, and robustness results from Section 4.3. We furthermore showed how the structure of the sparse formulation can be exploited in order to speed up the computation of the required Newton search directions, resulting in a reduced numerical complexity that scales only linearly with the prediction horizon. Finally, the presented benchmark study demonstrated the effectiveness of the proposed sparsity-exploiting MPC algorithm and showed that its current C implementation is in fact competitive with existing fast MPC solvers.

Table 4.1. Benchmarking results for the oscillating masses example for different numbers of masses M and different horizon lengths N. Listed are, averaged over 100 random initial states, the run times for executing ten iterations of the respective optimization algorithm. Also given are the numbers of optimization variables $p = N(n+m)$, equality constraints $r = Nn$, and inequality constraints $q = 2(N(n+m)+n)$ in the sparse formulation.

Problem dimensions					Computation times [ms]			
M	N	p	r	q	qpOASES	FORCES Pro	rbMPC$_d$	rbMPC$_s$
2	10	54	40	108	0.16	0.12	0.17	0.46
4	10	118	80	236	0.72	0.40	1.17	1.18
6	10	182	120	364	1.92	0.87	2.52	1.81
10	10	310	200	620	6.24	3.35	7.12	3.19
15	10	470	300	940	14.72	9.34	16.10	6.72
30	10	950	600	1900	79.04	64.64	134.56	24.12
6	30	522	360	1044	13.23	2.43	13.57	3.34
10	30	890	600	1780	67.14	9.96	86.56	6.57
15	30	1350	900	2700	205.51	27.92	348.38	15.38
30	30	2730	1800	5460	1677.93	198.61	2059.02	61.33

Table 4.2. Further benchmarking results for the oscillating masses example. Listed are, again averaged over 100 random initial states, the run times that are required by each of the respective optimization algorithms to converge fully to the optimal solution (residual tolerance 10^{-3}). The problem dimensions p, r, and q are defined analogously to Table 4.1. For large problem dimensions, the sparsity exploiting relaxed barrier function based approach is up to 30 times faster than the corresponding condensation-based algorithm.

Problem dimensions					Computation times [ms]			
M	N	p	r	q	qpOASES	FORCES Pro	rbMPC$_d$	rbMPC$_s$
2	10	54	40	108	0.15	0.11	0.13	0.41
4	10	118	80	236	0.72	0.44	1.44	1.13
6	10	182	120	364	2.02	0.97	3.83	2.04
10	10	310	200	620	10.24	3.86	9.61	3.72
15	10	470	300	940	35.84	11.20	26.66	7.22
30	10	950	600	1900	461.80	81.62	373.20	33.68
6	30	522	360	1044	17.53	2.69	14.35	3.10
10	30	890	600	1780	100.50	11.24	82.93	6.23
15	30	1350	900	2700	503.61	34.05	313.75	12.55
30	30	2730	1800	5460	8215.69	259.78	2158.95	66.12

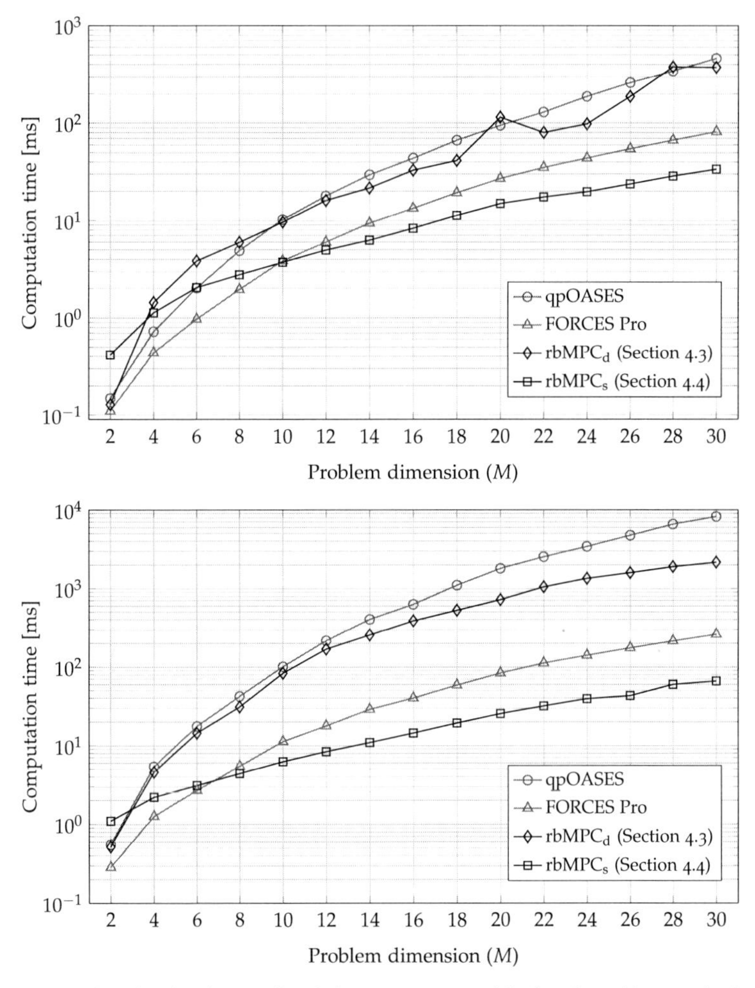

Figure 4.7. Graphical and more detailed representation of the benchmarking results from Table 4.2. Depicted are the average runtimes required for full convergence of the respective optimization algorithms for a varying number of masses M when considering a fixed prediction horizon of $N = 10$ (top), respectively $N = 30$ (bottom) .

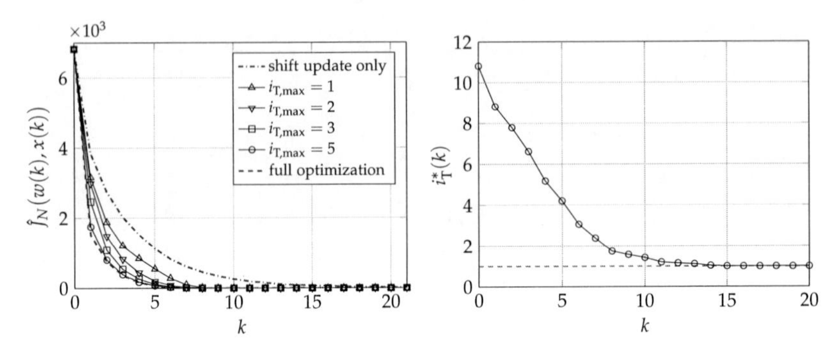

Figure 4.8. *Left:* Averaged evolution of the cost function for 100 random initial conditions $x_{0,i}$ when applying the sparsity exploiting anytime MPC algorithm with different upper bounds on the maximal number of Newton iterations $i_{\mathrm{T}}(k) \leq i_{\mathrm{T,max}}$. Already for five Newton iterations, the behavior is almost identical to that when applying the exact optimal solution. *Right:* Averaged evolution of the number of Newton iterations $i_{\mathrm{T}}^*(k)$ that is required to arrive at the optimal solution. Due to the applied warm-starting strategy (and the convergence of the closed-loop system to the origin), the required number of iterations decreases rapidly over time and converges to $i_{\mathrm{T}}(k) = 1$.

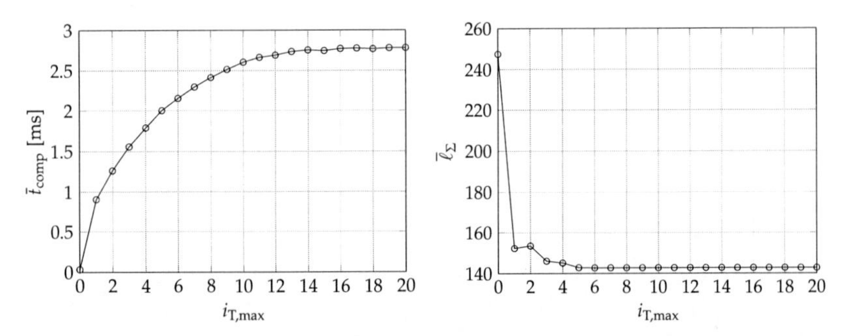

Figure 4.9. Behavior of the resulting numerical and closed-loop performance depending on the maximal number of Newton-iterations that are performed per sampling step. *Left:* Averaged evolution of the computation time per sampling step when executing the proposed anytime MPC algorithm with $i_{\mathrm{T}}(k) \leq i_{\mathrm{T,max}}$. *Right:* Corresponding behavior of the resulting control performance in terms of the averaged cumulative stage cost. Applying only one Newton iteration per sampling step already leads to good closed-loop performance while allowing for an average computation time of below one millisecond.

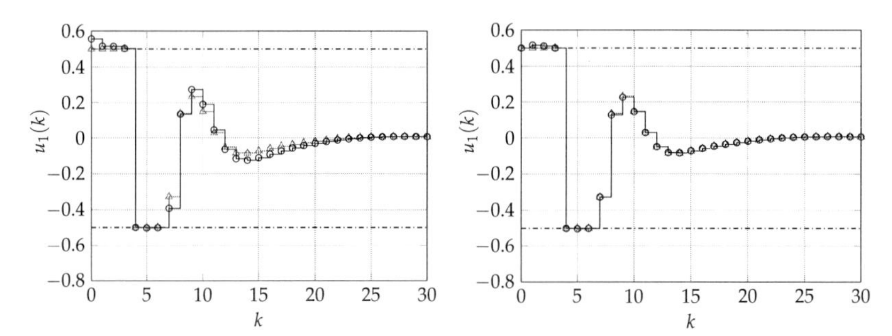

Figure 4.10. Applied control input $u_1(k)$ for the oscillating masses example for initial condition $x_1(0) = 2$, $x_M(0) = 2$ and $x_i(0) = 0$ for $i = 2, \ldots, M-1$. Shown is the resulting control for the proposed anytime MPC algorithm when performing only one optimizing Newton step at each sampling instant (○). In the left plot, the underlying optimizer iteration is initialized with $w_0 = \bar{K}_w x_0$. In the right plot, the iteration is initialized with the associated optimal solution that is returned by FORCES Pro. For comparison, the input trajectory resulting from applying the FORCES Pro solver in closed-loop operation is depicted in gray (△). Although the proposed anytime algorithm performs only one Newton iteration at each sampling step, the resulting closed-loop behavior is almost identical.

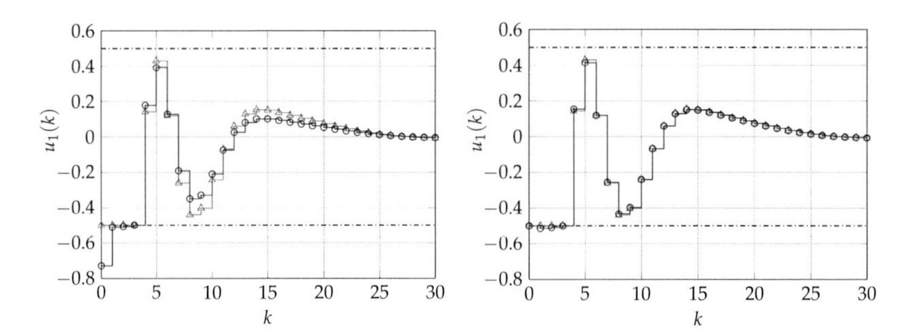

Figure 4.11. Applied control input $u_1(k)$ for the oscillating masses example for a random initial condition $x(0) \in \mathbb{R}^n$ with $-2 \cdot \mathbb{1}_n \leq x(0) \leq 2 \cdot \mathbb{1}_n$. Shown are again the resulting control trajectories associated to the proposed anytime MPC algorithm with only one optimizing Newton step at each sampling instant (○) and the conventional MPC approach based on FORCES Pro (△). While there are more pronounced deviations in the case of suboptimal initialization (left), the resulting controls are again almost identical when initializing the underlying iteration scheme with the initial FORCES Pro solution.

4.5 Continuous-time MPC algorithms

As discussed in the previous sections, one of the main advantages of the proposed relaxed barrier function based MPC framework is that it allows one to reformulate the underlying constrained optimal control problem in terms of a strongly convex and *smooth* (that is, twice continuously differentiable) cost function. Assuming that a continuous-time representation of the underlying system dynamics is known, this smooth reformulation allows us to design continuous-time optimization algorithms that asymptotically track the optimal solution of the associated barrier function based MPC problem. The result is a novel class of relaxed barrier function based *continuous-time MPC algorithms* that, as will be discussed briefly in the following, allow one to characterize a stabilizing feedback law via the solution of a suitably defined ordinary differential equation, eliminating thereby in principle completely the need for an iterative online-optimization. Similar approaches for the case of nonrelaxed logarithmic barrier functions have for example been presented by Ohtsuka (2004) and DeHaan and Guay (2007) as well as in (Feller and Ebenbauer, 2013). As continuous-time optimization approaches and the related concept of continuous-time MPC algorithms are not the focus of this thesis and can rather be seen as a topic for possible future research, we in the following only provide a short introduction and explain briefly some of the main ideas and preliminary results. More details on the discussed approach can be found in (Feller and Ebenbauer, 2014a).

We again consider the problem of controlling a discrete-time linear system of the form (4.1) to the origin while satisfying polytopic state and input constraints of the type (4.2) at each sampling instant. Based on the MPC design procedure described in Section 4.1, we can approach this problem by setting up the corresponding relaxed barrier function based MPC formulation and then solving the resulting unconstrained optimization problem at each sampling instant. In particular, we know by now that a stabilizing feedback law with desirable constraint satisfaction and robustness properties can be characterized via the minimizer of a globally defined, twice continuously differentiable, and strongly convex cost function of the form

$$\hat{J}_N(U,x) = \frac{1}{2}U^\top H U + x^\top F U + \frac{1}{2}x^\top Y x + \varepsilon \hat{B}_{\mathrm{xu}}(U,x)\,. \tag{4.83}$$

In contrast to the previous discussions, we now assume in addition that the system matrices (A,B) of the discrete-time representation (4.1) have been obtained by discretizing an associated continuous-time system of the form

$$\dot{x}(t) = A_c x(t) + B_c u(t)\,, \tag{4.84}$$

with a sampling time T_s. It is assumed that (4.84) describes the underlying continuous-time dynamics of the controlled plant exactly and that $A_c \in \mathbb{R}^{n\times n}$, $B_c \in \mathbb{R}^{n\times m}$, and $T_s \in \mathbb{R}_{++}$ are given. The main idea is now to use the given continuous-time plant dynamics (4.84) for designing a continuous-time optimization algorithm (i.e., a suitably defined continuous-time dynamical system) whose solution asymptotically tracks the optimal solution of the relaxed barrier function based MPC problem when treating the system state $x(t)$ as a continuously evolving parameter. More precisely, we are going to use the knowledge on the future evolution of the controlled system state in order to asymptotically follow the parametric minimizer $\hat{U}^*(x(t))$ of (4.83) by means of a suitably defined direct continuation method (Richter and DeCarlo, 1983). Towards this goal, we propose to consider the following continuous-time dynamical system

$$\dot{U}(t) = -\left(\nabla^2_{U}\hat{J}_N(U(t),x(t))\right)^{-1}\left(\kappa\nabla_U\hat{J}_N(U(t),x(t)) + \nabla^2_{Ux}\hat{J}_N(U(t),x(t))\,\dot{x}(t)\right), \quad (4.85)$$

with given initial condition $U(t_0) = U_0 \in \mathbb{R}^{Nm}$ and scaling $\kappa \in \mathbb{R}_{++}$ with unit 1/s. Note that (4.85) can be interpreted as a continuous-time Newton method with an additional correction term that takes the evolution of the underlying system state into account via the dynamics (4.84). In order to show that the solution of (4.85) in fact converges to $\hat{U}^*(x(t))$, we consider the Lyapunov function candidate $W : \mathbb{R}^{Nm} \times \mathbb{R}^n \to \mathbb{R}_+$ defined as

$$W(U,x) = \frac{1}{2}\nabla_U\hat{J}_N(U,x)^\top\nabla_U\hat{J}_N(U,x)\,. \quad (4.86)$$

Per definition, $W(U,x)$ is globally defined and continuously differentiable. Moreover, it satisfies $W(U,x) \geq 0\ \forall(U,x) \in \mathbb{R}^{Nm} \times \mathbb{R}^n$ as well as $W(U,x) = 0 \Leftrightarrow \nabla_U\hat{J}_N(U,x) = 0$. Computing the derivative of $W(U,x)$ along the respective continuous-time dynamics for $U = U(t)$ and $x = x(t)$, we obtain

$$\dot{W}(U,x) = \nabla_U\hat{J}_N(U,x)^\top\left(\nabla^2_{U}\hat{J}_N(U,x)\dot{U} + \nabla^2_{Ux}\hat{J}_N(U,x)\dot{x}\right). \quad (4.87)$$

Inserting now the expression for $\dot{U} = \dot{U}(t)$ from (4.85) immediately reveals that

$$\dot{W}(U,x) = -\kappa\nabla_U\hat{J}_N(U,x)^\top\nabla_U\hat{J}_N(U,x) = -2\kappa W(U,x)\,. \quad (4.88)$$

Thus, independently of the respective initial conditions, the function $W(U(t),x(t))$ converges exponentially to zero as $t \to \infty$. As this implies $\nabla_U\hat{J}_N(U(t),x(t)) \to 0$ and the underlying relaxed barrier function based cost function is strongly convex, this reveals that for any initial state $x(t_0) \in \mathbb{R}^n$, any initialization $U(t_0) \in \mathbb{R}^{Nm}$, and any measurable function of time $u(t)$, the solution $U(t)$ of (4.85) converges asymptotically to $\hat{U}^*(x(t))$, that is, $U(t) \to \hat{U}^*(x(t))$ whenever $t \to \infty$. Furthermore, $U(t) \equiv \hat{U}^*(x(t))$ for all $t \geq t_0$ whenever $U(t_0) = \hat{U}^*(x(t_0))$, which shows that the solution $U(t)$ of (4.85) tracks the parametric minimizer of (4.83) *exactly* whenever it is initialized with the optimal solution associated to the initial state $x(t_0)$. For more details on the existence and boundedness of the respective solutions, we refer the reader to the more rigorous proofs given in (Feller and Ebenbauer, 2013) and (Feller and Ebenbauer, 2014a).

The above continuous-time asymptotic tracking algorithm can in a quite intuitive way be used to design a sample-and-hold control input that can be shown to stabilize the considered discrete-time system dynamics of the plant. In particular, we propose to choose the control input $u(t)$ for any time interval $t \in [t_k, t_k + T_s]$ as $u(t) = \Pi_0 U(t_k)$, where $U(t_k)$ refers to the solution of system (4.85) at time $t_k = kT_s$. Consequently, the associated discrete-time dynamics of the plant state are given by

$$x(k+1) = Ax(k) + B\Pi_0\, U(kT_s)\,. \tag{4.89}$$

Introducing the error between the continuous-time solution and the exact parametric optimizer as $w_U(k) := U(t_k) - \hat{U}^*(x(t_k))$, system (4.89) can be rewritten as

$$x(k+1) = Ax(k) + B\Pi_0\, \hat{U}^*(x(k)) + w_U(k)\,, \tag{4.90}$$

revealing that the resulting discrete-time closed-loop dynamics are identical to those of the relaxed barrier function based MPC scheme discussed in Chapter 3, perturbed by the unknown but bounded additive disturbance $w_U(k)$. However, from the foregoing arguments we know that $w_U(k) \to 0$ for $k \to \infty$ and that $w_U(k) = 0$ for all $k \in \mathbb{N}$ whenever the underlying continuous-time tracking algorithm is initialized with $U(0) = \hat{U}^*(x(0))$. From the latter it follows immediately that in the case of an optimal initialization, the proposed continuous-time MPC algorithm applies at each sampling instant the exact optimal control input $u(k) = \Pi_0\hat{U}^*(x(k))$, which is nothing else than the control input of the associated discrete-time MPC scheme discussed in Chapter 3. Consequently, the closed-loop behavior of both approaches will be identical, which in particular reveals that also the constraint satisfaction properties discussed in Section 3.5 will in this case hold true for the proposed continuous-time MPC algorithm. In the case of a potentially suboptimal initialization $U(0) = U_0 \in \mathbb{R}^{Nm}$, the ISS results presented in Section 3.6 still ensure that the state of system (4.90) remains bounded and converges to the origin for $k \to \infty$. However, note that it is in this case not immediately possible to give guarantees on the violation of state and input constraints as the underlying barrier function based cost function might increase along trajectories of the closed-loop system as long as $U(t_k)$ is not sufficiently close to the optimal solution $\hat{U}^*(x(t_k))$.

Summarizing our short discussion on continuous-time MPC algorithms, we conclude that the barrier function based reformulation allows us to set up and employ continuous-time optimization algorithms that asymptotically track the state-dependent optimal solution of the respective relaxed barrier function based MPC problem. By making use of a sample-and-hold strategy and exploiting the robust stability properties of the underlying MPC approach, this leads to a novel class of MPC algorithms which, instead of iteratively solving an optimization problem at each sampling instant, require to integrate an ordinary differential equation in parallel to the evolving plant dynamics. Note that while the proposed continuous-time tracking algorithm explicitly relies on the underlying continuous-time plant dynamics, both the design and the stability analysis of the relaxed barrier function based MPC scheme are based on a discretized model of the plant.

Besides offering an alternative approach for the implementation of relaxed barrier function based MPC schemes, the presented continuous-time approach is in particular also interesting from a conceptual point of view and opens up several directions for possible future research. For example, it appears to be both interesting and intuitive to investigate the relation of the presented continuous-time approach to the Newton-based MPC iteration scheme discussed in Section 4.3 of this thesis. In particular, we might ask whether the continuous-time approach can be linked to the respective discrete-time iteration scheme by means of a suitable discretization or integration procedure and, vice versa, to which extent the continuous-time tracking formulation can be interpreted as the limiting case of a one-step iteration scheme with arbitrarily small sampling time. Furthermore, from a more application oriented perspective, the continuous-time approach connects the relaxed barrier function based MPC framework to the broad field of *analogue optimization* (Dennis, 1959; Vichik and Borrelli, 2014), in which the continuous-time dynamics of fast parametric programming algorithms are physically realized in form of analogue electrical circuits. In this context it would also be interesting to investigate whether the sample-and-hold strategy proposed above is actually required or whether the stability of the resulting closed-loop dynamics may also be analyzed completely in the continuous-time framework.

4.6 Chapter reflection

In this chapter, we introduced and discussed as the second main part and contribution of this thesis a novel framework for the design, analysis, and numerical implementation of relaxed barrier function based model predictive control algorithms with important systems theoretic and algorithmic guarantees. In particular, we presented a novel class of anytime MPC algorithms that guarantee important stability, constraint satisfaction, and robustness properties of the closed-loop system after each internal iteration of the underlying optimization algorithm and, in addition, allow us to execute the underlying optimizer iterations in a highly efficient manner.

After introducing the considered problem setup in Section 4.1, we presented and discussed structural properties of the underlying relaxed barrier function based optimization problem in Section 4.2, revealing in particular that the barrier function based MPC framework from Chapter 3 results in a twice continuously differentiable, strongly convex, and self-concordant cost function that in addition satisfies important Lipschitz conditions on the associated gradient and Hessian. As a consequence, many efficient and well-known line-search methods like (accelerated) gradient methods or the Newton method can be applied, including in particular also the convergence results and complexity estimates that are available for these methods.

In Section 4.3, we then introduced a novel class of MPC iteration schemes that exploit many of the characteristic properties of the relaxed barrier function based formulation and allow one to analyze the dynamics of both the controlled system and the underlying optimization algorithm in an integrative fashion. The main result of this section was that by combining a tailored warm-starting technique with a suitable class of line-search methods, we can design barrier function based MPC algorithms that recover many of the desired systems theoretic properties from the previous chapter – and this independently of the number of performed optimization algorithm iterations. In fact, the presented simulation results indicate that a good control performance can in many cases already be achieved when performing only one optimizing Newton iteration per sampling step.

Inspired by the well-known fact that the numerical complexity of the condensed formulation typically scales rather badly with the dimension of the considered MPC problem, we extended the proposed MPC iteration scheme approach in Section 4.4 to the associated non-condensed and sparse MPC problem formulation with equality constraints. The main result of this section was that, while essentially allowing for the same systems theoretic guarantees, the sparse formulation leads to a reduced numerical complexity in the underlying search direction computation, leading therefore to a very similar class of anytime MPC algorithms with significantly improved numerical performance. As shown by the presented benchmarking example, the resulting MPC algorithm is in terms of both speed and overall control performance competitive with existing fast MPC solvers.
Finally, in Section 4.5 we briefly discussed how the relaxed barrier function based MPC approach can be used for designing and analyzing continuous-time MPC algorithms which obtain a stabilizing control input by integrating a suitably defined ordinary differential equation in parallel to the evolving plant dynamics.

Recurring themes within the presented algorithmic MPC framework and the accompanying theoretical analysis were to analyze the dynamics of controlled plant and underlying optimization algorithm together in an integrative fashion – acknowledging thereby the fact that the required online optimization typically needs to be performed in parallel to the evolving system dynamics – and to exploit the inherent properties of the relaxed barrier function based MPC formulation for nevertheless deriving hard systems theoretic guarantees. The link between theory and practice was in particular formed by the presented MPC iteration schemes, which combine the ideas of a one-step-ahead prediction and a warm-starting shift operator with suitable descent-direction line-search methods. Furthermore, an approach that was crucial to many of the presented theoretical results was to use the relaxed barrier function based cost function as a Lyapunov function and to exploit the fact that the underlying MPC design inherently provides a very natural and intuitive warm-starting strategy that leads to a guaranteed cost function decrease.

Chapter 5

Experimental Case Study: Predictive Control of a Self-Driving Car

In this chapter, we present an experimental case study in which the presented relaxed barrier function based MPC approach is applied in a real-time environment to a challenging practical application – namely, the predictive control of a self-driving car. Although the nonlinearity of the underlying system dynamics as well as the formulation as a time-varying trajectory tracking control problem do not allow us to directly apply the theoretical results and stability guarantees from the previous chapters, the obtained results demonstrate in a proof-of-concept fashion the efficiency, robustness, and versatility of the presented relaxed barrier function based MPC framework, illustrating in particular also its potential for more general and challenging control application scenarios.
The autonomous driving experiments reported in Section 5.5 of this chapter have been conducted in August 2015 in cooperation with the Model Predictive Control Lab and the Hyundai Center of Excellence at the University of California, Berkeley, USA.

5.1 Problem setup and chapter outline

We consider in the following the problem of controlling a self-driving, autonomous car in such a way that it follows a given reference trajectory while satisfying at the same time constraints on the respective control inputs. This problem as well as different predictive control approaches towards its solution have already been discussed by several researchers, see for example (Falcone, 2007; Falcone et al., 2007, 2008; Beal and Gerdes, 2013; Carvalho et al., 2013, 2015; Kong et al., 2015) to only name a few. Inspired by the promising results from the previous two chapters, we now want to investigate whether the underlying constrained trajectory tracking control problem can also be handled by means of the relaxed barrier function based MPC framework presented in this thesis. Towards this end, we extend the proposed class of barrier function based anytime MPC algorithms from the previous chapter in the following to the required nonlinear tracking formulation, and investigate the performance of the resulting predictive controller both by means of numerical simulations and an experimental implementation in real-time. The considered Hyundai Azera test car as well as a high-level illustration of the implemented control architecture are shown in Figure 5.1 on the next page.

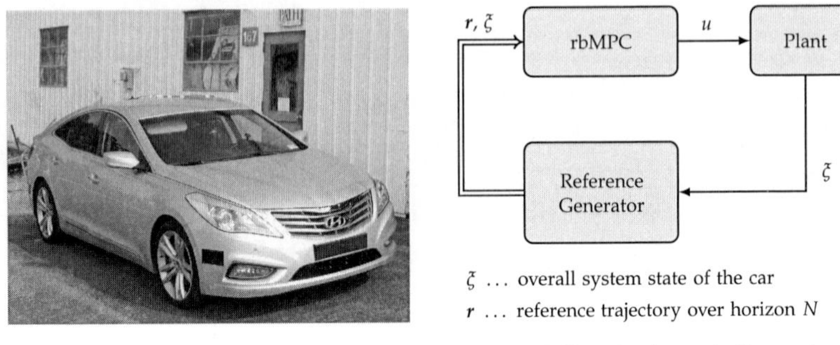

Figure 5.1. Photograph of the Hyundai Azera test car (left) and schematic illustration of the implemented control architecture with state-dependent reference generator (right).

The chapter is structured as follows. In Section 5.2, we briefly introduce and discuss the simplified kinematic vehicle model that is used for describing the relevant nonlinear dynamics of the car in a planar inertial frame. In Section 5.3, we then present a modified relaxed barrier function based MPC algorithm that, besides making use of a suitable state-dependent reference generator, is based on successively linearizing the underlying system dynamics at each sampling instant around the current system state. In Section 5.4, we briefly discuss some preliminary numerical simulations, before presenting in Section 5.5 experimental test results and real-time data from the conducted autonomous-driving experiments. A short conclusion of the chapter is given in Section 5.6.

5.2 Kinematic vehicle model

Following the works of Carvalho et al. (2015) and Kong et al. (2015), we make use of a simplified kinematic bicycle model, for which a graphical illustration is given in Figure 5.2. Here, (x, y) refer to the coordinates of the center of mass in an inertial frame (X, Y), ψ is the inertial heading angle, and v is the speed of the vehicle. Furthermore, β is the velocity heading angle with respect to the longitudinal axis of the car, and δ_f and δ_r denote the front and rear steering angles. The only characteristic parameters of the car are given by l_f and l_r, which represent the distance of the center of mass to the front and rear axles. The control inputs that can be used to control the car are the steering angles δ_f, δ_r and the acceleration a in the given velocity direction. A detailed derivation of both kinematic and dynamic system equations for the considered bicycle model can for example be found in (Rajamani, 2011, Chapter 2). As most cars can typically only apply steering at the front wheels, we assume in the following that $\delta_r = 0$ (Kong et al., 2015). In this case, the continuous-time dynamics of the kinematic bicycle model are given by

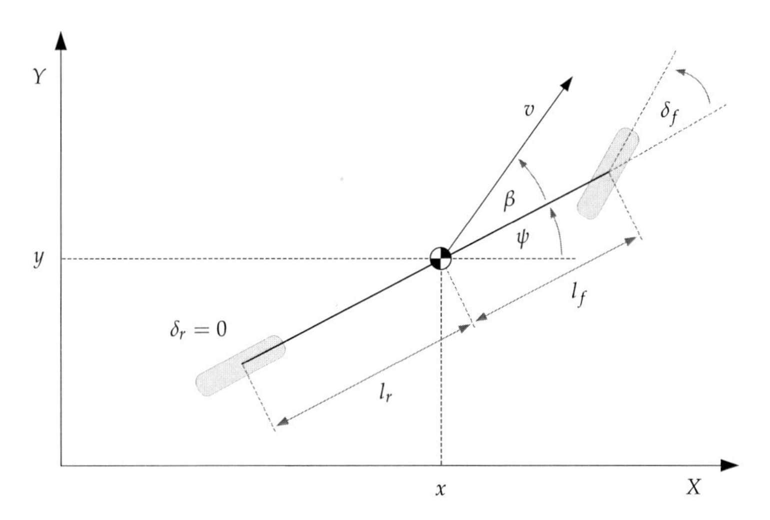

Figure 5.2. Kinematic bicycle model with position (x, y), inertial heading angle ψ, velocity heading angle β, and the front and rear steering angles δ_f and δ_r. We assume $\delta_r = 0$.

$$\dot{x} = v \cos(\psi + \beta) \tag{5.1a}$$
$$\dot{y} = v \sin(\psi + \beta) \tag{5.1b}$$
$$\dot{\psi} = \frac{v}{l_r} \sin(\beta) \tag{5.1c}$$
$$\dot{v} = a \tag{5.1d}$$
$$\beta = \tan^{-1}\left(\frac{l_r}{l_f + l_r} \tan(\delta_f)\right), \tag{5.1e}$$

which reveals that β can be seen as a virtual control input from which the required front steering angle δ_f can be computed directly by means of (5.1e). Thus, the overall system state is given by $\xi = [x, y, \psi, v]^\top \in \mathbb{R}^4$, while the considered control input vector is $u = [\beta, a]^\top \in \mathbb{R}^2$. By discretizing system (5.1), e.g. with an Euler forward method, one directly obtains a discrete-time model representation of the form

$$\xi(k+1) = f(\xi(k), u(k)). \tag{5.2}$$

Note that although we focus in the following on the simplified kinematic bicycle model discussed above, the modified barrier function based MPC algorithm presented in Section 5.3 can in principle be applied to any discrete-time model of the form (5.2).

5.3 A modified barrier function based MPC algorithm

As the considered vehicle dynamics are nonlinear and the goal is to track a given state reference (instead of stabilizing a desired equilibrium point), the relaxed barrier function based MPC problem formulation from the previous chapters needs to be adapted. Following the linear time-varying MPC approach presented by Falcone et al. (2007, 2008), we make in the following use of a tracking MPC formulation that employs at each sampling instant a linear time-varying (LTV) prediction model that is obtained by linearizing the nonlinear system dynamics around the current predicted system state trajectory. We first present the general idea for the conventional MPC case without barrier functions.

Assuming that the current system state $\xi(k)$ and the current control input $u(k)$ are given and that the control input is kept constant over a given prediction horizon $N \in \mathbb{N}_+$, the associated future evolution of the system state can for $i \geq k$ be predicted as

$$\hat{\xi}(i+1|k) = f\left(\hat{\xi}(i|k), u(k)\right), \qquad \hat{\xi}(k|k) = \xi(k). \tag{5.3}$$

The variation of the original nonlinear system (5.2) around the predicted state and input trajectories associated to (5.3) can then be approximated by the following LTV system:

$$\Delta\xi(i+1|k) = A_{i,k}\, \Delta\xi(i|k) + B_{i,k}\, \Delta u(i|k), \tag{5.4}$$

where the time-varying matrices $A_{i,k}$ and $B_{i,k}$ are defined as

$$A_{i,k} = \left.\frac{\partial f}{\partial \xi}\right|_{(\hat{\xi}(i|k),u(k))}, \qquad B_{i,k} = \left.\frac{\partial f}{\partial u}\right|_{(\hat{\xi}(i|k),u(k))}, \tag{5.5}$$

and $\Delta\xi(i|k) := \xi(i) - \hat{\xi}(i|k)$, $\Delta u(i|k) := u(i) - u(k)$. Obviously, the above LTV system describes the deviations of the future trajectories of the original nonlinear system (5.2) from the predicted state trajectory (5.3) which assumes that the constant input $u(k)$ is applied over the whole prediction horizon. Returning to absolute coordinates, we therefore obtain the following approximate prediction model for the future evolution of the system state:

$$\xi(i+1|k) = A_{i,k}\, \xi(i|k) + B_{i,k}\, u(i) + d_{i,k} \tag{5.6}$$

where $d_{i,k} := \hat{\xi}(i+1|k) - A_{i,k}\,\hat{\xi}(i|k) - B_{i,k}\, u(k)$. Note that (5.4) and (5.6) represent equivalent first order approximations of the original system (5.2) around the nominal predicted trajectory $\hat{\xi}(i|k)$. In the latter formulation, the nominal trajectory is hidden in the time-varying term $d_{i,k}$. In the following, we will use the LTV model (5.6) as time-varying, approximate prediction model within a suitably defined tracking MPC formulation. In order to reduce the numerical complexity, we assume from now on that the time-varying system matrices are kept constant over the prediction horizon, that is, $A_{i,k} = A_{k,k} =: A_k$ and $B_{i,k} = B_{k,k} =: B_k$. However, time-varying prediction matrices over the prediction horizon can without further modifications directly be used within the following MPC approach and may in certain driving maneuvers lead to an improved control performance.

Based on the above LTV prediction model and the well-known Δu-formulation for tracking MPC problems, we consider the following open-loop optimal control problem

$$\min_{\Delta u} \sum_{i=0}^{N-1} \|\xi_i - r_i(k)\|_Q^2 + \|\Delta u_i\|_R^2 + \|\xi_N - r_N(k)\|_P^2 \tag{5.7a}$$

$$\text{s.t. } \xi_{i+1} = A_k \xi_i + B_k u_i + d_{i,k}, \quad \xi_0 = \xi(k) \tag{5.7b}$$

$$u_{\max} \leq u_i \leq u_{\min}, \ i = 0, \ldots, N-1 \tag{5.7c}$$

$$\Delta u_{\min} \leq \Delta u_i \leq \Delta u_{\max}, \ i = 0, \ldots, N-1 \tag{5.7d}$$

$$u_i = u_{i-1} + \Delta u_i, \ i = 0, \ldots, N-1 \tag{5.7e}$$

$$u_{-1} = u(k-1), \tag{5.7f}$$

where $Q \in \mathbb{S}_{++}^4$, $R \in \mathbb{S}_{++}^2$, and $P \in \mathbb{S}_{++}^4$ are suitably chosen weighting matrices. Furthermore, $\boldsymbol{r}(k) = \{r_0(k), r_1(k), \ldots, r_N(k)\}$ represents a given state reference sequence over the considered prediction horizon, and the time-varying matrices A_k and B_k are obtained by linearizing the nonlinear dynamics (5.2) at each sampling instant around $(\xi(k), u(k-1))$. In particular, assuming that (5.2) has been obtained by a Euler forward discretization with sampling time $T_s \in \mathbb{R}_{++}$, the linearization matrices are for given $\xi(k) = [x, y, \psi, v]^\top$ and $u(k-1) = [\beta, a]^\top$ given by

$$A_k = \begin{bmatrix} 1 & 0 & -v\sin(\psi+\beta)T_s & \cos(\psi+\beta)T_s \\ 0 & 1 & v\cos(\psi+\beta)T_s & \sin(\psi+\beta)T_s \\ 0 & 0 & 1 & \frac{1}{l_r}\sin(\beta)T_s \\ 0 & 0 & 0 & 1 \end{bmatrix}, \quad B_k = \begin{bmatrix} -v\sin(\psi+\beta)T_s & 0 \\ v\cos(\psi+\beta)T_s & 0 \\ \frac{v}{l_r}\cos(\beta)T_s & 0 \\ 0 & T_s \end{bmatrix}. \tag{5.8}$$

The input constraints as well as additionally imposed input rate constraints are characterized by the given vectors $u_{\min}, u_{\max} \in \mathbb{R}^m$ and $\Delta u_{\min}, \Delta u_{\max} \in \mathbb{R}^m$, respectively. The terms $d_{i,k}$ are at each sampling instant computed according to the procedure described above. The optimization variables are given by the sequence of control input variations $\boldsymbol{\Delta u} = \{\Delta u_0, \ldots, \Delta u_{N-1}\}$, and the applied control input is given by

$$u(k) = u(k-1) + \Delta u_0^*\Big(\xi(k), \boldsymbol{r}(k), u(k-1)\Big). \tag{5.9}$$

As usual in the Δu-formulation, the time varying reference sequence as well as the previous control input need to be considered as additional parameters in the above parametrized optimal control problem. Throughout this chapter, we assume that a suitable reference is given at each sampling instant and do not discuss the actual realization of the reference generator in detail, see Remark 5.1 below. Similar to the conventional linear MPC case, the above tracking MPC formulation can be rewritten as a parametric QP in the optimization variable $\Delta U = [\Delta u_0^\top, \ldots, \Delta u_{N-1}^\top]^\top \in \mathbb{R}^{Nm}$ (Bemporad et al., 2002).

In the following, we eliminate again all the inequality constraints from the above formulation by making use of suitably defined relaxed logarithmic barrier functions. As we will only consider rather small prediction horizons, we make use of the condensed, unconstrained formulation and eliminate therefore also the underlying linear dynamic equations (5.7b), leading to an unconstrained problem formulation of the form

$$\min_{\Delta U} \frac{1}{2} \Delta U^\top H(\zeta) \Delta U + \zeta^\top F(\zeta) U + \frac{1}{2} \zeta^\top Y(\zeta) \zeta + \varepsilon \hat{B}_{\zeta u}(\Delta U, \zeta) \,. \tag{5.10}$$

Here, the extended parameter vector $\zeta \in \mathbb{R}^{4(N+2)+2}$ is at each sampling instant given by $\zeta = [\xi(k)^\top, R(k)^\top, u(k-1)^\top]^\top$, where $R(k) \in \mathbb{R}^{4(N+1)}$ refers to the stacked vector of state references over the considered prediction horizon. Note that the problem matrices depend on the current realization of $\zeta = \zeta(k)$ and therefore need to be computed at each sampling instant. The construction of the matrices $H(\zeta)$, $F(\zeta)$, and $Y(\zeta)$ as well as of the overall relaxed barrier function $\hat{B}_{\zeta u}(\cdot)$ follows along the lines of Appendix B.9 and is therefore not discussed here in detail. Note, however, that it is still guaranteed that the above cost function will be twice continuously differentiable and strongly convex in ΔU.

Based on the proposed unconstrained problem formulation and the ideas presented in Chapter 4 of this thesis, we now aim to compute the required control input at each sampling instant by means of a suitably defined, iterative optimization algorithm operator. In particular, the applied control input is given by

$$u(k) = u(k-1) + \Pi_0 \, \Delta U(k) \,, \tag{5.11}$$

where the current optimizer state $\Delta U(k)$ is at each sampling instant computed by performing only a finite number of Newton-based optimization algorithm iterations, cf. Section 4.3.5. More precisely, the proposed iterative MPC algorithm operates as follows:

Step 1. Measure the current system state and construct $\zeta = \zeta(k)$. Furthermore, read out the current optimizer state $\Delta U(k)$ and apply the control input given in (5.11); Construct the matrices of the condensed problem formulation (5.10).

Step 2. Start the Newton-based optimizer iteration based on the current extended parameter vector $\zeta(k)$, using the current optimizer $\Delta U(k)$ as a warm-start solution.

Step 3. After $i_T(k)$ optimizer iterations, set $\Delta U(k+1) = \Phi^{i_T(k)}(\Delta U(k), \zeta(k))$.

Step 4. Set $k \leftarrow k+1$ and go to Step 1.

As proposed in the context of the anytime MPC algorithm presented in Section 4.3, the underlying Newton-based optimization algorithm makes use of a backtracking line-search that enforces at each of its internal iteration a decay in the cost function (5.10). More details are given in Section 5.4 below. Note however that no guarantees on convergence or constraint satisfaction can be given due to the nonlinearity of the underlying system dynamics and the use of an approximate LTV prediction model. Furthermore, we do not apply the proposed shift-update from the linear MPC case but use in the applied warm-starting procedure simply the optimizer state from the previous sampling instant.

Remark 5.1. At each sampling instant, the proposed barrier function based tracking MPC algorithm requires for each of the underlying system states a reference trajectory over the prediction horizon, cf. Figure 5.1. For the results presented in the following, this is ensured by making use of a state-dependent reference generator. At each sampling instant, the current position of the car is projected onto the desired reference track, and a suitable reference path is placed ahead of the vehicle. The distance between the generated reference points is thereby determined by the desired reference velocity, cf. (Kong et al., 2015). However, although the underlying reference generator is of crucial importance for the presented predictive controller and often heavily influences the resulting closed-loop performance, its actual realization shall not be discussed here in detail.

5.4 Numerical simulations

Before we discuss the actual practical implementation of the above relaxed barrier function based tracking MPC algorithm, we present some preliminary simulation results that have been used for tuning the underlying problem parameters. For all the results presented in the following, we make use of the diagonal weight matrices

$$Q = \operatorname{diag}(10, 10, 10, 1)\,, \qquad R = \operatorname{diag}(1000, 5)\,, \qquad P = Q\,. \tag{5.12}$$

The sampling time is chosen as $T_s = 0.1$ s, the prediction horizon as $N = 10$, and the barrier function parameters are $\varepsilon = \delta = 10^{-3}$. The box constraints on the control input and control input rates are given by

$$\begin{bmatrix} -0.6458 \\ -2 \end{bmatrix} \leq u(k) \leq \begin{bmatrix} 0.6458 \\ 2 \end{bmatrix}, \quad \begin{bmatrix} -0.0175 \\ -0.2 \end{bmatrix} \leq \Delta u(k) \leq \begin{bmatrix} 0.0175 \\ 0.2 \end{bmatrix}, \tag{5.13}$$

which corresponds to $|\beta| \leq 37°$, $|\dot{\beta}| \leq 10°/\text{s}$, $|a| \leq 2$ m/s², and $|\dot{a}| \leq 2$ m/s³ in the continuous-time system formulation. The underlying Newton-based optimization algorithm makes use of a backtracking line search with parameters $\rho = 0.5$, $c_1 = 0.01$, and performs only one optimizing Newton update per sampling step, that is, $i_T(k) = 1$ for all $k \in \mathbb{N}$. A the start of each simulation, the control input and optimizer iteration are initialized with $u(-1) = 0 \in \mathbb{R}^m$, $\Delta U_0 = 0 \in \mathbb{R}^{Nm}$. The resulting closed-loop behavior of the controlled vehicle is simulated based on the continuous-time system model (5.1). The relevant parameters for the Hyundai Azera test car are given by $l_f = 1.105$ m, $l_r = 1.738$ m.

ISO 3888-2 lane change

As a first simulation example, we consider the ISO 3888-2 lane change scenario (Hansen and Hedrick, 2015; VEHICO GmbH, 2017), in which the vehicle is supposed to perform a double lane change at a constant speed of 12 m/s. The lane change is characterized by an entry and an exit lane with a length of 12 m and a side lane with a length of 11 m. All the lanes have a width of 3 m and the lateral offset of the side lane with respect to the entry and exit lanes is 1 m. The longitudinal offset between entry and side lane is 13.5 m, whereas it is 12.5 m between side and exit lane, see Figure 5.3 on the following page.

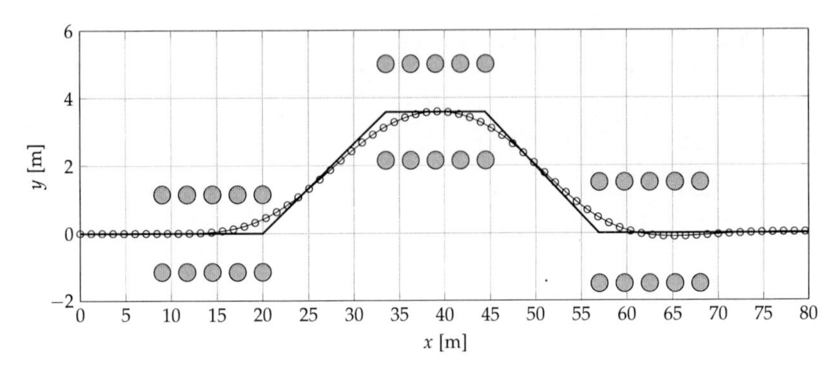

Figure 5.3. Simulated closed-loop behavior for the ISO 3888-2 lane change scenario with constant reference speed $v_r = 12$ m/s. Depicted are the piecewise defined reference trajectory (thick black line) and the resulting closed-loop trajectory (○) in the (x, y)-space.

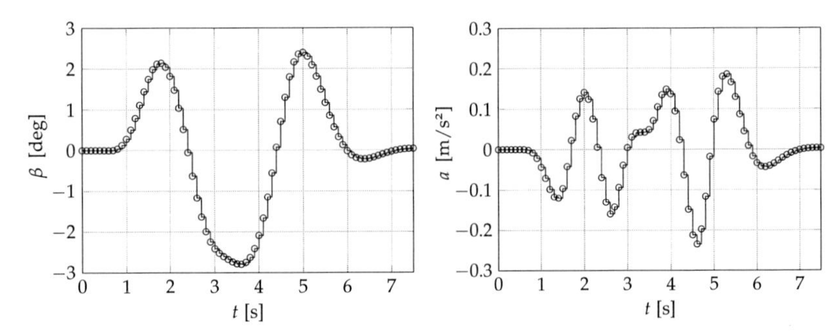

Figure 5.4. Evolution of the applied control inputs (virtual steering angle and longitudinal acceleration) over time. The inputs stay far away from the respective input constraints.

It is assumed that upon initialization the vehicle is approaching the entry lane with the desired speed of 12 m/s. Some selected simulation results are depicted in Figures 5.3 and 5.4 above. Both plots show that the proposed MPC algorithm is able to track the desired reference trajectory in a satisfactory way. Note also that the applied control inputs are actually anticipative due to the fact that they exploit the knowledge of the future reference trajectory over the prediction horizon. As the speed is supposed to be kept constant and the required steering angles are rather small, the applied control inputs stay in this scenario quite far away from the respective input and input rate constraints. Plots of the remaining system states ψ and v are omitted here for the sake of brevity.

Winding track

As a second example we consider a scenario in which the car is supposed to follow a given reference path that has been recorded on the so-called *winding track* at the Hyundai California Proving Grounds (Kong et al., 2015). A graphical illustration of the reference track can be found in Figure 5.5 at the bottom of this page. The winding track is a road course with many differently shaped turns and straight line paths, which makes it well suited for testing the performance of the proposed tracking MPC algorithm. The track coordinates and heading angles were extracted from GPS data, while the desired speed reference has been obtained by recording the velocity measurements of a human driver. In order to demonstrate the potential robustness and constraint satisfaction properties of our MPC algorithm, we consider both a scenario in which the car is at initialization placed directly on the track and a scenario in which the car starts quite far away from the track. In particular, the considered initial conditions are $x_{0,1} = [416.8617, -224.7493, 0.5841, 12.8371]$ as well as $x_{0,2} = [400, -100, -0.5236, 0]$. The obtained simulation results and a short discussion of the resulting closed-loop behavior are presented in the Figures 5.5 and 5.6 below. Although only one Newton step is carried out at each sampling instant, both a good tracking performance and a satisfactory constraint satisfaction behavior are achieved.

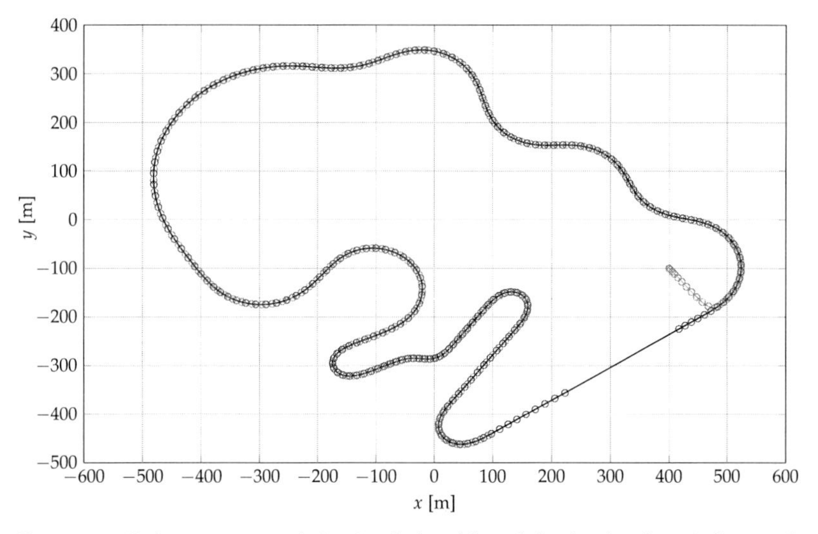

Figure 5.5. Reference map and simulated closed-loop behavior for the winding track scenario in the (x, y)-space. Depicted are the reference trajectory (black line) and the resulting closed-loop trajectory for initialization on the track (○) and off the track (○).

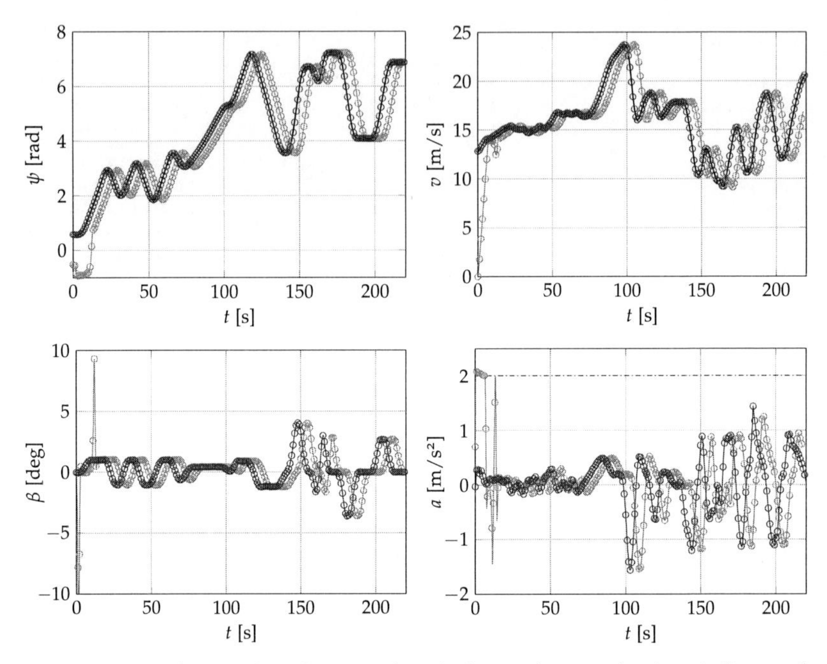

Figure 5.6. Heading angle, velocity, and applied control inputs for the winding track scenario and the two different initial conditions $x_{0,1}$ and $x_{0,2}$. Both the angle and velocity references (black line) are tracked almost perfectly if the car is initialized on the track and with the desired velocity $v(0) = v_r(0)$ ($\circ$). If the car is initialized far away from the track and with $v(0) = 0$, the tracking behavior is again very good after a short transition phase ($\circ$). The observed time delay is due to the fact that the vehicle acceleration hits the upper bound $a_{\max} = 2$ m/s² during the initial transition phase in which the underlying controller tries to catch up with the given (infeasible) reference trajectory.

5.5 Experimental results

The experimental results presented in the following are based on autonomous-driving experiments that were conducted in cooperation with the Model Predictive Control Lab and the Hyundai Center of Excellence at the University of California, Berkeley, USA.
The position, heading angle, and velocity of the considered Hyundai Azera test car are measured by means of a GPS base station, comprising a differential GPS, an inertial measurement unit (IMU) and a digital signal processor. All measurements are recorded with

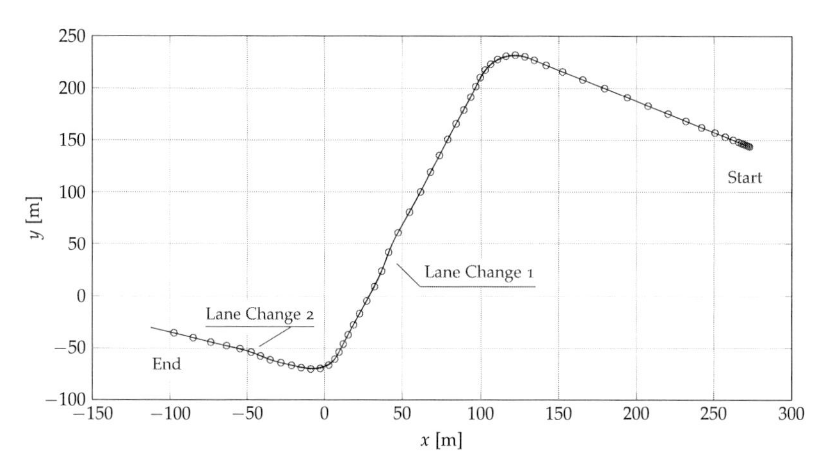

Figure 5.7. Reference map and recorded tracking behavior of the self-driving car for the considered experimental test scenario with double lane change in the (x, y)-plane.

a sampling rate of ten milliseconds. The proposed MPC algorithm as well as the required localization and low-level controllers are all run in real-time on a dSPACE MicroAutobox. In order to achieve a better control performance, the MPC is executed every 50 milliseconds (that is, twice as fast as assumed by the underlying system discretization). All the parameters are chosen exactly as in the numerical simulations, with the exception that the vehicle acceleration constraints are now $|a| \leq 3\ \mathrm{m/s^2}$, and $|\dot{a}| \leq 3\ \mathrm{m/s^3}$ and that the maximal number of performed Newton iterations is limited to $i_T(k) \leq 10$.

The considered experimental test scenario is to let the car follow a given reference map which consists of three straight lines, two sharp turns, and two lane changes, see Figure 5.7 above. The vehicle speed during the two sharp turns and the second lane change is approximately 25 km/h, while the first lane change is performed at a speed of approximately 55 km/h. The underlying reference speed map has again be obtained from human driver experiments, and the state-dependent reference generator outlined above is used for placing at each sampling instant a suitable reference path ahead of the vehicle. The proposed MPC algorithm is activated manually in the dSPACE Control Desk while the car is rolling at slow speed along the desired reference map. The experiment is manually stopped before the car reaches the end of the given reference track.

Plots of the respective reference trajectories, the measured state and input variables, and the number of actually performed Newton iterations are given in the Figures 5.7–5.10. Note that for the sake of an improved graphical representation not every data point has been marked with a circle. Note also that a measurement-induced time offset between the reference and the recorded closed-loop data has been removed manually.

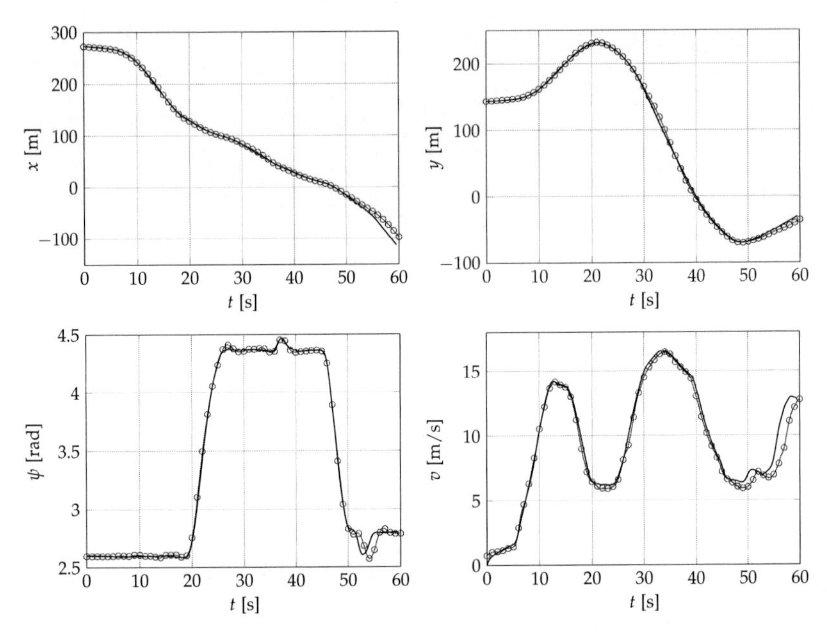

Figure 5.8. Reference trajectories (black line) and measured realizations of the vehicle position, heading angle, and velocity (○) for the considered autonomous-driving experiment with two sharp turns and two lane changes.

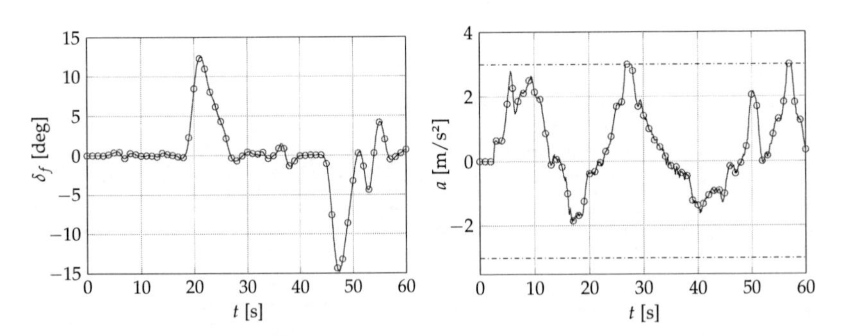

Figure 5.9. Measured steering angle and vehicle acceleration. Note that the control input changes every 50 ms, but that only every 20th data point is marked with a circle.

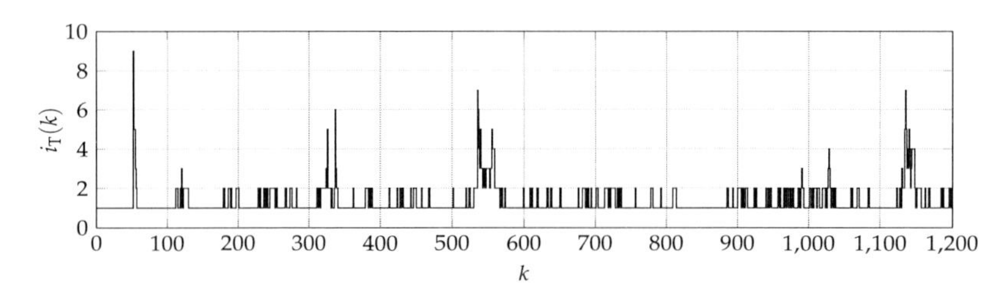

Figure 5.10. Number of performed Newton iterations at each execution of the underlying optimization algorithm (iteration until gradient residual less than 10^{-3}). Note that the spikes correspond nicely to the initialization, the two turns, and the two lane changes.

On the whole, the obtained experimental results demonstrate that the proposed MPC algorithm can indeed be implemented successfully in a real-time application scenario. As in the presented numerical simulations, the data reveals that already one or two Newton iterations per sampling step are in many cases sufficient for achieving a good overall tracking performance under almost exact constraint satisfaction. From a more implementation-oriented point of view, a further result from experimental case study was that it clearly confirmed the robustness and numerical simplicity of the proposed relaxed barrier function based MPC algorithm. In particular, the underlying Newton-based iteration scheme was implemented in a straightforward fashion as an Embedded MATLAB Function with only 150 lines of code, and a first real-time implementation with satisfactory tracking performance was compiling and running successfully after only a few hours of testing. More extensive experimental tests at the Hyundai California Proving Grounds were planned, but could unfortunately not be realized due to time constraints.

5.6 Chapter reflection

We presented in this chapter a proof-of-concept case study in which we extended and applied the relaxed barrier function based MPC framework introduced in this thesis to the predictive path-following control of an autonomous vehicle. The modified MPC algorithm presented in Section 5.3 combines a linearization-based LTV formulation from the literature with the relaxed barrier function based MPC concept that has been proposed in the previous two chapters. Both the simulation results presented in Section 5.4 and the conducted real-time experiments discussed in Section 5.5 show that a good tracking performance as well as a satisfactory level of constraint satisfaction can be achieved despite the nonlinearity and complexity of the underlying system dynamics – even when performing only a small number of optimization algorithm iterations at each sampling step. The experimental results therefore confirm and underline the relevance and potential of the proposed MPC approach for challenging real-world control applications.

Chapter 6

Conclusions

In this chapter, we summarize the main results of the thesis and give an outlook on some interesting directions for possible future research. Concerning the summarizing conclusions, we deliberately focus on the main conceptual results and their role in the bigger picture of the presented relaxed barrier function based MPC framework. More detailed summaries can be found in the chapter reflections at the end of each chapter.

Summary and Conclusions

The key motive and overall goal of this thesis was the design, analysis, and implementation of model predictive control approaches and algorithms based on relaxed logarithmic barrier functions. In particular, we introduced the novel concept of relaxed barrier function based MPC, presented a comprehensive systems theoretical framework for the design and analysis of relaxed barrier function based linear MPC schemes, and proposed and tested different classes of numerically efficient MPC iteration schemes and algorithms that allow one to analyze the resulting overall closed-loop system consisting of controlled plant and underlying optimization algorithm in an integrative fashion. In all cases, important systems theoretic properties such as asymptotic stability, constraint satisfaction, and robustness of the closed-loop system were investigated in detail. Besides the overall conclusion that the concept of relaxed barrier functions offers many inherent advantages for the design of novel, conceptually simple MPC schemes with strong (robust) stability properties, one of the main results of the thesis was that the relaxed barrier function based MPC formulation in particular also lends itself directly to the analysis and implementation of numerically efficient *anytime MPC algorithms*. Such algorithms, which guarantee asymptotic stability of the overall closed-loop system independently of the number of performed optimization algorithm iterations, possess many desirable real-time capabilities and are therefore of particular interest for applications in which the underlying optimization operates on the same time-scale as the controlled plant dynamics.

Recurring and important themes in the thesis were the unconstrained (or equality constrained), smooth, and convex reformulation of the underlying MPC problem based on relaxed barrier functions as well as the principle approach of analyzing the overall closed-loop dynamics in an integrative fashion as an interconnection of both plant and optimization algorithm dynamics. Moreover, we tried to always discuss the obtained results both from a systems theoretic and a more practical, algorithmic perspective.

In the first main part of the thesis, consisting of Chapter 3, we introduced the novel concept of relaxed barrier function based MPC and presented an accompanying systems theoretic framework for the design and analysis of relaxed barrier function based MPC schemes for discrete-time linear systems. In particular, we showed that asymptotic stability of the closed-loop system can be guaranteed based on conceptually different choices of the employed terminal cost function term – including both terminal set based and terminal set free approaches. Moreover, we showed that, despite the underlying barrier function relaxation, exact or approximate satisfaction of input and state constraints can still be guaranteed by choosing the corresponding relaxation parameter in a suitable way. An additional interesting result, in particular with respect to many practical applications, was that the relaxed barrier function based MPC formulation lends itself via the presented separation principle also directly to the design and analysis of novel, estimation-based output feedback MPC schemes with guaranteed robust stability properties.

In Chapter 4, the second main part of the thesis, we introduced and discussed a novel framework for the design, the analysis, and the efficient numerical implementation of relaxed barrier function based MPC algorithms with important systems theoretic guarantees. In particular, Chapter 4 extended and complemented the more theory-oriented Chapter 3 by providing tailored optimization algorithms and iteration schemes that allow for an efficient and stability preserving implementation of the respective relaxed barrier function based MPC approaches. As a first main result, we discussed structural properties of the resulting MPC problem formulation and showed that it allows one to characterize the control input via the minimization of a twice continuously differentiable, strongly convex, and self-concordant cost function. Furthermore, by exploiting properties of the relaxed barrier function based formulation, analyzing the interconnected dynamics of plant and optimization algorithm in an integrative fashion, and combining tailored warm-starting techniques with a suitable class of line-search methods, we showed that it is possible to derive barrier function based *anytime MPC algorithms* that guarantee many desirable systems theoretic properties *independently* of the number of performed optimization algorithm iterations. Moreover, we also extended the MPC iteration scheme approach to the non-condensed MPC problem formulation and showed that the numerical complexity of the required search-direction computation can be reduced significantly by exploiting the sparsity structure of the resulting equality constrained optimization problem.

In Chapter 5, we presented an experimental case study in which the proposed class of barrier function based MPC algorithms was applied in a real-time environment to the predictive path-following control of a self-driving car. A satisfactory overall closed-loop performance was achieved despite the nonlinearity and complexity of the underlying system dynamics – even if performing only a small number of optimizing Newton iterations at each sampling step. In total, the obtained experimental results therefore demonstrated in a proof-of-concept fashion the efficiency, robustness, and versatility of the presented relaxed barrier function based MPC framework, highlighting in particular also its potential for more general and challenging predictive control applications.

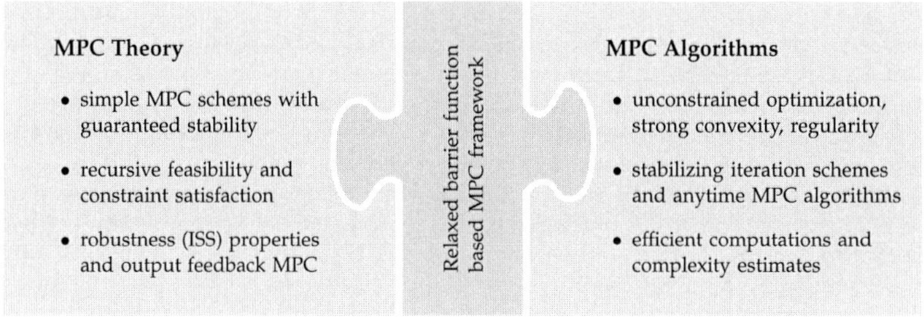

MPC Theory

- simple MPC schemes with guaranteed stability
- recursive feasibility and constraint satisfaction
- robustness (ISS) properties and output feedback MPC

Relaxed barrier function based MPC framework

MPC Algorithms

- unconstrained optimization, strong convexity, regularity
- stabilizing iteration schemes and anytime MPC algorithms
- efficient computations and complexity estimates

Figure 6.1. The presented relaxed barrier function based MPC framework as a unifying building block for bridging the gap between MPC theory and MPC algorithms.

Summarizing, we can conclude that the relaxed barrier function based MPC approach presented in this thesis provides a novel and self-contained MPC framework that allows one to design, analyze, and implement conceptually novel classes of MPC schemes and algorithms in a holistic and integrated fashion – taking algorithmic aspects into account in the MPC design and, conversely, allowing for a rigorous systems theoretic analysis of the actually implemented MPC algorithms. In total, it thus enables us to derive structurally simple and stabilizing MPC implementations with important systems theoretic and real-time guarantees, contributing thereby to the overall goal of bridging the gap between MPC theory and MPC algorithms. A schematic illustration of this fact, reflecting and summarizing in particular many of the different aspects that were discussed in this thesis, is given in Figure 6.1. The price to pay, and one main disadvantage of the relaxed barrier function based approach, is that both input and state constraints may in principle be violated in closed-loop operation (see also the discussion below).

Outlook

We conclude the thesis by briefly pointing out a selection of possible directions for future research. While some of the following aspects can be seen as rather straightforward extensions or generalizations of the material presented in this thesis, others take a much broader perspective and sketch more ambitious topics and goals of a potential long-term research vision.

Related to the theoretical framework presented in Chapter 3, one quite intuitive direction for future research would be to investigate whether potential alternative MPC design approaches can be derived by studying alternative choices for the underlying terminal cost function term. In this context, it might also be worthwhile to search for less conservative upper bounds for the quadratically relaxed barrier functions or to improve the practical applicability of the proposed zero terminal cost function approach by deriving less conservative estimates for the required stabilizing prediction horizon. Furthermore, although revealing important structural properties and dependencies, the constraint satisfaction results presented in this thesis tend to be quite conservative (compare Section 3.5). The practical relevance of the presented results may be increased significantly by deriving less conservative bounds on the maximal constraint satisfaction.

All the stability results presented in this thesis are based on the assumption that the resulting relaxed barrier function based control input is applied to plant – even if it does not satisfy the underlying input constraints. Obviously, this might not be possible when considering practical applications with hard input constraints and input saturations, e.g. due to physically motivated actuator limits. Thus, although the results presented in this thesis revealed that the resulting constraint violation is often very small and can, in particular, be tuned directly via the underlying relaxation parameter, the practical applicability of the respective MPC schemes may be improved by analyzing the closed-loop behavior in the presence of hard input constraints and introducing, if possible, suitable measures that allow one to derive stability guarantees also in this more challenging scenario. In this context, it might also be worthwhile to investigate potential advantages of relaxed barrier function based MPC schemes that adapt the underlying relaxation parameter online or that make use of different relaxation parameters for different types of constraints or for different stages within the prediction horizon (e.g., enforcing the constraints at the beginning of the prediction horizon with a higher priority than those at the end).

Furthermore, while this thesis mainly focused on the problem of stabilizing the origin of a given dynamical system, one possible topic for future research could be to extend the presented relaxed barrier function based MPC approach to more general control problem formulations, including in particular also tracking of a given reference signal. A first step into this direction has been discussed in Chapter 5, however, in a rather hands-on fashion and without providing a detailed analysis of the associated theoretical properties. From a more general perspective, it might also be interesting to apply the relaxed barrier function based approach to other, quite recent concepts from the area of predictive control, such as distributed MPC (Christofides et al., 2013; Pannocchia, 2015; Müller and Allgöwer, 2017) or MPC based on economic performance criteria (Rawlings et al., 2012; Müller and Allgöwer, 2017). In particular, the relaxation of coupling constraints may in the context of distributed MPC allow ensuring recursive feasibility of the underlying local optimization problems independently of the applied distributed optimization or synchronization procedures as well as in the presence of selfish and conflicting agent behavior.

Finally, a both important and very natural next step, which would also be a key enabler for an increased practical applicability, is to investigate whether and how the relaxed barrier function based MPC approach may be extended to more general system classes, such as, e.g., linear time-varying systems, bilinear system, or even more general nonlinear systems. While some ideas into this direction have been outlined in Section 3.7 of this thesis, and the presented experimental case study illustrated the potential of the relaxed barrier function based approach for more challenging control scenarios, a detailed analysis still needs to be done and many open research questions remain.

Concerning the algorithmic framework presented in Chapter 4, we can on the one hand basically formulate the same future research challenges as discussed above for the theoretical MPC framework, i.e., investigating alternative stabilizing MPC formulations, deriving less conservative stability and constraint satisfaction guarantees, and extending the approach to more general system classes. On the other hand, possible future research may in particular also focus on the design and analysis of tailored optimization algorithms for the resulting class of relaxed barrier function based optimization problems. Besides extending and improving the Newton-based approaches that were proposed and discussed in this thesis, this may in particular include the design of tailored first order methods which explicitly exploit structural properties of the barrier function based gradient and Hessian. First promising results in this direction have already been obtained as part of ongoing research, allowing for example for the constructive design of tailored first order methods which have a significantly faster guaranteed convergence rate than Nesterov's accelerated gradient method. In addition, it might also be worthwhile to study the convergence behavior and potential of so-called *composite* or *multi-step* Newton methods (Tapia et al., 1996; Ortega and Rheinboldt, 1970) in the context of the proposed relaxed barrier function based MPC formulations. Widening the focus to nonlinear systems, a further interesting research direction is to develop also in this context suitable anytime iteration schemes and study their conceptual relations to both sequential quadratic programming (SQP) approaches and the one-step NMPC iteration scheme proposed by Diehl et al. (2005, 2007). In particular, it could be investigated whether the NMPC iteration scheme and the associated convergence analysis developed in the aforementioned works may be extended to the constrained case by making use of the relaxed barrier function based MPC concept proposed in this thesis. Going even further into the direction of embedded MPC implementations, it might moreover be worthwhile to study in more detail the influence of the underlying number representation and to aim for a co-design of software and hardware (Kerrigan et al., 2012).

Finally, following a quite different and broader strand of research, one could aim to extend or transfer the relaxed barrier function based approach presented in this thesis to the area of constrained state or parameter estimation. In particular, one could investigate whether and in which way the concept of relaxed barrier functions may also be useful in the context of optimization-based estimation procedures such as *moving horizon estimation* (Rao, 2000; Alessandri et al., 2003; Diehl et al., 2009; Sui and Johansen, 2014). However, although moving horizon estimation (MHE) is often referred to as the "dual" to model predictive control, it has to be noted that the underlying stabilization problem (related now to the estimation error dynamics) is of a quite different nature and requires in general also different analysis and design tools than in the MPC case. Thus, extending the results obtained in this thesis to the MHE setup is not necessarily straightforward. Some preliminary results which can be seen as a promising first step towards the design and analysis of MHE schemes based on relaxed barrier functions can be found in (Gharbi, Feller, and Ebenbauer, 2017).

Concerning further developments into the direction of *MHE algorithms*, e.g., in the spirit of the anytime MPC algorithms presented in this thesis, it remains to be seen whether the relaxed barrier function based approach will in the future turn out to be as useful in the MHE setting as it did in the MPC case. However, if this enterprise is successful, it may pave the way to a novel approach for the integrated design, analysis, and implementation of *estimation-based anytime MPC algorithms* that combine the anytime MPC algorithms presented in this thesis in an overall relaxed barrier function based framework with potential anytime MHE algorithms. In this context, the underlying estimation may then not only be used for the reconstruction of the system state but also for the online identification of system parameters, the estimation of external disturbances, or the iterative learning of periodic reference trajectories.

Appendix A

Stability of Discrete-Time Systems

In the following, we briefly recall some fundamental stability concepts that are used throughout the thesis. In particular, we focus on the concepts of (global) asymptotic stability and input-to-state stability as well as on their characterizations in terms of Lyapunov and comparison functions.

A.1 Stability of autonomous systems

Most of the material in this section is rather standard and can be found in the references (LaSalle, 1986), (Jiang and Wang, 2002), and (Rawlings and Mayne, 2009). Comparable definitions for the continuous-time case can be found in (Khalil, 2002).

Consider a time-invariant discrete-time system of the form

$$x(k+1) = f(x(k)), \quad x(0) = \xi, \tag{A.1}$$

where $x(k) \in \mathbb{R}^n$ denotes the system state at time $k \in \mathbb{N}$ and $f : \mathbb{R}^n \to \mathbb{R}^n$ describes the transition to the associated successor state. We assume in the following that the function $f(\cdot)$ is continuous and we use $x(k, \xi)$ to denote the solution of (A.1) at time k when starting from initial condition $\xi \in \mathbb{R}^n$.

Definition A.1 (Equilibrium point). *A point $\bar{x} \in \mathbb{R}^n$ is an* equilibrium point *of the system dynamics (A.1) if $\bar{x} = f(\bar{x})$, implying that $x(k, \xi) = \bar{x}$ for all $k \in \mathbb{N}$ whenever $\xi = \bar{x}$.*

Definition A.2 (Positively invariant set). *A closed set $\mathcal{S}$ is* positively invariant *under the system dynamics (A.1) if $x \in \mathcal{S}$ implies $f(x) \in \mathcal{S}$.*

We assume without loss of generality that $\bar{x} = 0$ is an equilibrium point of system (A.1) and define the following stability notions with respect to the origin. However, as discussed in the aforementioned references, the following concepts can be easily generalized to arbitrary, nonzero equilibrium points as well as to considering positively invariant sets instead of isolated equilibria.

Definition A.3 (Asymptotic stability). *The origin $\bar{x} = 0$ is* stable *under the system dynamics (A.1) if for every $\varepsilon \in \mathbb{R}_{++}$, there exists a $\delta \in \mathbb{R}_{++}$ such that $\|\xi\| \leq \delta$ implies $\|x(k, \xi)\| \leq \varepsilon$ for all $k \in \mathbb{N}$. It is* asymptotically stable with region of attraction $\mathcal{X}_s$ *if it is stable, $\mathcal{X}_s$ is positively invariant under (A.1), and* $\lim_{k\to\infty} \|x(k, \xi)\| = 0$ *for any $\xi \in \mathcal{X}_s$. It is* globally asymptotically stable *if it is asymptotically stable with region of attraction $\mathcal{X}_s = \mathbb{R}^n$.*

Definition A.4 (Exponential stability). *The origin $\bar{x} = 0$ is* exponentially stable with region of attraction $\mathcal{X}_s$ *under the dynamics (A.1) if $\mathcal{X}_s$ is positively invariant under (A.1) and there exist $c \in \mathbb{R}_{++}$ and $\sigma \in (0,1)$ such that $\|x(k,\xi)\| \le c\|\xi\|\sigma^k$ for all $k \in \mathbb{N}$ for any $\xi \in \mathcal{X}_s$. It is said to be* globally exponentially stable *if it is exponentially stable with region of attraction $\mathcal{X}_s = \mathbb{R}^n$.*

With a slight abuse of terminology, we call a system of the form (A.1) asymptotically (exponentially) stable if the origin is an asymptotically (exponentially) stable equilibrium point under the corresponding system dynamics. Furthermore, under the assumption that the function $f(\cdot)$ is continuous, equivalent stability definitions may also be given in terms of comparison functions.

Definition A.5 ($\mathcal{K}$-, $\mathcal{K}_\infty$-, and $\mathcal{KL}$-function). *A function $\alpha : \mathbb{R}_+ \to \mathbb{R}_+$ with $\alpha(0) = 0$ is a* $\mathcal{K}$-function, *or $\alpha \in \mathcal{K}$, if it is continuous and strictly increasing. It is a* $\mathcal{K}_\infty$-function *if $\alpha \in \mathcal{K}$ and if in addition $\alpha(r) \to \infty$ for $r \to \infty$. A continuous function $\beta : \mathbb{R}_+ \times \mathbb{R}_+ \to \mathbb{R}_+$ is a* $\mathcal{KL}$-function *if $\beta(\cdot, s) \in \mathcal{K}$ for fixed $s \in \mathbb{R}_+$ while for fixed $r \in \mathbb{R}_+$, the function $\beta(r,\cdot)$ is decreasing and $\beta(r,s) \to 0$ whenever $s \to \infty$.*

Proposition A.1. *Consider system (A.1) with $f : \mathbb{R}^n \to \mathbb{R}^n$ being a continuous function. Then, the origin $\bar{x} = 0$ is globally asymptotically stable (GAS) under the system dynamics (A.1) if and only if there exists a $\mathcal{KL}$-function β such that for any $\xi \in \mathbb{R}^n$ it holds that $\|x(k,\xi)\| \le \beta(\|\xi\|, k)$ for all $k \in \mathbb{N}$. Moreover, it is globally exponentially stable (GES) if and only if the aforementioned relation holds for a $\mathcal{KL}$-function of the form $\beta(r,s) = cr\sigma^k$ with $c \in \mathbb{R}_{++}$ and $\sigma \in (0,1)$.*

An energy-related approach for characterizing and analyzing stability properties of system (A.1) is provided by the framework of Lyapunov stability theory and, in particular, by the concept of Lyapunov functions.

Definition A.6 (Lyapunov function). *A continuous function $V : \mathbb{R}^n \to \mathbb{R}_+$ is said to be a* Lyapunov function *for system (A.1) if there exist $\mathcal{K}_\infty$-functions α_1, α_2 and a continuous and positive definite function $\alpha_3 : \mathbb{R} \to \mathbb{R}_+$ such that the following holds for any $x \in \mathbb{R}^n$:*

$$\alpha_1(\|x\|) \le V(x) \le \alpha_2(\|x\|), \tag{A.2a}$$

$$V(f(x)) - V(x) \le -\alpha_3(\|x\|). \tag{A.2b}$$

A smooth Lyapunov function is a Lyapunov function $V(\cdot)$ satisfying $V \in \mathcal{C}^\infty$. Moreover, if the function $V(\cdot)$ satisfies the above inequalities only for all x in a positively invariant set $\in \mathcal{X}_s \subset \mathbb{R}^n$, then $V(\cdot)$ is said to be a Lyapunov function for system (A.1) on the set $\mathcal{X}_s$. As stated by the following result, see (Jiang and Wang, 2002, Theorems 1 and 2), the existence of a (smooth) Lyapunov function is a a necessary and sufficient condition for global asymptotic, respectively global exponential, stability.

Theorem A.1. *Consider system (A.1) with $f : \mathbb{R}^n \to \mathbb{R}^n$ being a continuous function. Then, the system is GAS if and only if it admits a smooth Lyapunov function $V(\cdot)$. Moreover, it is GES if and only if it admits a continuous Lyapunov function $V(\cdot)$ satisfying for some fixed $c \in \mathbb{R}_{++}$ and any $x \in \mathbb{R}^n$ the inequalities*

$$\|x\|^2 \le V(x) \le c\|x\|^2, \tag{A.3a}$$

$$V(f(x)) - V(x) \le -\|x\|^2. \tag{A.3b}$$

Remark A.1. Note that the more commonly used conditions $c_1\|x\|^2 \leq V(x) \leq c_2\|x\|^2$, $V(f(x)) - V(x) \leq -c_3\|x\|^2$ with $c_1, c_2, c_3 \in \mathbb{R}_{++}$ can always be rewritten as (A.3) by dividing all inequalities by c_1 and defining $c := c_2/c_1 \in \mathbb{R}_{++}$. In particular, satisfaction of (A.3b) can then be ensured by choosing $c_1 \leq c_3$, which is always possible.

Remark A.2. It is worth noting that, in the discrete-time setting, continuity is not a necessary requirement for $V(\cdot)$ to serve as a Lyapunov function for system (A.1). As shown by Theorems B.11 and B.14 in (Rawlings and Mayne, 2009), the existence of $V : \mathbb{R}^n \to \mathbb{R}_+$ satisfying the respective inequalities is sufficient to ensure global asymptotic (exponential) stability even if $V(\cdot)$ is not a continuous function (assuming that it is at least locally bounded). This is especially important in the context of (nonlinear) MPC, where the value function might not necessarily be continuous in the system state.

A.2 Input-to-state stability

As in the continuous-time framework introduced by Sontag (1989), the concept of input-to-state stability allows to characterize the stability properties of a dynamical system that is affected by an external input signal. Most of the following material is based on (Jiang and Wang, 2001) and (Angeli, 1999), to which we also refer the reader for more details.

By slightly extending the autonomous system dynamics (A.1), we consider in the following time-invariant discrete-time systems of the form

$$x(k+1) = f(x(k), w(k)), \quad x(0) = \xi, \tag{A.4}$$

where $x(k) \in \mathbb{R}^n$ and $w(k) \in \mathbb{R}^m$ denote the realizations of the system state and the external input at time $k \in \mathbb{N}$, respectively. Again, $f : \mathbb{R}^n \times \mathbb{R}^m \to \mathbb{R}^n$ describes the transition to the respective successor state and we assume that $f(\cdot,\cdot)$ is continuous and that $f(0,0) = 0$. The solution to the system dynamics (A.4) at time $k \in \mathbb{N}$ given an initial condition $\xi \in \mathbb{R}^n$ is denoted as $x(k, \xi, w)$, where $w := \{w(0), w(1), \ldots\}$ refers to the sequence of external inputs. Then, roughly speaking, a system of the form (A.4) is called input-to-state stable if $x(k, \xi, w)$ remains bounded for any bounded input sequence w and becomes eventually small whenever the input signal is small, i.e., if there exists a well-defined gain between the norm of the input signal and the norm of the resulting future states. A more rigorous definition is the following.

Definition A.7 (Input-to-state stability). *A system of the form (A.4) is said to be* input-to-state stable (ISS) *if there exist $\beta \in \mathcal{KL}$ and $\gamma \in \mathcal{K}$ such that for any bounded input sequence w and any $k \in \mathbb{N}_+$ it holds that*

$$\|x(k, \xi, w)\| \leq \beta(\|\xi\|, k) + \gamma(\|w_{[k-1]}\|). \tag{A.5}$$

Here, $w_{[k-1]}$ denotes the *truncation* $w_{[k-1]} = \{w(0), w(1), \ldots, w(k-1)\}$ of the input sequence, while $\|w_{[k-1]}\| := \max_{i \in \{1,\ldots,k-1\}} \|w(i)\|$. Note that ISS implies 0-GAS, i.e., that the origin of the unforced system with $w(k) = 0\ \forall\, k \in \mathbb{N}$ is globally asymptotically stable. However, as shown by Jiang and Wang (2001), the converse is not true in general.

As in the context of asymptotic stability of autonomous systems, ISS properties of system (A.4) can be characterized by means of a suitable class of Lyapunov functions.

Definition A.8. *A continuous function $V : \mathbb{R}^n \to \mathbb{R}_+$ is said to be an* ISS Lyapunov function *for system (A.4) if there exist $\mathcal{K}_\infty$-functions α_1, α_2, α_3 and a $\mathcal{K}$-function σ such that for all $x \in \mathbb{R}^n$ and all $w \in \mathbb{R}^m$*

$$\alpha_1(\|x\|) \le V(x) \le \alpha_2(\|x\|) \tag{A.6a}$$

$$V(f(x,w)) - V(x) \le -\alpha_3(\|x\|) + \sigma(\|w\|) \,. \tag{A.6b}$$

As shown by Jiang and Wang (2001), the existence of a (smooth) ISS Lyapunov function is a necessary and sufficient condition for system of the form (A.4) to be ISS.

Theorem A.2 (Jiang and Wang, 2001). *A system of the form (A.4) with $f : \mathbb{R}^n \times \mathbb{R}^m \to \mathbb{R}^n$ being a continuous function is ISS if and only if it admits a smooth ISS Lyapunov function.*

A slightly weaker notion of input-to-state stability is given by the concept of *integral input-to-state stability* (Angeli, 1999), see also (Sontag, 1989, 1998) as well as (Angeli et al., 2000) for the original definition in the continuous-time case.

Definition A.9 (Integral input-to-state stability). *A system of the form (A.4) is said to be* integral input-to-state stable (IISS) *if there exist $\beta \in \mathcal{KL}$ and $\sigma \in \mathcal{K}_\infty$ such that for any bounded input sequence w and any $k \in \mathbb{N}$ it holds that*

$$\|x(k,\xi,w)\| \le \beta(\|\xi\|,k) + \sum_{i=0}^{k-1} \sigma(\|w(i)\|) \,. \tag{A.7}$$

Note that the above IISS definition involves an upper bound for the maximal magnitude of the system state that depends on the usual fading term due to the initial condition plus an additional term that captures the *energy* of the input signal. Thus, roughly speaking, the IISS property implies that the future system state will eventually be small if the *integrals* of the inputs are small. It can be easily verified that every ISS system is also IISS, while the converse it not true in general. An example for a physical system that is IISS but not ISS has been presented by Angeli et al. (2000) in the continuous-time setting. Again, we can also give an equivalent Lyapunov function based characterization.

Definition A.10. *A function $V : \mathbb{R}^n \to \mathbb{R}_+$ is said to be an* IISS Lyapunov function *for system (A.4) if there exist $\mathcal{K}_\infty$-functions α_1, α_2, σ and a positive definite function $\rho : \mathbb{R} \to \mathbb{R}_+$ such that for all $x \in \mathbb{R}^n$ and all $w \in \mathbb{R}^m$*

$$\alpha_1(\|x\|) \le V(x) \le \alpha_2(\|x\|) \tag{A.8a}$$

$$V(f(x,w)) - V(x) \le -\rho(\|x\|) + \sigma(\|w\|) \,. \tag{A.8b}$$

Theorem A.3 (Angeli, 1999). *A system of the form (A.4) with $f : \mathbb{R}^n \times \mathbb{R}^m \to \mathbb{R}^n$ being a continuous function is IISS if and only if it admits an IISS Lyapunov function.*

It is evident from the definition that IISS implies 0-GAS of the associated autonomous system. However, unlike to the ISS case, there is no gap between 0-GAS and IISS. In particular, the following equivalence result has been proven by Angeli (1999).

Theorem A.4 (Angeli, 1999). *A system of the form (A.4) with $f : \mathbb{R}^n \times \mathbb{R}^m \to \mathbb{R}^n$ being a continuous function is IISS if and only if it is 0-GAS, i.e., GAS whenever $w(k) = 0 \; \forall \, k \in \mathbb{N}$.*

Thus, every globally asymptotically stable dynamical system of the above form also possesses certain inherent robustness properties which may be characterized by the concept of integral input-to-state stability. An important consequence of IISS that also has a nice physical interpretation is given by the *bounded energy converging state* property (Sontag, 1998; Angeli, 1999), which directly follows from (A.7) and ensures that

$$\sum_{i=0}^{\infty} \sigma(\|w(i)\|) < \infty \quad \Rightarrow \quad \lim_{k \to \infty} \|x(k, \xi, w)\| = 0 \,. \tag{A.9}$$

Thus, as long as the energy that is supplied to the system through the external input signal is finite, the system state will eventually converge to the desired set point, which is in this case given by the origin.

Appendix B

Auxiliary Results

B.1 An example for a self-concordant barrier function

Consider the logarithmic barrier function $f(x) = -\sum_{i=1}^{n} \ln(x_i)$ for the nonnegative orthant with $\mathcal{D}_f = \mathbb{R}^n_{++}$. The gradient and Hessian are easily obtained as

$$\nabla f(x) = -\begin{bmatrix} \frac{1}{x_1} & \frac{1}{x_2} & \cdots & \frac{1}{x_n} \end{bmatrix}^\top, \quad \nabla^2 f(x) = \operatorname{diag}\left(\frac{1}{x_1^2}, \frac{1}{x_2^2}, \cdots, \frac{1}{x_n^2}\right). \tag{B.1}$$

We first verify that $f(x)$ is a self-concordant function according to Definition 2.3. Concerning condition $i)$, we note that $y \in \mathcal{B}(x, 1)$ is equivalent to $\sum_{i=1}^{n} \left((y_j - x_j)/(x_j)\right)^2 < 1$, which directly implies that $y_i > 0$. Thus, $y \in \mathbb{R}^n_{++} = \mathcal{D}_f$ whenever $y \in \mathcal{B}(x, 1)$, which shows that condition $i)$ of Definition 2.3 is satisfied. Moreover, assuming $y \in \mathcal{B}(x, 1)$, it holds that

$$\|v\|^2_{\nabla^2 f(y)} = v^\top \nabla^2 f(y) v = \sum_{i=1}^{n} \left(\frac{v_i}{y_i}\right)^2 = \sum_{i=1}^{n} \left(\frac{v_i}{x_i}\right)^2 \left(\frac{x_i}{y_i}\right)^2 \le \|v\|^2_{\nabla^2 f(x)} \max_i \left(\frac{x_i}{y_i}\right)^2 \tag{B.2}$$

for any arbitrary vector $v \in \mathbb{R}^n$. On the other hand, however,

$$1 - \|y - x\|_{\nabla^2 f(x)} = 1 - \sqrt{\sum_{i=1}^{n} \left(\frac{y_i - x_i}{x_i}\right)^2} \le 1 - \left|\frac{y_i}{x_i} - 1\right| \le \frac{y_i}{x_i} \quad \forall i \in \{1, \ldots, n\}, \tag{B.3}$$

which shows that $x_i/y_i \le 1/(1 - \|y - x\|_{\nabla^2 f(x)})$ for all $i \in \{1, \ldots, n\}$. Inserting the latter relation into (B.2) directly proves satisfaction of the rightmost inequality in (2.19). The leftmost inequality can be shown in a similar way and is, in fact, redundant (Renegar, 2001). Thus, also condition $ii)$ from Definition 2.3 is satisfied, and f is a self-concordant function. Furthermore, Equation (B.1) immediately implies that

$$\nabla f(x)^\top \left(\nabla^2 f(x)\right)^{-1} \nabla f(x) = n, \tag{B.4}$$

which shows that f is a ϑ-self-concordant barrier function with complexity value $\vartheta = n$, i.e., it is an n-self-concordant barrier function according to Definition 2.4. Note that as f is in this case three times continuously differentiable, the same result may also be derived based on the more common definition due to Nesterov and Nemirovskii (1994).

B.2 Proof of the quadratic upper bound for B_K

In the following, we verify the quadratic upper bound given in (2.31). Let us first consider a general self-concordant function $f : \mathcal{D}_f \to \mathbb{R}$. Then, Definition 2.6 implies that for any $\bar{x} \in \mathcal{D}_f$ and any $r \in [0,1)$, the Dikin ellipsoid $\mathcal{W}_f(\bar{x}, r)$ is a subset of the set $\mathcal{B}_f(\bar{x}, 1) = \{x \in \mathbb{R}^n : \|x - \bar{x}\|_{\nabla^2 f(\bar{x})} < 1\} \subseteq \mathcal{D}_f$ introduced in Definition 2.3. Furthermore, (2.19) holds, implying that for any $\bar{x} \in \mathcal{D}_f$ and any $v \in \mathbb{R}^n$,

$$\|v\|_{\nabla^2 f(x)} \leq \frac{1}{1 - \|x - \bar{x}\|_{\nabla^2 f(\bar{x})}} \|v\|_{\nabla^2 f(\bar{x})} \leq \frac{1}{1-r} \|v\|_{\nabla^2 f(\bar{x})} \quad \forall\, x \in \mathcal{W}_f(\bar{x}, r)\,. \tag{B.5}$$

Consider now the function $B_K : \mathcal{X}_K^\circ \to \mathbb{R}_+$ as defined in (2.29), which is obviously self-concordant due to Assumption 2.4 and the fact that the property of self-concordance is preserved under affine transformations. Performing a Taylor expansion of $B_K(\cdot)$ around the origin, see for example Theorem 2.1 in (Nocedal and Wright, 1999), reveals that

$$B_K(x) = B_K(0) + [\nabla B_K(0)]^\top x + \frac{1}{2}\|x\|^2_{\nabla^2 B_K(sx)} = \frac{1}{2}\|x\|^2_{\nabla^2 B_K(sx)} \tag{B.6}$$

for some $s \in (0,1)$, where the first two terms vanish due to the recentering. Consider now an arbitrary $x \in \mathcal{W}_{B_K}(0, r)$, i.e., $\bar{x} = 0$, and note that $\mathcal{W}_{B_K}(0, r) \subseteq \mathcal{B}_{B_K}(0, 1) \subseteq \mathcal{X}_K^\circ$. As $s \in (0,1)$, and $\mathcal{W}_{B_K}(0, r)$ is a convex set, it holds that $x' = sx \in \mathcal{W}_{B_K}(0, r)$. However, then (B.5) and (B.6) together imply that

$$B_K(x) = \frac{1}{2}\|x\|^2_{\nabla^2 B_K(x'=sx)} \leq \frac{1}{2(1-r)^2}\|x\|^2_{\nabla^2 B_K(0)} \quad \forall\, x \in \mathcal{W}_{B_K}(0, r)\,. \tag{B.7}$$

This is nothing else than the bound given in (2.31), which completes the proof.

B.3 Weight recentered logarithmic barrier functions

In the following, we summarize the main aspects of the concept of weight recentered logarithmic barrier functions and present results on the existence and computation of suitable weighting vectors for polytopic constraint sets.

Definition B.1 (Weight recentered log barrier function). *Let $\mathcal{P} = \{z \in \mathbb{R}^{n_z} : Cz \leq d\}$ with $C \in \mathbb{R}^{q \times n_z}$, $d \in \mathbb{R}^q_{++}$ be a polytope containing the origin. Furthermore, let $w_z \in \mathbb{R}^q_{++}$ be a given weighting vector with nonzero elements, satisfying $\bar{C} w_z = 0$, where $\bar{C} := C^\top \mathrm{diag}(\frac{1}{d^1}, \ldots, \frac{1}{d^q})$. Then, the strictly convex function $B_{z,W} : \mathcal{P}^\circ \to \mathbb{R}_+$ defined as*

$$B_{z,W}(z) = \sum_{i=1}^{q} w_z^i \Big(-\ln(-C^i z + d^i) + \ln(d^i) \Big) \tag{B.8}$$

is called the weight recentered logarithmic barrier function for the set $\mathcal{P}$. As $B_{z,W}(0) = 0$ and $\nabla B_{z,W}(0) = \bar{C} w_z = 0$, this function is by construction positive definite with respect to the origin.

While a suitable weighting vector typically may be derived rather easily for small problems with box constraints, it is not immediately clear whether such a weighting always exists and how it may be obtained in a constructive fashion when considering general polytopic sets. Without loss of generality, we consider in the following directly the polytope in normalized representation, i.e., after dividing each row of the constraint matrix with the respective right-hand side element d^i. Concerning the question of existence of a suitable weighting vector, we can then state the following.

Lemma B.1. *Let $\mathcal{P} = \{z \in \mathbb{R}^{n_z} : Cz \leq \mathbb{1}\}$ with $C \in \mathbb{R}^{q \times n_z}$ describe a polytope containing the origin. Then, there always exists a vector $w_z \in \mathbb{R}^q_{++}$ satisfying $C^\top w_z = 0$.*

Proof. The proof is done by contradiction and uses the fact that $\mathcal{P}$ is a polytope and hence bounded. In order to avoid the strict inequality $w_z^i > 0$, we rewrite the weighting vector as $w_z = \mathbb{1} + \bar{w}$, where we require that $\bar{w} \in \mathbb{R}^q_+$. Consequently, we have to show that there exists always a vector $\bar{w} \in \mathbb{R}^q_+$ with the property that $C^\top \bar{w} = -C^\top \mathbb{1}$. Assume for the sake of contraction that this is not the case. Then Farkas' Lemma states that there exists a vector $\bar{z} \in \mathbb{R}^r$ satisfying $C\bar{z} \geq 0$ as well as $-\mathbb{1}^\top C\bar{z} < 0$, i.e., there exists $\bar{z} \neq 0$ with $C\bar{z} \geq 0$. However, this would immediately imply that the line $g(\kappa) := -\kappa\bar{z}$ satisfies $g(\kappa) \in \mathcal{P}$ for any $\kappa \in \mathbb{R}_+$, which clearly is a contradiction to the fact $\mathcal{P}$ as a polytope is bounded. Thus, there always exists $\bar{w} \in \mathbb{R}^q_+$ satisfying $C^\top \bar{w} = -C^\top \mathbb{1}$ and hence also $w_z \in \mathbb{R}^q_{++}$ satisfying $C^\top w_z = 0$. This completes the proof. ∎

As any polytope containing the origin can be cast into an equivalent normalized representation, the above Lemma directly implies that a suitable (not necessarily unique) weighting vector w_z according to Definition B.1 always exists.
Furthermore, by considering the normalized representations of the polytopes $\mathcal{X}$ and $\mathcal{U}$, the above Lemma can be applied directly to our problem setup. In particular, we might choose the weighting vectors $w_x \in \mathbb{R}^{q_x}_{++}$ and $w_u \in \mathbb{R}^{q_u}_{++}$ as $w_x = 1 + \bar{w}_x$ and $w_u = 1 + \bar{w}_u$, where $\bar{w}_x$, $\bar{w}_u$ are solutions to the quadratic program

$$\min_{\bar{w}_x, \bar{w}_u} \frac{1}{2}\|\bar{w}_x\|_2^2 + \frac{1}{2}\|\bar{w}_u\|_2^2 \tag{B.9a}$$

$$\text{s.t.} \quad \bar{C}_x \bar{w}_x = -\bar{C}_x \mathbb{1}, \ \bar{w}_x \geq 0 \tag{B.9b}$$

$$\bar{C}_u \bar{w}_u = -\bar{C}_u \mathbb{1}, \ \bar{w}_u \geq 0\,. \tag{B.9c}$$

While Lemma B.1 ensures that problem (B.9) always has a feasible solution, the strongly convex cost function ensures uniqueness of the resulting vectors $\bar{w}_x$ and $\bar{w}_u$. Note that the above design procedure inherently guarantees in addition that $w_x \geq \mathbb{1}$ and $w_u \geq \mathbb{1}$, which is crucial in order to ensure self-concordance of the resulting recentered barrier function, see (Renegar, 2001; Boyd and Vandenberghe, 2004) as well as Appendix C.18 of this thesis.

B.4 Barrier function based MPC with generalized terminal sets

As it reveals some novel results that may be of interest in their own right, we briefly revisit the concept of *nonrelaxed* barrier function based MPC for the more general problem setup introduced in Section 3.4.1. In particular, we consider the following logarithmic barrier function based open-loop optimal control problem

$$\tilde{J}_N^*(x) = \min_{u} \sum_{k=0}^{N-1} \tilde{\ell}(x_k, u_k) + \tilde{F}(x_N) \tag{B.10a}$$

$$\text{s.t. } x_{k+1} = Ax_k + Bu_k,\ x_0 = x\,. \tag{B.10b}$$

As before, $N \in \mathbb{N}_+$ denotes the given, finite prediction horizon and the barrier function based cost function terms are given by

$$\tilde{\ell}(x,u) = \|x\|_Q^2 + \|u\|_R^2 + \varepsilon B_{\mathrm{x}}(x) + \varepsilon B_{\mathrm{u}}(u)\,, \quad \tilde{F}(x) = x^\top Px + \varepsilon B_{\mathrm{f}}(x)\,, \tag{B.11}$$

where $B_{\mathrm{x}} : \mathcal{X}^\circ \to \mathbb{R}_+$ and $B_{\mathrm{u}} : \mathcal{U}^\circ \to \mathbb{R}_+$ are suitably recentered (nonrelaxed) logarithmic barrier functions for the sets $\mathcal{X}$ and $\mathcal{U}$, respectively, and $B_{\mathrm{f}} : \mathcal{X}_f^\circ \to \mathbb{R}_+$ is a logarithmic barrier function for the associated terminal set. In particular, $B_{\mathrm{x}}(\cdot)$ and $B_{\mathrm{u}}(\cdot)$ may be defined analogously to (3.18) by simply replacing the primordial relaxed barrier function $\hat{B}(\cdot)$ with the natural logarithm, while $B_{\mathrm{f}}(\cdot)$ is for the considered problem setup defined according to (3.21), which we repeat here for the sake of convenience:

$$B_{\mathrm{f}}(x) = -\ln(1 - \varphi_{\mathrm{f}}(x))\,. \tag{B.12}$$

Furthermore, $P \in \mathbb{S}_{++}^n$ denotes a suitably chosen terminal weight matrix and $\varepsilon \in \mathbb{R}_{++}$ the barrier function weighting parameter. The associated optimal open-loop control input sequence $\tilde{\boldsymbol{u}}^*(x) = \{\tilde{u}_0^*(x), \ldots, \tilde{u}_{N-1}^*(x)\}$ then leads to the closed-loop system

$$x(k+1) = Ax(k) + B\tilde{u}_0^*(x(k))\,. \tag{B.13}$$

The key task in the corresponding MPC design is to choose the terminal set $\mathcal{X}_f$, i.e., the function $\varphi_{\mathrm{f}}(\cdot)$, as well as the matrix P in such a way that the origin of system (B.13) is asymptotically stable. Let us first recall the definition of the associated feasible set

$$\mathcal{X}_N := \left\{x \in \mathcal{X} : \exists\, \boldsymbol{u} = \{u_0, \ldots, u_{N-1}\} \text{ s. t. } u_k \in \mathcal{U},\, x_k(\boldsymbol{u}, x) \in \mathcal{X},\, x_N(\boldsymbol{u}, x) \in \mathcal{X}_f\right\}\,, \tag{B.14}$$

with $k = 0, \ldots, N-1$, as well as

$$\mathcal{X}_K := \{x \in \mathcal{X} : Kx \in \mathcal{U}\}\,, \qquad B_K(x) = B_{\mathrm{x}}(x) + B_{\mathrm{u}}(Kx)\,, \tag{B.15}$$

where $K \in \mathbb{R}^{m \times n}$ is an auxiliary stabilizing control gain that is assumed to be given. As discussed in Section 2.3, the state and input constraint barrier function $B_K(\cdot)$ plays a central role when investigating the stability properties of the closed-loop system with standard MPC stability concepts that use the associated value function as a Lyapunov function candidate. In particular, a crucial ingredient in the design of both the terminal set and the terminal weight was the existence of a region in the state space in which the function $B_K(\cdot)$ can be upper bounded by a quadratic function. In the approach of Wills and Heath (2004), this was achieved by exploiting characteristic properties of the considered class of self-concordant barrier function on their respective Dikin ellipsoid. In the following, we present and discuss a slightly more general set of sufficient conditions that may be used for ensuring asymptotic stability of the closed-loop system in a similar way.

Assumption B.1. *For (A, B) stabilizable, given $Q \in \mathbb{S}^n_{++}$, $R \in \mathbb{S}^m_{++}$, $\varepsilon \in \mathbb{R}_{++}$, and a stabilizing local control gain $K \in \mathbb{R}^{m \times n}$ with $A_K := A + BK$, $|\lambda_i(A_K)| < 1$, the following holds:*

A1: *The functions $B_x : \mathcal{X}^\circ \to \mathbb{R}_+$ and $B_u : \mathcal{U}^\circ \to \mathbb{R}_+$ are suitably recentered logarithmic barrier functions for the sets $\mathcal{X}$ and $\mathcal{U}$, attaining their minima at the origin.*

A2: *There exist a compact set $\mathcal{N}_K \subseteq \mathcal{X}_K := \{x \in \mathcal{X} : Kx \in \mathcal{U}\}$ with $0 \in \mathcal{N}_K^\circ$ and an associated matrix $M \in \mathbb{S}^n_+$ such that the quadratic upper bound $B_K(x) \leq x^\top M x$ holds for any $x \in \mathcal{N}_K$.*

A3: *The terminal set $\mathcal{X}_f$ is given by (3.20), where $\varphi_f : \mathbb{R}^n \to \mathbb{R}_+$ is positive definite, convex, and radially unbounded and satisfies $\varphi_f(A_K x) \leq \varphi_f(x)$ for any $x \in \mathcal{X}_f$. Moreover, $\varphi_f(\cdot)$ is chosen in such a way that $\mathcal{X}_f \subseteq \mathcal{N}_K$ for the set $\mathcal{N}_K$ is given in A2.*

A4: *The terminal cost matrix $P \in \mathbb{S}^n_{++}$ is chosen as the positive definite solution to the Lyapunov equation $P = A_K^\top P A_K + K^\top R K + Q + \varepsilon M$, where the matrix $M \in \mathbb{S}^n_+$ is given in A2.*

One way to ensure satisfaction of the conditions in Assumption B.1 in a constructive way is given by the original approach presented by Wills and Heath (2004), in which the role of the set $\mathcal{N}_K$ is played by the Dikin ellipsoid $\mathcal{W}_{B_K}(0, r)$ with adjustable radius $r \in (0, 1)$ and the terminal set is chosen as a scaled sublevel set of the quadratic terminal cost, that is, $\varphi_f(x) = \frac{1}{\alpha} x^\top P x$ with suitably chosen $\alpha \in \mathbb{R}_{++}$, see Section 2.3. Moreover, several alternative constructive design approaches, which may in many cases lead to an increased region of attraction of the closed-loop system and therefore represent an interesting contribution in their own right, have been presented in (Feller and Ebenbauer, 2013, 2014b, 2015a). For the sake of a simplified discussion, these alternative approaches are summarized and discussed in Appendix B.5.

Theorem B.1. *Consider the barrier function based MPC problem (B.10)–(B.11) and let Assumption B.1 hold. Moreover, let the associated feasible set $\mathcal{X}_N$ defined according to (B.14) have nonempty interior. Then, the origin of the resulting closed-loop system (B.13) is asymptotically stable with region of attraction $\mathcal{X}_N^\circ$. Furthermore, the barrier function based feedback $u(k) = \tilde{u}_0^*(x(k))$ ensures strict satisfaction of the pointwise state and input constraints (3.2) for all $k \in \mathbb{N}$.*

Proof. The proof uses standard MPC stability arguments and comprises some of the main ideas presented in (Wills and Heath, 2004; Feller and Ebenbauer, 2013, 2014b). We first show recursive feasibility of problem (B.10). For any $x_0 \in \mathcal{X}_N^\circ$, there exists by definition an optimal input sequence $\tilde{\boldsymbol{u}}^*(x_0) = \{\tilde{u}_0^*, \ldots, \tilde{u}_{N-1}^*\}$ that guarantees strict satisfaction of all input, state, and terminal set constraints and results in a feasible open-loop state sequence $\boldsymbol{x}^*(x_0) = \{x_0, x_1^*, \ldots, x_N^*\}$ with $x_N^* \in \mathcal{X}_f^\circ$. The associated successor state is given by $x_0^+ = Ax_0 + B\tilde{u}_0^* = x_1^*$. Due to the linear dynamics and the properties of the terminal set $\mathcal{X}_f$, see A3 of Assumption B.1, $\tilde{\boldsymbol{u}}^+(x_0) = \{\tilde{u}_1^*, \ldots, \tilde{u}_{N-1}^*, Kx_N^*\}$ is a suboptimal but feasible input sequence for the initial state x_0^+, resulting in the feasible open-loop state sequence $\boldsymbol{x}(\tilde{\boldsymbol{u}}^+(x_0), x_0^+) = \{x_0^+, x_2^*, \ldots, x_N^*, A_K x_N^*\}$ with $A_K x_N^* \in \mathcal{X}_f^\circ$. This shows that for any $x_0 \in \mathcal{X}_N^\circ$, there exists a feasible input sequence that ensures that the successor state $x_0^+ = Ax_0 + B\tilde{u}_0^*$ lies again in the interior of the feasible set $\mathcal{X}_N$, which guarantees recursive feasibility of the open-loop optimal control problem (B.10). In the following, we show that the value function satisfies in addition

$$\tilde{J}_N^*(x_0^+) - \tilde{J}_N^*(x_0) \leq -\tilde{\ell}(x_0, \tilde{u}_0^*(x_0)) \;\; \forall\, x_0 \in \mathcal{X}_N^\circ \,. \tag{B.16}$$

First, due to suboptimality of the candidate solution $\tilde{u}^+(x_0)$, it holds that $\tilde{J}_N^*(x_0^+) - \tilde{J}_N^*(x_0) \leq \tilde{J}_N(\tilde{u}^+(x_0), x_0^+) - \tilde{J}_N^*(x_0)$, where $\tilde{J}_N(\tilde{u}^+(x_0), x_0^+)$ denotes the value of the cost function evaluated for the suboptimal input sequence $\tilde{u}^+(x_0)$. Moreover, $\tilde{J}_N(\tilde{u}^+(x_0), x_0^+) - \tilde{J}_N^*(x_0) = \tilde{F}(A_K x_N^*) - \tilde{F}(x_N^*) + \tilde{\ell}(x_N^*, K x_N^*) - \tilde{\ell}(x_0, \tilde{u}_0^*) \leq -\tilde{\ell}(x_0, \tilde{u}_0^*)$ for any $x_0 \in \mathcal{X}_N^\circ$ since

$$\tilde{F}(A_K x_N^*) - \tilde{F}(x_N^*) + \tilde{\ell}(x_N^*, K x_N^*) \tag{B.17a}$$

$$= \|A_K x_N^*\|_P^2 - \|x_N^*\|_P^2 + \|x_N^*\|_Q^2 + \|K x_N^*\|_R^2 + \varepsilon B_K(x_N^*) + \varepsilon B_f(A_K x_N^*) - \varepsilon B_f(x_N^*) \tag{B.17b}$$

$$\leq \varepsilon \left(B_f(A_K x_N^*) - B_f(x_N^*)\right) = \varepsilon \, \frac{\ln\left(1 - \varphi_f(x_N^*)\right)}{\ln\left(1 - \varphi_f(A_K x_N^*)\right)} \leq 0 \;\; \forall \;\; x_N^* \in \mathcal{X}_f^\circ \,. \tag{B.17c}$$

Here, the first inequality follows from the quadratic bound $B_K(x) \leq x^\top M x \; \forall\, x \in \mathcal{X}_f \subseteq \mathcal{N}_K$ and the choice of the terminal cost matrix P, see A2–A4 of Assumption B.1. As $\varepsilon \in \mathbb{R}_{++}$, the second inequality holds due to the assumption that $\varphi_f(A_K x) \leq \varphi_f(x)$ for any $x \in \mathcal{X}_f$, see A3 of Assumption B.1.
Finally, the considered problem setup and the design of the barrier functions according to A1 of Assumption B.1 ensure that $\tilde{J}_N^* : \mathcal{X}_N^\circ \to \mathbb{R}_+$ is a well-defined and positive definite function with $\tilde{J}_N^*(x) \to \infty$ whenever $x \to \partial \mathcal{X}_N$. Thus, in combination with property (B.16), it can be used as a Lyapunov function that proves asymptotic stability of the origin of system (B.13) with region of attraction $\mathcal{X}_N^\circ$. ∎

From a conceptual point of view, Assumption B.1 and Theorem B.1 can be seen as a generalization of the barrier function based MPC approach due to Wills and Heath (2004), allowing in principle for novel quadratic barrier function bounds and more general terminal set formulations. Moreover, as discussed in Section 3.4.1, many of the underlying key ideas can also be used in the relaxed barrier function based framework.

B.5 Constructive design of ellipsoidal and polytopic terminal sets for relaxed and nonrelaxed barrier function based MPC

The results presented in Section 3.4.1 (and Appendix B.4) are all based on the assumption that the terminal set and the quadratic terminal weight satisfy the conditions stated in Assumption 3.2, respectively Assumption B.1. In particular, this requires the knowledge of a local quadratic upper bound for the barrier function $B_K(\cdot)$, e.g, as described by the pair $(\mathcal{N}_K, M)$. As the relaxed barrier function $\hat{B}_K(\cdot)$ is on the set $\mathcal{X}_K$ always upper bounded by its nonrelaxed counterpart $B_K(\cdot)$, the following quadratic upper bounds for the nonrelaxed formulation can also be applied in the relaxed case.

As proposed by Wills and Heath (2004) and as summarized in Section 2.3, one way to derive such a quadratic upper bound is by exploiting the self-concordance of the underlying logarithmic barrier functions, leading to

$$\mathcal{N}_K = \mathcal{W}_{B_K}(0,r)\,, \qquad M = \frac{1}{2(1-r)^2}\nabla^2 B_K(0)\,, \tag{B.18}$$

where $\mathcal{W}_{B_K}(0,r)$ refers to the Dikin ellipsoid of the function $B_K(\cdot)$ with radius $r \in (0,1)$. Furthermore, the conditions A3 and A4 can then be enforced easily by choosing the terminal set as the maximal ellipsoidal subset of the quadratic terminal penalty term, i.e.,

$$\varphi_{\mathrm{f}}(x) = \frac{1}{\alpha}x^\top P x\,, \qquad P = (A+BK)^\top P(A+BK) + Q + K^\top RK + \varepsilon M\,. \tag{B.19}$$

with a suitably defined scaling factor $\alpha \in \mathbb{R}_{++}$, see Section 2.3. The main disadvantage of this approach is that the shape and size of the resulting terminal set are limited by the choice of P and the size of the associated Dikin ellipsoid, which may be considerably smaller than the set $\mathcal{X}_K$. In the following, we present not only a novel quadratic upper bound that is valid on a polytopic set of adjustable size but also different novel approaches that allow us to choose the corresponding terminal set as a maximal volume ellipsoid or polytope within the set $\mathcal{X}_K$.

A) A quadratic barrier function bound on a scalable polytopic set

Let us first recall the definition of the set $\mathcal{X}_K$ as well as of the corresponding barrier function $B_K : \mathcal{X}_K^\circ \to \mathbb{R}_+$ given in (B.15), which we repeat here for the sake of convenience

$$\mathcal{X}_K := \{x \in \mathcal{X} : Kx \in \mathcal{U}\}\,, \qquad B_K(x) = B_{\mathrm{x}}(x) + B_{\mathrm{u}}(Kx)\,. \tag{B.20}$$

As discussed in Section 3.4.1, $B_{\mathrm{x}} : \mathcal{X}^\circ \to \mathbb{R}_+$ and $B_{\mathrm{u}} : \mathcal{U}^\circ \to \mathbb{R}_+$ are assumed to be weight recentered logarithmic barrier functions of the form

$$B_{\mathrm{x}}(x) = \sum_{i=1}^{q_{\mathrm{x}}} w_{\mathrm{x}}^i \left(-\ln(-C_{\mathrm{x}}^i x + d_{\mathrm{x}}^i) + \ln(d_{\mathrm{x}}^i)\right)\,, \qquad w_{\mathrm{x}} \in \mathbb{R}_{++}^{q_{\mathrm{x}}}\,, \tag{B.21a}$$

$$B_{\mathrm{u}}(u) = \sum_{i=1}^{q_{\mathrm{u}}} w_{\mathrm{u}}^i \left(-\ln(-C_{\mathrm{u}}^i u + d_{\mathrm{u}}^i) + \ln(d_{\mathrm{u}}^i)\right)\,, \qquad w_{\mathrm{u}} \in \mathbb{R}_{++}^{q_{\mathrm{u}}}\,. \tag{B.21b}$$

Lemma B.2. *Consider the linear constraints (3.2) and let the barrier function $B_K : \mathcal{X}_K^\circ \to \mathbb{R}_+$ be defined according to (B.20)–(B.21) for a given $K \in \mathbb{R}^{m\times n}$. Then, it holds that*

$$B_K(x) \le \gamma\, x^\top \left(M_{\mathrm{x}} + K^\top M_{\mathrm{u}} K\right) x \quad \forall\, x \in \mathcal{P}_K(\gamma)\,, \tag{B.22}$$

where $\gamma \in \mathbb{R}_{++}$ is a scalar design parameter, the matrices $M_{\mathrm{x}} \in \mathbb{S}_+^n$ and $M_{\mathrm{u}} \in \mathbb{S}_+^m$ are defined as $M_{\mathrm{x}} := C_{\mathrm{x}}^\top \mathrm{diag}(w_{\mathrm{x}}) C_{\mathrm{x}}$ and $M_{\mathrm{u}} := C_{\mathrm{u}}^\top \mathrm{diag}(w_{\mathrm{u}}) C_{\mathrm{u}}$, respectively, and $\mathcal{P}_K(\gamma) \subset \mathcal{X}_K$ refers to the polytopic set

$$\mathcal{P}_K(\gamma) := \left\{x \in \mathbb{R}^n \,\middle|\, C_{\mathrm{x}}\, x \le d_{\mathrm{x}} - \frac{1}{\sqrt{\gamma}}\mathbb{1},\; C_{\mathrm{u}} K x \le d_{\mathrm{u}} - \frac{1}{\sqrt{\gamma}}\mathbb{1}, \right\}. \tag{B.23}$$

Proof. We first consider the state constraints and show that

$$f(x) := \gamma x^\top M_x x - B_x(x) \geq 0 \quad \forall x \in \mathcal{P}_K(\gamma), \tag{B.24}$$

which then obviously implies that $B_x(x) \leq \gamma x^\top M_x x$ for all $x \in \mathcal{P}_K(\gamma)$. Due to the recentering of $B_x(\cdot)$, the function $f : \mathcal{X}_K^\circ \to \mathbb{R}$ satisfies $f(0) = 0$ and $\nabla f(0) = 0$, which reveals that it attains a local minimum with function value zero at the origin. Thus, $f(x) \geq 0$ will at least hold in that region around the origin in which the function $f(\cdot)$ is convex, i.e., for which its Hessian is positive semi-definite viz.

$$\nabla^2 f(x) = C_x^\top \left(\gamma \operatorname{diag}(w_x) - \operatorname{diag}\left(\frac{w_x^1}{(-C_x^1 x + d_x^1)^2}, \dots, \frac{w_x^{q_x}}{(-C_x^{q_x} x + d_x^{q_x})^2} \right) \right) C_x \in \mathbb{S}_+^n . \tag{B.25}$$

This is obviously the case whenever $C_x^i x \leq d_x^i - 1/\sqrt{\gamma}$ for all $i = 1, \dots, q_x$, leading to the result that

$$B_x(x) \leq \gamma x^\top M_x x \quad \forall x \in \mathcal{P}_x(\gamma) := \left\{ x \in \mathbb{R}^n \mid C_x x \leq d_x - \frac{1}{\sqrt{\gamma}} \mathbb{1} \right\} \tag{B.26}$$

A similar result can be derived for the input constraint barrier function $B_u(\cdot)$ by making use of exactly the same arguments. Inserting $u = Kx$ and combining both results finally yields the claimed quadratic bound for the barrier function $B_K(\cdot)$. ∎

Concerning condition A2 of Assumption 3.2 as well as Assumption B.1, Lemma B.2 provides the following choice for the pair $(\mathcal{N}_K, M)$:

$$\mathcal{N}_K = \mathcal{P}_K(\gamma), \qquad M = \gamma \left(M_x + K^\top M_u K \right) . \tag{B.27}$$

Note that the set $\mathcal{P}_K$ given in (B.23) will be empty when choosing $\gamma < 1/(\min(d_x^\top, d_u^\top))^2$. On the other hand, it can be easily seen that $\mathcal{P}_K \to \mathcal{X}_K$ for $\gamma \to \infty$, which reveals that the set $\mathcal{N}_K$ on which the quadratic upper bound holds can now be scaled in such a way that it approximates the set $\mathcal{X}_K$ arbitrarily close. In particular, analogous to the radius r in (B.18), the scalar factor $\gamma \in \mathbb{R}_{++}$ can be seen as a design parameter that allows to find a trade-off between the norm of the resulting quadratic bound matrix M and the size of the associated set $\mathcal{N}_K$. However, unlike as in the approach of Wills and Heath (2004), the region $\mathcal{N}_K$, and thus the size of the terminal set $\mathcal{X}_f \subseteq \mathcal{N}_K$, is now not limited by the size of the unit radius Dikin ellipsoid.

As the above arguments provide separate quadratic bounds for the state and input constraint barrier function, a suitable auxiliary controller gain $K \in \mathbb{R}^{m \times n}$ and a corresponding terminal weight $P \in \mathbb{S}_{++}^n$ satisfying condition A4 can be obtained as solutions to the following modified discrete-time algebraic Riccati equation

$$K = -\left(R + B^\top P B + \varepsilon\gamma M_u \right)^{-1} B^\top P A \tag{B.28a}$$

$$P = (A + BK)^\top P (A + BK) + K^\top (R + \varepsilon\gamma M_u) K + Q + \varepsilon\gamma M_x . \tag{B.28b}$$

The last step in the MPC design, which is then applicable to both nonrelaxed and relaxed barrier function based MPC schemes, is to find a suitable terminal set of the form

$$\mathcal{X}_f = \{x \in \mathbb{R}^n \mid \varphi_f(x) \leq 1\}, \tag{B.29}$$

where the function $\varphi_f : \mathbb{R}^n \to \mathbb{R}_+$ needs to be chosen in such a way that all the conditions in A3 of Assumption 3.2/B.1 are satisfied. In the following two subsection, we are going to present different approaches for choosing the function φ_f based on both ellipsoidal and polytopic terminal set formulations.

Remark B.1. Note that a quadratic upper bound that is very similar to the one provided in Lemma B.2 can also be derived for the case of gradient recentered logarithmic barrier functions, see Lemma 2 in (Feller and Ebenbauer, 2013). Consequently, all the results that will be presented in the following still apply when making use of gradient recentered instead of weight recentered barrier functions.

B) Ellipsoidal terminal sets

Recalling the conditions stated in A3 , we no want to choose the function $\varphi_f : \mathbb{R}^n \to \mathbb{R}_+$ in (3.20) in such a way that the resulting terminal set is the largest positively invariant ellipsoid that can be fitted into the set $\mathcal{P}_K$ given in Lemma B.2. In particular, we propose to choose

$$\varphi_f(x) = x^\top P_f x, \quad P_f \in \mathbb{S}^n_{++}, \tag{B.30}$$

where P_f is a design parameter referring to the *shape matrix* of the associated ellipsoidal terminal set. Obviously, $\varphi_f(\cdot)$ is in this case positive definite, strictly convex, and radially unbounded. In order to ensure that it also satisfies the additional invariance condition $\varphi_f(A_K x) \leq \varphi(x)\ \forall, x \in \mathcal{X}_f$, we demand that

$$(A + BK)^\top P_f (A + BK) \preceq P_f, \tag{B.31}$$

which is a linear matrix inequality in the design parameter P_f. Furthermore, the resulting ellipsoidal terminal set has to be as subset of the polytope $\mathcal{P}_K(\gamma)$ derived above, i.e.,

$$\mathcal{X}_f = \{x \in \mathbb{R}^n \mid x^\top P_f x \leq 1\} \subseteq \mathcal{P}_K(\gamma), \qquad \mathcal{P}_K(\gamma) = \{x \in \mathbb{R}^n \mid C_K x \leq \mathbb{1}\} \tag{B.32}$$

where we assume without loss of generality that $\mathcal{P}_K(\gamma)$ is, for a fixed $\gamma \in \mathbb{R}_{++}$, given in normalized representation with $C_K \in \mathbb{R}^{\bar{q} \times n}$, $\bar{q} = q_x + q_u$, cf. (B.23). It can be shown, for example by applying the S-procedure to (B.32), that the aforementioned set containment condition is fulfilled whenever

$$C_K^i P_f^{-1} C_K^{i\top} \leq 1, \quad i = 1, \dots, \bar{q}. \tag{B.33}$$

Finally, it is a well-known fact that the volume of the ellipsoidal set $\mathcal{X}_f$ as defined above strongly depends on the eigenvalues of the shape matrix and will, in fact, always be proportional to the determinant of P_f^{-1}. Thus, putting it all together, a suitable shape matrix may be computed as

$$P_f = X^{*-1}, \qquad X^* := \arg\min_{X} \; -\log\det X \tag{B.34a}$$

$$\text{s.t.} \begin{bmatrix} X & (A+BK)X \\ X(A+BK)^\top & X \end{bmatrix} \succeq 0, \;\; X \succ 0, \tag{B.34b}$$

$$C_K^i X C_K^{i\top} \leq 1, \;\; i = 1, \ldots, \bar{q}. \tag{B.34c}$$

Note that we made use of the Schur complement in order to reformulate (B.31) in the optimization variable $X = P_f^{-1}$. Furthermore, we applied the usual monotonic transformation based on the logarithm in order to render the objective function convex (recall that the determinant is a log-concave function). As the underlying constraints are linear, (B.34) is a convex optimization problem that can be solved by efficient and reliable algorithms.

Thus, the overall result is a constructive and numerically tractable approach to compute an ellipsoidal terminal set that satisfies all the necessary conditions and has maximal size with respect to the set $\mathcal{P}_K(\gamma)$, cf. Figure B.1. As the set $\mathcal{P}_K(\gamma)$ can be used to approximate the set $\mathcal{X}_K$ arbitrary close, the resulting terminal set can in principle also be interpreted as the maximal suitable ellipsoidal terminal set for the given choice of K.

Remark B.2. As the shape and size of the set $\mathcal{P}_K$ heavily depends on the choice of the auxiliary controller gain $K \in \mathbb{R}^{m\times n}$, one might ask whether K can also be treated as an additional design parameter when constructing the maximal volume terminal set. In fact, this is possible and suitable choices for K and P_f can be computed as

$$P_f = X^{*-1}, \qquad (X^*, Y^*) := \arg\min_{X,Y} \; -\log\det X \tag{B.35a}$$

$$K = Y^* X^{*-1}, \qquad \text{s.t.} \begin{bmatrix} X & AX + BY \\ XA^\top + Y^\top B^\top & X \end{bmatrix} \succeq 0, \;\; X \succ 0, \tag{B.35b}$$

$$\begin{bmatrix} 1 & \bar{C}_\mathrm{u}^i Y \\ Y^\top \bar{C}_\mathrm{u}^{i\top} & X \end{bmatrix} \succeq 0, \;\; i = 1, \ldots, q_\mathrm{u} \tag{B.35c}$$

$$\bar{C}_\mathrm{x}^i X \bar{C}_\mathrm{x}^{i\top} \leq 1, \;\; i = 1, \ldots, q_\mathrm{x}, \tag{B.35d}$$

where $\bar{C}_\mathrm{x}$, $\bar{C}_\mathrm{u}$ are the matrices corresponding to the normalized state and input constraints in the definition of $\mathcal{P}_K(\gamma)$, i.e., $P_K(\gamma) = \{x \in \mathbb{R}^n \,|\, \bar{C}_\mathrm{x} x \leq \mathbb{1}, \bar{C}_\mathrm{u} K x \leq \mathbb{1}\}$, which now have to be treated separately. Moreover, $Y \in \mathbb{R}^{m\times n}$ denotes a suitably defined auxiliary optimization variable that is required in order to formulate (B.35c) as a linear constraint. The main advantage of this approach is that the size of the terminal set will scale proportionally with the choice of the design parameter γ. However, as K is now solely designed with the aim to achieve the largest possible terminal set, this may potentially be bought at the cost of a significantly increased terminal weight P.

Remark B.3. Note that the size of the terminal set directly influences the size of the resulting feasible set $\mathcal{X}_N$ given in (3.29), respectively (B.14). Thus, in the context of the nonrelaxed and relaxed barrier function based MPC schemes addressed by Theorem B.1 and Theorem 3.1 (respectively Corollary 3.1) a larger terminal set will typically result in a larger region of attraction. However, regarding the closed-loop performance, also the globally stabilizing relaxed barrier function based MPC approach addressed in Theorem 3.2 will usually benefit from a less restrictive terminal set constraint.

C) Polytopic terminal sets

In contrast to the previous paragraph and the approach of Wills and Heath (2004), we now want to turn our attention to polytopic terminal sets of the form

$$\mathcal{X}_f = \{x \in \mathbb{R}^n \,|\, C_f x \leq \mathbb{1}\} \,, \qquad \mathcal{X}_f \subseteq \mathcal{P}_K(\gamma) \,, \tag{B.36}$$

where $C_f \in \mathbb{R}^{q_f \times n}$ is a given matrix and we assume without loss of generality a normalized representation. Especially in the context of linear systems, such polytopic terminal sets are typically a natural choice as they are directly compatible with the usual polytopic representation of (maximal) positively invariant sets. Furthermore, compared for example to ellipsoids, polytopic representations are more flexible and might allow to formulate larger terminal sets, particularly in the presence of asymmetric state and input constraints. However, an inherent conceptual disadvantage of polytopic terminal sets is that, instead of a *one single* (quadratic) inequality constraint, they are described by a finite number of linear inequalities. As shown in the following discussion, this necessitates the use of novel and more elaborate formulations for the function $\varphi_f(\cdot)$ in (B.29).

Remark B.4. Note that an intuitive and quite natural barrier function formulation for a polytopic terminal set of the form (B.36) would be given by

$$B_f(x) = -\sum_{i=1}^{q_f} w_f^i \left(\ln(1 - C_f^i x)\right) \,, \qquad w_f \in \mathbb{R}_{++}^{q_f} \,, \tag{B.37}$$

see also Definition 3.3 in Section 3.2 as well as Wills (2003), where the authors make use of a similar gradient recentered formulation. However, in this case it is not clear whether and how it can be ensured that the terminal set barrier function (or a suitably relaxed version of it) will decrease under the auxiliary local system dynamics $x^+ = A_K x$ as it is required by the proofs of Theorems B.1, 3.1, and 3.2. As a consequence, asymptotic stability of the resulting closed-loop system will in general not be guaranteed when making use of a terminal set barrier function of the form (B.37).

The basic idea of the approaches presented in the following is to choose the function $\varphi_f(\cdot)$ in such a way that it penalizes only the most important of the linear constraints in the underlying terminal set formulation, i.e., the constraint which is the closest to being violated. As a first step, we derive a non-smooth formulation of $\varphi_f(\cdot)$ based on the so-called Minkowski functional (Blanchini and Miani, 2007).

Definition B.2 (C-set)**.** *A convex and compact subset of* $\mathbb{R}^n$ *that contains the origin as an interior point is called a C-set.*

Definition B.3 (Minkowski functional)**.** *Let* $\mathcal{S} \subset \mathbb{R}^n$ *be a given a C-set. Then, the Minkowski functional* $\varphi_{\mathcal{S}} : \mathbb{R}^n \to \mathbb{R}_+$ *is defined as*

$$\psi_{\mathcal{S}}(x) = \inf \{\mu \in \mathbb{R}_+ \,|\, x \in \mu\mathcal{S}\} \ . \tag{B.38}$$

Note that for any given C-set $\mathcal{S} \subset \mathbb{R}^n$, the Minkowski functional is a continuous, positive definite, and convex function in the variable $x \in \mathbb{R}^n$. Moreover, the Minkowski functional satisfies $\psi_{\mathcal{S}}(x) = 1 \Leftrightarrow x \in \partial\mathcal{S}$ as well as $\psi_{\mathcal{S}}(x) < 1 \; \forall x \in \mathcal{S}^\circ$, which reveals that $x \in \mathcal{S} \Leftrightarrow \psi_{\mathcal{S}}(x) \leq 1$, cf. Figure B.1. Thus, the Minkowski functional is a natural candidate for describing the above polytopic terminal set by means of one single inequality viz.

$$x \in \mathcal{X}_f = \{x \in \mathbb{R}^n \,|\, C_f x \leq \mathbb{1}\} \;\Leftrightarrow\; \psi_{\mathcal{X}_f}(x) \leq 1 \, . \tag{B.39}$$

Based on these observations, we propose to choose the function $\varphi_f(\cdot)$ as the Minkowski functional of the polytopic set $\mathcal{X}_f$, which is in this case nothing else than the maximum over the underlying linear constraint values, i.e.,

$$\varphi_f(x) = \psi_{\mathcal{X}_f}(x) = \max_{i=1,\dots,r} \{C_f^i x\} \, . \tag{B.40}$$

Obviously, this function is positive definite, radially unbounded, and convex, and thus satisfies the first part of condition A3. Furthermore, under rather mild assumptions on the given polytopic terminal set, it is rather straightforward to show that it also satisfies the required decrease condition $\varphi_f(A_K x) \leq \varphi_f(x)$.
To this end, let us assume that the polytopic terminal set $\mathcal{X}_f$ given in (B.36) is positively invariant under the local system dynamics, i.e., $x^+ = A_K x \in \mathcal{X}_f \; \forall x \in \mathcal{X}_f$ for a given stabilizing auxiliary control gain $K \in \mathbb{R}^{m \times n}$ as discussed above. Now let $\bar{x} \in \mathcal{X}_f$ be arbitrary and define $\psi_{\mathcal{X}_f}(\bar{x}) := \mu$, that is, $\bar{x} \in \mu\mathcal{X}_f$ with $\mu \leq 1$. Due to the positive invariance of $\mathcal{X}_f$ and the linearity of the underling system dynamics, it holds that $A_K\bar{x} \in \mu\mathcal{X}_f$, and hence $\psi_{\mathcal{X}_f}(A_K\bar{x}) \leq \mu$. As $\varphi_f(x) := \psi_{\mathcal{X}_f}(x)$ and since $\bar{x} \in \mathcal{X}_f$ was arbitrary, this implies that $\varphi_f(A_K x) \leq \varphi_f(x)$ for any $x \in \mathcal{X}_f$ as desired (in fact, this holds for any $x \in \mathbb{R}^n$).
In order to satisfy also the last requirement of condition A3, we need to assume that the given polytopic terminal set satisfies $\mathcal{X}_f \subseteq \mathcal{P}_K(\gamma)$, where $\mathcal{P}_K(\gamma)$ is defined as in Lemma B.2. However, note this can always be ensured by scaling a given positively invariant set $\mathcal{P} \subseteq \mathcal{X}_K$ with a scaling factor $\kappa \in (0,1)$ until $\mathcal{X}_f := \kappa\mathcal{P}$ satisfies the above set containment condition. In particular, recalling the above definitions of the sets $\mathcal{X}_K$ and $\mathcal{P}_K(\gamma) \subset \mathcal{X}_K$, it is straightforward to show that any $\kappa \leq 1 - 1/(\sqrt{\gamma}\, d_{\min})$ with $d_{\min} := \min(d_x^\top, d_u^\top)$ will actually do the job, cf. (B.23). A suitable positively invariant set $\mathcal{P}$ that is admissible under the given state and input constraints may for example be computed based on the methods presented in (Gilbert and Tan, 1991).

Thus, by choosing the function $\varphi_f(\cdot)$ as the Minkowski functional (B.40), we can guarantee asymptotic stability for the barrier function based MPC approaches discussed above

also in the case of polytopic terminal sets. However, as revealed by the above discussion, the Minkowski functional of a polytopic terminal set is inherently non-smooth. As this renders the resulting optimization problem inaccessible to standard nonlinear programming techniques, we are interested in a suitable approximation that still satisfies all the properties of Assumption 3.2/B.1 and is at least twice continuously differentiable.
To this end, let us for a general polytopic C-set $\mathcal{P} := \{x \in \mathbb{R}^n : Cx \leq \mathbb{1}\}$ with $C \in \mathbb{R}^{q\times n}$ consider the following function

$$\tilde{\psi}^p_{\mathcal{P}}(x) := \sqrt[p]{\sum_{i=1}^{q} \sigma_p(C^i x)}\,, \qquad \sigma_p(\xi) = \begin{cases} 0 & \text{if } \xi < 0 \\ \xi^p & \text{if } \xi \geq 0 \end{cases}, \tag{B.41}$$

where $p \in \mathbb{N}_+$ is a positive integer and $\sigma_p : \mathbb{R}^n \to \mathbb{R}_+$ is use in order to allow for asymmetric sets (Blanchini and Miani, 2007). As discussed in the aforementioned reference, the function $\tilde{\psi}^p_{\mathcal{P}} : \mathbb{R}^n \to \mathbb{R}_+$ allows for a "smooth" approximation of the Minkowski functional associated with the set $\mathcal{P}$. In particular, we can state the following.

Lemma B.3. *Let $\mathcal{P} := \{x \in \mathbb{R}^n : Cx \leq \mathbb{1}\}$ be a polytopic C-set with associated Minkowski functional $\psi_{\mathcal{P}} : \mathbb{R}^n \to \mathbb{R}_+$ according to Definition B.3 and let $\tilde{\psi}^p_{\mathcal{P}} : \mathbb{R}^n \to \mathbb{R}_+$ be given by (B.41). Then it holds that $\lim_{p\to\infty} \tilde{\psi}^p_{\mathcal{P}}(x) = \psi_{\mathcal{P}}(x)$ uniformly on every compact subset of $\mathbb{R}^n$. Furthermore, $\tilde{\psi}^p_{\mathcal{P}} : \mathbb{R}^n \to \mathbb{R}_+$ is a positive definite, radially unbounded, and convex function.*

Proof. We easily see that $\max_i\{C^i x\} \leq \sqrt[p]{\sum_{i=1}^{q} \sigma_p(C^i x)} \leq \sqrt[p]{q}\max_i\{C^i x\}$ and, hence, $\psi_{\mathcal{P}}(x) \leq \tilde{\psi}^p_{\mathcal{P}}(x) \leq \sqrt[p]{q}\,\psi_{\mathcal{P}}(x)$ for any $x \in \mathbb{R}^n$, $p \in \mathbb{N}_+$, cf. Blanchini and Miani (2007, p. 95). Let $\mathcal{D}$ be a compact subset of $\mathbb{R}^n$ with $\sup_{x\in\mathcal{D}} \psi_{\mathcal{P}}(x) =: L < \infty$. Then, it holds that $\lim_{p\to\infty} \sup_{x\in\mathcal{D}} |\tilde{\psi}^p_{\mathcal{P}}(x) - \psi_{\mathcal{P}}(x)| \leq \lim_{p\to\infty} |\sqrt[p]{r} - 1|L = 0$ since $\lim_{p\to\infty} \sqrt[p]{r} = 1$ for any $r \in \mathbb{N}_+$. This proves the claimed uniform convergence for all $x \in \mathcal{D}$. As $\mathcal{D}$ was arbitrary, this implies that $\lim_{p\to\infty} \tilde{\psi}^p_{\mathcal{P}}(x) = \psi_{\mathcal{P}}(x)$ uniformly on every compact subset of $\mathbb{R}^n$.
Furthermore, $\tilde{\psi}^p_{\mathcal{P}}(x)$ is by design nonnegative and satisfies $\tilde{\psi}^p_{\mathcal{P}}(x) = 0 \Leftrightarrow x = 0$ due to the definition of $\mathcal{P}$. To see this, note that the existence of $\bar{x} \neq 0$ such that $\tilde{\psi}^p_{\mathcal{P}}(\bar{x}) = 0$ would imply $C^i\bar{x} \leq 0$ for all $i = 1,\ldots,q$. However, then $\kappa\bar{x} \in \mathcal{P}\ \forall \kappa \geq 0$, which would contradict our assumption that $\mathcal{P}$ as a polytope is bounded. Finally, convexity follows from the fact that any function of the form $f(x) = \left(\sum_{i=1}^{q} g_i(x)^p\right)^{1/p}$ with $p \geq 1$ is convex if the functions $g_1(\cdot),\ldots,g_q(\cdot)$ are convex and nonnegative, see Boyd and Vandenberghe (2004, p. 87). ∎

Based on Lemma B.3, let us now consider the following smoothed polytopic terminal set approximation

$$\mathcal{X}^p_f = \left\{x \in \mathbb{R}^n \mid \tilde{\psi}^p_{\mathcal{X}_f}(x) \leq 1\right\}, \tag{B.42}$$

which satisfies $\mathcal{X}^p_f \subset \mathcal{X}_f \subseteq \mathcal{P}_K(\gamma)$ for any $p \in \mathbb{N}_+$ as well as $\lim_{p\to\infty} \mathcal{X}^p_f = \mathcal{X}_f$. However, unfortunately the underlying approximation of the Minkowski functional is not smooth at the origin and even the first derivative fails to exist for $x = 0$. In order to circumvent

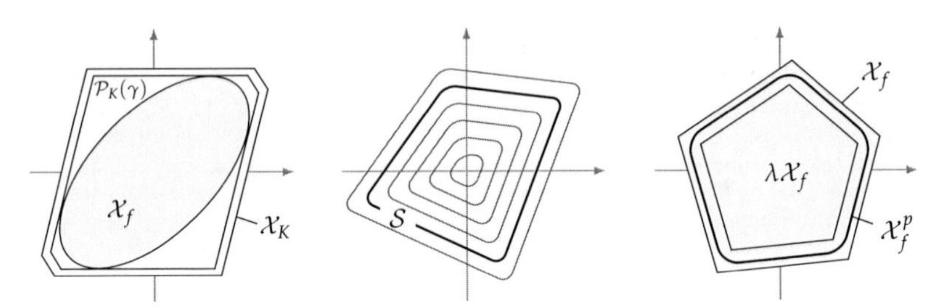

Figure B.1. *Left:* Principle idea of choosing the terminal set as the maximal volume ellipsoid contained in the set $\mathcal{P}_K(\gamma) \subset \mathcal{X}_K$. *Middle:* Level sets of the Minkowski functional for a given C-set $\mathcal{S}$. *Right:* Smooth approximation of a λ-contractive terminal set $\mathcal{X}_f$.

this problem and ensure sufficient smoothness of the resulting overall cost function, we propose to choose the associated characterizing function $\varphi_\mathrm{f} : \mathbb{R}^n \to \mathbb{R}_+$ as

$$\varphi_\mathrm{f}(x) = \left(\tilde{\psi}^p_{\mathcal{X}_f}(x)\right)^p = \sum_{i=1}^{q_\mathrm{f}} \sigma_p(C^i_\mathrm{f}\, x)\,. \tag{B.43}$$

Note that this function is $(p-1)$-times continuously differentiable for any $x \in \mathbb{R}^n$ while still resulting in the same terminal set approximation, i.e., $\mathcal{X}_f^p = \{x \in \mathbb{R}^n \,|\, \varphi_\mathrm{f}(x) \leq 1\}$. A direct consequence of Lemma B.3 is that $\varphi_\mathrm{f} : \mathbb{R}^n \to \mathbb{R}_+$ as defined above is a positive definite, radially unbounded, and convex function, and thus satisfies the first part of condition A3. Furthermore, the following result shows that if we assume that the original terminal set is *contractive* instead of only positively invariant, then there always exists a finite but large enough choice for the parameter p such that, in addition, $\varphi_\mathrm{f}(A_K x) \leq \varphi(x)$ holds for any $x \in \mathcal{X}_f^p$.

Definition B.4. *Given a positive scalar $\lambda \in (0,1)$, a C-set $\mathcal{S}$ is called* λ-contractive *with respect to a linear discrete-time system of the form $x^+ = Ax$ if it satisfies $x^+ = Ax \in \lambda\mathcal{S}$ for all $x \in \mathcal{S}$.*

Assumption B.2. *Let $K \in \mathbb{R}^{m\times n}$ be a given stabilizing auxiliary control gain and let the set $\mathcal{P}_K$ be defined according to Lemma B.2. We assume that the polytopic terminal set $\mathcal{X}_f$ given in (B.36) satisfies $\mathcal{X}_f \subseteq \mathcal{P}_K$ and is in addition λ-contractive with respect to the local system dynamics $x^+ = A_K x$ for a given $\lambda \in (0,1)$, where $A_K := A + BK$.*

Lemma B.4. *Let $\mathcal{X}_f$ in (B.36) be a λ-contractive polytopic terminal set according to Assumption B.2 and consider $\varphi_\mathrm{f} : \mathbb{R}^n \to \mathbb{R}_+$ given in (B.43), resulting in the smooth polytopic terminal set approximation $\mathcal{X}_f^p$ in (B.42). Then, for any $\lambda \in (0,1)$, there exists an integer $p_\lambda \in [1,\infty)$ such that $\varphi_\mathrm{f}(A_K x) \leq \varphi_\mathrm{f}(x)\ \forall\, x \in \mathcal{X}_f$ holds for any $p \geq p_\lambda$.*

Proof. The proof consists of two parts. First, we show that $\varphi_\mathrm{f}(A_K x) \leq \varphi_\mathrm{f}(x)\ \forall x \in \mathcal{X}_f^p$ whenever the parameter p is chosen large enough to ensure

$$\lambda\mathcal{X}_f \subset \mathcal{X}_f^p \subset \mathcal{X}_f\,. \tag{B.44}$$

Then, we show in a second step that for any $\lambda \in (0,1)$ there exists a finite $p_\lambda \in \mathbb{N}_+$ such that (B.44) will be satisfied for any $p \geq p_\lambda$.

Part 1. Assume that (B.44) holds for given p and λ and a λ-contractive set $\mathcal{X}_f$. Consider an arbitrary $x \in \mathcal{X}_f^p$ and define $\mu := \tilde{\psi}^p_{\mathcal{X}_f}(\bar{x})$ as well as $\mu_2 := \psi_{\mathcal{X}_f}(\bar{x})$. As $\psi_{\mathcal{X}_f}(x) \leq \tilde{\psi}^p_{\mathcal{X}_f}(x)$ for all $x \in \mathbb{R}^n$ and any p, it obviously always holds that $\mu_2 \leq \mu$. Thus, $\bar{x} \in \mu_2\mathcal{X}_f \subseteq \mu\mathcal{X}_f$, which by the contraction property of the set $\mathcal{X}_f$ implies that $A_K\bar{x} \in \lambda\mu\mathcal{X}_f$. However, then $\psi_{\mathcal{X}_f}(A_K\bar{x}) \leq \lambda\mu$ due to the definition of the Minkowski functional. Let us now exploit the fact that $\lambda\mathcal{X}_f \subset \mathcal{X}_f^p$. As this trivially implies that $\mu\lambda\mathcal{X}_f \subset \mu\mathcal{X}_f^p$ for any $\mu \in \mathbb{R}_{++}$, we can state that $\tilde{\psi}^p_{\mathcal{X}_f}(x) \leq \mu$ whenever $\psi_{\mathcal{X}_f}(x) \leq \mu\lambda$. Together with our previous result that $\psi_{\mathcal{X}_f}(A_K\bar{x}) \leq \lambda\mu$ we thus get that $\tilde{\psi}^p_{\mathcal{X}_f}(A_K\bar{x}) \leq \mu$. Noting that we defined $\mu := \tilde{\psi}^p_{\mathcal{X}_f}(\bar{x})$, this finally yields that $\tilde{\psi}^p_{\mathcal{X}_f}(A_K\bar{x}) \leq \tilde{\psi}^p_{\mathcal{X}_f}(\bar{x})$. By recalling the definition of $\varphi_f(\cdot)$ and raising both sides of the previous inequality to the power of p, it follows immediately that $\varphi_f(A_K\bar{x}) \leq \varphi_f(\bar{x})$. As $\bar{x} \in \mathcal{X}_f^p$ was arbitrary, this yields the desired decrease property and completes the first part of the proof.

Part 2. As outlined above, it always holds that $\mathcal{X}_f^p \subset \mathcal{X}_f$ as $\psi_{\mathcal{X}_f}(x) \leq \tilde{\psi}^p_{\mathcal{X}_f}(x)\ \forall x \in \mathbb{R}^n$ for any $p \in \mathbb{N}_+$. However, in combination with Lemma B.3, this implies that the set $\mathcal{X}_f^p$ uniformly converges to $\mathcal{X}_f$ from the inside as $p \to \infty$. Hence, if $\lambda \in (0,1)$, we can always find a finite p_λ such that $\lambda\mathcal{X}_f \subset \mathcal{X}_f^p$ for any $p \geq p_\lambda$. ■

Remark B.5. Let $\{v_1, \dots, v_{n_v}\}$ be the vertices of the λ-contractive polytopic set $\mathcal{X}_f$ for a given $\lambda \in (0,1)$. Then, due to convexity of set $\mathcal{X}_f^p$, it holds that $\lambda\mathcal{X}_f \subset \mathcal{X}_f^p$ if and only if $\tilde{\psi}^p_{\mathcal{X}_f}(\lambda v_j) = \lambda\tilde{\psi}^p_{\mathcal{X}_f}(v_j) < 1, j = 1, \dots, n_v$. Hence, for given $\lambda \in (0,1)$, a sufficiently large $p \geq p_\lambda$ can be determined by iteratively increasing p until this condition is satisfied.

Essentially, the proof of Lemma B.4 reveals that the desired decrease condition will be satisfied whenever the parameter p is large enough to ensure $\lambda\mathcal{X}_f \subset \mathcal{X}_f^p \subset \mathcal{X}_f$ for the given contraction factor $\lambda \in (0,1)$, i.e., whenever the original polytopic terminal set is approximated close enough. Due to the convergence result presented in Lemma B.3, such a p always exist and may be computed by iteratively increasing p until $\varphi_f(\lambda v_j) < 1$ is satisfied for all vertices $\{v_1, \dots, v_{n_v}\}$ of the original polytope $\mathcal{X}_f$. A graphical illustration of this idea is given in Figure B.1. Furthermore, note that a suitable λ-contractive terminal set $\mathcal{X}_f$ can be constructed by computing a polytope $\mathcal{P}$ that is positively invariant under the dynamics $x^+ = \frac{1}{\lambda}A_K x$ and then scale this set with a factor $\kappa \in (0,1)$ until $\mathcal{X}_f = \kappa\mathcal{P} \subseteq \mathcal{P}_K(\gamma)$, see the discussion above.

Thus, all the conditions stated in Assumption 3.2/B.1 can still be ensured when replacing the non-smooth Minkowski function from (B.40) with the suitably defined smooth approximation given in (B.43). In this way, we can also in the context of polytopic terminal sets not only ensure asymptotic stability of the resulting closed-loop system but also achieve a smooth formulation of the underlying overall cost function.

Remark B.6. As an alternative to the approximation given in (B.43) one might also make use of a smooth reformulation based on

$$\varphi_{\mathrm{f}}(x) = \frac{1}{p} \ln \left(\sum_{i=1}^{r} w_{\mathrm{f}}^{i} e^{p C_{\mathrm{f}}^{i} x} \right), \tag{B.45}$$

where $p \in \mathbb{N}_+$ is again a positive integer and $w \in \mathbb{R}_+^r$ denotes a suitable weighting vector that ensures that the gradient of the above function vanishes at the origin. Having been presented in (Feller and Ebenbauer, 2015a), this alternative approach is based on the function $g(z) = \frac{1}{p} \ln \left(\sum_{i=1} e^{p z_i} \right)$, which is generally referred to as (scaled) *log-sum-exp function* or *Kreisselmeier-Steinhauser function* and which is also well-known to provide a smooth approximation of the maximum function, see (Kreisselmeier and Steinhauser, 1982) as well as (Boyd and Vandenberghe, 2004, p. 72). While this alternative formulation is smooth for every $x \in \mathbb{R}^n$, it does not automatically ensure that the function $\varphi_{\mathrm{f}}(\cdot)$ will be positive definite. To achieve nevertheless a suitable recentering, the following alternative terminal set barrier function has been proposed in (Feller and Ebenbauer, 2015a):

$$B_{\mathrm{f}}(x) = -\ln \left(1 - \tilde{\psi}_{\mathcal{X}_f}^{p}(x) \right) + \ln \left(1 - \mathbb{1}^\top w_{\mathrm{f}} \right). \tag{B.46}$$

As before, a corresponding relaxed formulation may be obtained by replacing the natural logarithm with the primordial relaxed barrier function (3.4). More details on this alternative approach as well as on the design of the parameters w_{f} and p can be found in (Feller and Ebenbauer, 2015a).

Summarizing, we presented several conceptually different and constructive design approaches that allow to choose the function $\varphi_{\mathrm{f}}(\cdot)$ characterizing the terminal set and hence the terminal set barrier function in such a way that the sufficient stability conditions in Assumption B.1 are satisfied. In the case of an ellipsoidal formulation, the terminal set can in principle be computed as a suitably defined maximal volume inscribed ellipsoid within the set $\mathcal{X}_K$. While the presented polytopic terminal set formulations are particularly interesting from a conceptual point of view and may in many practical application lead to a further enlargement of the resulting terminal set, they require a slightly more sophisticated formulation of the characterizing function $\varphi_{\mathrm{f}}(\cdot)$ – especially when aiming for a smooth formulation of the overall optimization problem. In addition, all the presented approaches heavily exploit the novel quadratic upper bound for the barrier function $B_K(\cdot)$ presented in Lemma B.2.

D) Example

We briefly illustrate the different terminal set design approaches discussed above means of an academic numerical example. In particular, we consider a discretized double integrator system with the system model

$$x(k+1) = \begin{bmatrix} 1 & T_s \\ 0 & 1 \end{bmatrix} x(k) + \begin{bmatrix} T_s^2 \\ T_s \end{bmatrix} u(k), \qquad T_s = 0.1. \tag{B.47}$$

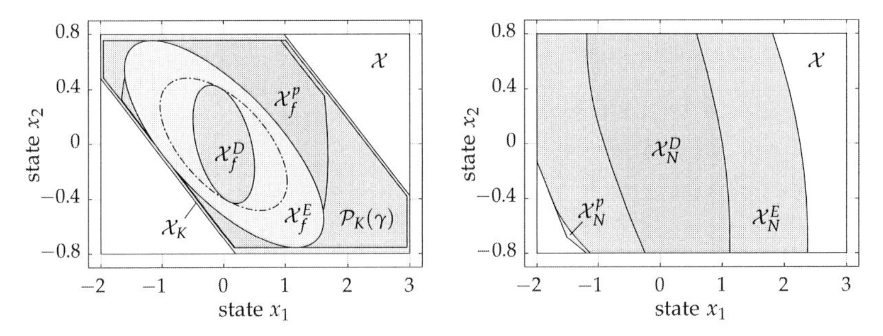

Figure B.2. Resulting terminal sets for the considered double integrator example (left). The largest region is achieved by the proposed smooth polytopic terminal set formulation, whereas the size of the ellipsoidal terminal set due to Wills and Heath (2004) is clearly limited by the surrounding Dikin ellipsoid (here indicated by the dashed line). The results are also reflected by the comparison of the associated feasible sets, depicted in the plot on the right-hand side. As can be seen, a much larger region of attraction (or strict constraint satisfaction) is obtained when using one of the novel design approaches proposed above.

The input and state constraints are assumed to be given by box constraints of the form $\mathcal{U} = \{u \in \mathbb{R} \,|\, -2 \leq u \leq 1\}$ and $\mathcal{X} = \{x \in \mathbb{R}^2 \,|\, -2 \leq x_1 \leq 3,\ -0.8 \leq x_2 \leq 0.6\}$, that is,

$$C_{\mathrm{x}} = \begin{bmatrix} 1 & 0 \\ 0 & 1 \\ -1 & 0 \\ 0 & -1 \end{bmatrix},\ d_{\mathrm{x}} = \begin{bmatrix} 3 \\ 0.8 \\ 2 \\ 0.8 \end{bmatrix}, \qquad C_{\mathrm{u}} = \begin{bmatrix} 1 \\ -1 \end{bmatrix},\ d_{\mathrm{u}} = \begin{bmatrix} 1 \\ 2 \end{bmatrix}. \tag{B.48}$$

The prediction horizon and the quadratic stage cost matrices are chosen as $N = 10$, $Q = \mathrm{diag}(1,\ 0.1)$, and $R = 1$, while the underlying barrier functions are formulated according to (B.21) with recentering weighting vectors $w_{\mathrm{x}} = [1,\ 1,\ 1.5,\ 1]^\top$ and $w_{\mathrm{u}} = [1,\ 2]^\top$. Based on this setup, the auxiliary control gain K, the terminal weight matrix P, and the set $\mathcal{P}_K(\gamma)$ can then be computed based on (B.23) and (B.28), respectively. Using for example $\gamma = 200$ and a barrier function weight $\varepsilon = 0.01$, this results in

$$K = \begin{bmatrix} -0.8983 & -1.3930 \end{bmatrix}, \qquad P = \begin{bmatrix} 23.2608 & 14.3725 \\ 14.3725 & 22.7877 \end{bmatrix}. \tag{B.49}$$

Based on the corresponding realization of the set $\mathcal{P}_K(\gamma)$ and the approaches discussed above, we can then construct the proposed ellipsoidal and polytopic terminal sets. In particular, a suitable ellipsoidal terminal set of the form $\mathcal{X}_f^E = \{x \in \mathbb{R}^2 \,|\, x^\top P_f\, x \leq 1\}$ can be computed based on solving the matrix valued optimization problem (B.34). Furthermore, a smooth polytopic terminal set approximation $\mathcal{X}_f^p = \{x \in \mathbb{R}^2 \,|\, \tilde{\psi}^p_{\mathcal{X}_f}(x) \leq 1\}$ can be obtained by computing a λ-contractive polytope $\mathcal{X}_f$ and then choosing $p \in \mathbb{N}_+$ large enough such that $\lambda\mathcal{X}_f \subset \mathcal{X}_f^p \subset \mathcal{X}_f$. For the considered example and $\lambda = 0.99$, it turns out that $p = 56$ is sufficient. The resulting terminal sets as well as the associated feasible sets $\mathcal{X}_N^E$ and $\mathcal{X}_N^p$ are depicted in Figure B.2 together with some comments. For comparison the plot also contains the respective terminal and feasible sets that would result from the more restrictive approach due to Wills and Heath (2004) when choosing the radius of the Dikin ellipsoid as $r = 0.95$, cf. Section 2.3.

In the above example, the Multi-Parametric Toolbox (Kvasnica et al., 2004) has been used to construct the required contractive and admissible polytopic set $\mathcal{X}_f$ for the considered problem setup, while the determinant maximization problem (B.34) for computing the maximum volume inscribed ellipsoid has been solved with YALMIP (Löfberg, 2004).

B.6 Parametrized solutions to the finite-horizon LQR problem with zero terminal state constraint

Consider a discrete-time linear system of the form (3.1) with (A, B) controllable and recall the finite-horizon LQR problem with zero terminal state constraint, i.e.,

$$v(x) = \arg\min_{v} \sum_{l=0}^{T-1} \|z_l\|_Q^2 + \|v_l\|_R^2 \tag{B.50a}$$

$$\text{s.t. } z_{l+1} = Az_l + Bv_l, \ l = 0, \ldots, T-1, \tag{B.50b}$$

$$z_T = 0, \ z_0 = x, \tag{B.50c}$$

for a given initial state $x \in \mathbb{R}^n$ and horizon $T \geq n$. By eliminating the predicted states z_l for $l = 1 \ldots, T$ based on (B.50b), we can derive the equivalent vectorized formulation

$$V^*(x) = \arg\min_{V} \frac{1}{2}V^\top HV + \frac{1}{2}x^\top FV + \frac{1}{2}V^\top F^\top x + \frac{1}{2}x^\top Yx \tag{B.51a}$$

$$\text{s.t. } A^T x + \begin{bmatrix} S_1 & S_2 \end{bmatrix} V = 0, \tag{B.51b}$$

where the vector $V \in \mathbb{R}^{Tm}$ is defined as $V = [v_0^\top, \cdots, v_{T-1}^\top]^\top$. Furthermore, the respective matrices are given as

$$S_1 = \begin{bmatrix} A^{T-1}B & \cdots & A^n B \end{bmatrix}, \ S_2 = \begin{bmatrix} A^{n-1}B & \cdots & B \end{bmatrix} \tag{B.52a}$$

$$H = 2\left(\tilde{R} + \Phi^\top \tilde{Q}\Phi\right), \ F = 2\Omega^\top \tilde{Q}\Phi, \ Y = 2\left(Q + \Omega^\top \tilde{Q}\Omega\right) \tag{B.52b}$$

$$\Omega = \begin{bmatrix} A \\ \vdots \\ A^T \end{bmatrix}, \ \Phi = \begin{bmatrix} B & \cdots & 0 \\ \vdots & \ddots & \vdots \\ A^{T-1}B & \cdots & B \end{bmatrix}, \ \begin{matrix} \tilde{Q} = I_{T-1} \otimes Q \\ \tilde{R} = I_T \otimes R. \end{matrix} \tag{B.52c}$$

Due to the assumed controllability of the system, there exists for any $x \in \mathbb{R}^n$ a feasible solution $V(x)$ to (B.51). Let us now partition $V \in \mathbb{R}^{Tm}$ as $V^\top = \begin{bmatrix} V_1^\top & V_2^\top \end{bmatrix}$ with $V_1 \in$

$\mathbb{R}^{(T-n)m}$ and $V_2 \in \mathbb{R}^{nm}$. Then, the terminal state constraint (B.51b) can be eliminated by choosing $V_2 = -S_2^+ \left(A^T x + S_1 V_1\right)$, where S_2^+ denotes the Moore-Penrose pseudoinverse of the matrix S_2. This leads to the following parametrization

$$V = \begin{bmatrix} I_{(T-n)m} \\ -S_2^+ S_1 \end{bmatrix} V_1 + \begin{bmatrix} 0 \\ -S_2^+ A^T \end{bmatrix} x = \Gamma_V V_1 + \Gamma_x x \,. \tag{B.53}$$

Inserting (B.53) into problem (B.51) then results in the following unconstrained optimization problem in the variable $V_1 \in \mathbb{R}^{(T-n)m}$:

$$V_1^*(x) = \arg\min_{V_1} \frac{1}{2} V_1^\top \tilde{H} V_1 + x^\top \tilde{F} V_1 + \frac{1}{2} x^\top \tilde{Y} x \,. \tag{B.54}$$

Here, the transformed problem matrices are given by $\tilde{H} = \Gamma_V^\top H \Gamma_V$, $\tilde{F} = \Gamma_x^\top H \Gamma_V + F^\top \Gamma_V$, and $\tilde{Y} = Y + \Gamma_x^\top H \Gamma_x + \Gamma_x^\top F^\top + F\Gamma_x$. Note that since $V(x) = \Gamma_x x$ is a feasible solution and $\mathrm{Im}(\Gamma_V) = \mathrm{Null}(\begin{bmatrix} S_1 & S_2 \end{bmatrix})$, the solution is invariant under this change of coordinates, which reveals that the problems (B.51) and (B.54) are in fact equivalent, see for example (Boyd and Vandenberghe, 2004, p. 132). Furthermore, the matrix $\tilde{H} = \Gamma_V^\top H \Gamma_V$ is guaranteed to be positive definite as the matrix Γ_V has full column rank and $H \in \mathbb{S}_{++}^{Tm}$. As a consequence, the unique optimal solution to the reduced problem can be computed easily as $V_1^*(x) = -\tilde{H}^{-1} \tilde{F}^\top x$ which results in

$$V^*(x) = \left(\Gamma_x - \Gamma_V \tilde{H}^{-1} \tilde{F}^\top\right) x = K_V\, x \,. \tag{B.55}$$

Inserting this solution into the quadratic cost function (B.51a) it can be shown that the associated cost-to-go is given by $J_{T,V}^*(x) = x^\top P_V x$ with $P_V = \frac{1}{2}\left(\tilde{Y} + \tilde{F}^\top \tilde{H}^{-1} \tilde{F}\right) \in \mathbb{S}_+^n$.

B.7 Continuity properties of nonrelaxed barrier function based MPC solutions with zero terminal state constraint

The following Lemma states an auxiliary result concerning the continuity properties of *nonrelaxed* barrier function based MPC solutions that are based on an additional zero terminal state constraint. In this thesis, this result is mainly used to prove the existence of a sufficiently small relaxation parameter that ensures strict constraint satisfaction for relaxed barrier function based MPC approaches with a purely quadratic terminal cost, see Theorem 3.7 in Section 3.5 as well as Theorem 4.5 in Section 4.3.3.

Lemma B.5. *Assume that* $\begin{bmatrix} A^{N-1}B & \cdots & AB & B \end{bmatrix}$ *has full row rank* n *and let* $\mathcal{X}_N := \{x \in \mathbb{R}^n | \exists\, U \in \mathbb{R}^{Nm}$ *s.t.* $u_k \in \mathcal{U}$, $x_k(U,x) \in \mathcal{X}$ *for* $k = 1, \ldots, N-1, x_N(U,x) = 0\}$ *be the set in the interior of which the problem*

$$\tilde{J}_N^*(x) = \min_U \sum_{k=0}^{N-1} \tilde{\ell}(x_k, u_k) \tag{B.56a}$$

$$s.t.\ x_{k+1} = Ax_k + Bu_k,\ x_N = 0,\ x_0 = x \,. \tag{B.56b}$$

has a solution. Here, $\tilde{\ell}(x,u) := \|x\|_Q^2 + \|u\|_R^2 + \varepsilon B_{\rm u}(u) + \varepsilon B_{\rm x}(x)$, *where* $B_{\rm u} : \mathcal{U}^\circ \to \mathbb{R}_+$ *and* $B_{\rm x} : \mathcal{X}^\circ \to \mathbb{R}_+$ *are recentered logarithmic barrier functions, e.g., functions of the form (3.18) with* $\hat{B}(\cdot)$ *being replaced by* $B(z) = -\ln(z)$. *Then, for any* $x_0 \in \mathcal{X}_N^\circ$, *the optimizer* $\tilde{U}^*(x_0)$ *associated to (B.56) is unique and* $\tilde{U}^* : \mathcal{X}_N^\circ \to \mathbb{R}^{Nm}$ *is a continuously differentiable function.*

Proof. Using the input vector notation, problem (B.56) is equivalent to

$$\tilde{J}_N^*(x) = \min_U \tilde{J}_N(U,x) \;\; \text{s.t.} \;\; S_1 U + S_2 x = 0 \tag{B.57}$$

where $\tilde{J}_N(\cdot,\cdot)$ is the nonrelaxed barrier function based cost function similar to (3.46), strongly convex in U, and $S_1 = \begin{bmatrix} A^{N-1}B & \cdots & B \end{bmatrix}$, $S_2 = A^N$ are used to encode the terminal equality constraint $x_N(U,x) = 0$ based on

$$x_k(U,x) = A^k x + \sum_{i=0}^{k-1} A^i B u_{k-1-i} \, . \tag{B.58}$$

For given $x_0 \in \mathcal{X}_N^\circ$, the KKT conditions for (B.56), which are in this case necessary and sufficient conditions for optimality, read

$$\nabla_U \tilde{J}_N(\tilde{U}^*(x_0), x_0) + S_1^\top \nu^*(x_0) = 0 \tag{B.59a}$$

$$S_1 \tilde{U}^*(x_0) + S_2 x_0 = 0 \, , \tag{B.59b}$$

with $\nu^*(x_0)$ denoting the associated Lagrange multiplier. The implicit function theorem states that unique and continuously differentiable solutions $\tilde{U}^* : \mathcal{X}_N^\circ \to \mathbb{R}^{Nm}$, $\nu^* : \mathcal{X}_N^\circ \to \mathbb{R}^n$ to (B.59) will exist if the corresponding Jacobian with respect to $(\tilde{U}^*, \nu^*)$, given by

$$\begin{bmatrix} \nabla_U^2 \tilde{J}_N(\tilde{U}^*(x_0), x_0) & S_1^\top \\ S_1 & 0 \end{bmatrix} \tag{B.60}$$

is nonsingular for all $x_0 \in \mathcal{X}_N^\circ$. Indeed, this follows from applying block inversion and noting that both the upper left corner block and the Schur complement are nonsingular. In fact, $\nabla_U^2 \tilde{J}_N(\tilde{U}^*(x_0), x_0) \in \mathbb{S}_{++}^{Nm} \; \forall x_0 \in \mathcal{X}_N^\circ$ due to strong convexity of the cost function. In addition, $S_1 \big(\nabla_U^2 \tilde{J}_N(\tilde{U}^*(x_0), x_0)\big)^{-1} S_1^\top \in \mathbb{S}_{++}^n \; \forall x_0 \in \mathcal{X}_N^\circ$ since S_1 has full row rank due to the assumed rank condition. For example, this assumption is always satisfied if (A,B) is controllable and $N \geq n$. Note furthermore that, based on this result, it could be shown easily that the value function $\tilde{J}_N^* : \mathcal{X}_N^\circ \to \mathbb{R}_+$ is twice continuously differentiable. ■

B.8 An algorithm to compute a suitable relaxation parameter with guaranteed constraint satisfaction properties

Assuming a given overall tolerance $\hat{z}_{\rm tol} \in \mathbb{R}_+$ and a set of initial condition $\mathcal{X}_0 \subset \mathcal{X}_N^\circ$, the algorithm on the next page may be used to compute a suitable relaxation parameter $\bar{\delta}_0$ that ensures a maximal constraint violation of $\hat{z}_{\rm x}(k) \leq \hat{z}_{\rm tol}\mathbb{1}$, $\hat{z}_{\rm u}(k) \leq \hat{z}_{\rm tol}\mathbb{1}$ for any $k \in \mathbb{N}$.

Algorithm *(Compute relaxation parameter $\bar{\delta}_0$ for given $\mathcal{X}_0$ and desired tolerance $\hat{z}_{\mathrm{tol}} \in \mathbb{R}_+$)*

Input: problem formulation, set $\mathcal{X}_0$, desired constraint satisfaction tolerance $\hat{z}_{\mathrm{tol}}$;
Output: choice for the relaxation parameter $\delta \in \mathbb{R}_{++}$ such that the maximal constraint violation in closed-loop operation is bounded by $\hat{z}_{\mathrm{tol}}$ for any initialization $x(0) \in \mathcal{X}_0$;
1: choose initial $\varepsilon, \delta \in \mathbb{R}_{++}$ and set up problem (3.56);
2: **repeat**
3: ∟ decrease the relaxation parameter: $\delta \leftarrow \gamma\delta$, $\gamma \in (0,1)$;
4: ∟ determine $\hat{\alpha}(\mathcal{X}_0, \delta) = \max_{x \in \mathcal{X}_0} \hat{\alpha}(x, \delta)$ based on (C.43);
5: ∟ compute the maximal violations $\hat{z}^i_{\mathrm{x}}(\mathcal{X}_0, \delta)$ and $\hat{z}^j_{\mathrm{u}}(\mathcal{X}_0, \delta)$ for the set $\mathcal{X}_0$ by inserting $\hat{\alpha}(0) = \hat{\alpha}(\mathcal{X}_0, \delta)$ into (C.38) and solving the resulting convex problems;
6: **until** $\hat{z}^i_{\mathrm{x}}(\mathcal{X}_0; \delta) \leq \hat{z}_{\mathrm{tol}}$ and $\hat{z}^j_{\mathrm{u}}(\mathcal{X}_0; \delta) \leq \hat{z}_{\mathrm{tol}}$ holds $\forall\, i = 1, \dots, q_{\mathrm{x}}$, $j = 1, \dots, q_{\mathrm{u}}$.

B.9 Vector formulation of the open-loop optimal control problem

We briefly illustrate in the following how the relaxed barrier function based open-loop optimal control problem (3.16) can be written as a strongly convex optimization problem in matrix/vector form. Both the condensed (that is, performing an elimination of the predicted system states) and the sparse formulation with equality constraints are discussed. While the following discussion focuses on the approach with a purely quadratic terminal cost, basically the same ideas can be used for the other approaches.

Condensed formulation

Note that for given $x \in \mathbb{R}^n$ and $U \in \mathbb{R}^{Nm}$, the predicted system states $x_k(U, x)$ can based on (3.16b) be expressed as

$$x_k(U, x) = A^k x + \sum_{i=0}^{k-1} A^i B u_{k-1-i}, \qquad k = 0, \dots, N. \tag{B.61}$$

Writing the open-loop optimal control problem with purely quadratic terminal cost in matrix form and eliminating the predicted system states by means of (B.61) reveals that the cost function matrices $H \in \mathbb{S}^{Nm}_{++}$, $F \in \mathbb{R}^{n \times Nm}$, and $Y \in \mathbb{S}^{Nm}_{++}$ in the condensed formulation as for example in (3.46) and (4.6) are given by

$$H = 2\left(\tilde{R} + \Gamma^\top \tilde{Q} \Gamma\right), \quad F = 2\Omega^\top \tilde{Q} \Gamma, \quad Y = 2\left(Q + \Omega^\top \tilde{Q} \Omega\right), \tag{B.62}$$

with the auxiliary matrices $\Omega \in \mathbb{R}^{Nn \times n}$, $\Gamma \in \mathbb{R}^{Nn \times Nm}$, $\tilde{Q} \in \mathbb{S}^{Nn}_{++}$, and $\tilde{R} \in \mathbb{S}^{Nm}_{++}$ being defined as

$$\Omega = \begin{bmatrix} A \\ A^2 \\ \vdots \\ A^N \end{bmatrix}, \quad \Gamma = \begin{bmatrix} B & 0 & \cdots & 0 \\ AB & B & \cdots & 0 \\ \vdots & \vdots & \ddots & \vdots \\ A^{N-1}B & A^{N-2}B & \cdots & B \end{bmatrix}, \quad \begin{aligned} \tilde{Q} &= \begin{bmatrix} I_{N-1} \otimes Q & 0 \\ 0 & P \end{bmatrix} \\ \tilde{R} &= I_N \otimes R\,. \end{aligned} \tag{B.63a}$$

Furthermore, the constraint matrices G, d, and E can be constructed in a very similar way and may (without a potentially possible redundancy reduction) be formulated as

$$G = \begin{bmatrix} 0 \\ ([I_{N-1}\ 0] \otimes C_{\mathrm{x}})\Gamma \\ I_N \otimes C_{\mathrm{u}} \end{bmatrix}, \quad E = -\begin{bmatrix} C_{\mathrm{x}} \\ ([I_{N-1}\ 0] \otimes C_{\mathrm{x}})\Omega \\ 0 \end{bmatrix}, \quad d = \begin{bmatrix} \mathbb{1}_N \otimes d_{\mathrm{x}} \\ \mathbb{1}_N \otimes d_{\mathrm{u}} \end{bmatrix}. \tag{B.64}$$

The associated relaxed logarithmic barrier function $\hat{B}_{\mathrm{xu}} : \mathbb{R}^{Nm} \times \mathbb{R}^n \to \mathbb{R}_+$ in (3.84) may then be written as

$$\hat{B}_{\mathrm{xu}}(U, x) = \sum_{i=1}^{q} w^i \left(\hat{B}(-G^i U + E^i x + d^i) - \ln\left(d^i\right) \right), \tag{B.65}$$

where $\hat{B} : \mathbb{R} \to \mathbb{R}$ refers to the primordial relaxed logarithmic barrier function defined according to (3.4). Furthermore, $w \in \mathbb{R}^q_{++}$ denotes a suitable recentering vector that may for example be constructed by stacking the vectors w_{x} and w_{u} in a suitable way.

Sparse formulation

As discussed in Section 4.1, the open-loop optimal control problem may also be formulated in a non-condensed, sparse form by retaining the underlying system dynamics as equality constraints. In particular, defining the optimization variable as

$$w = \begin{bmatrix} u_0^\top & x_1^\top & u_1^\top & x_2^\top & \cdots & u_{N-1}^\top & x_N^\top \end{bmatrix}^\top \in \mathbb{R}^{n_w}, \tag{B.66}$$

$n_w = N(m+n)$, problem (3.16) can be reformulated as

$$\hat{J}_N^*(x) = \min_w \ \frac{1}{2} w^\top H w + x^\top F w + \frac{1}{2} x^\top Y x + \varepsilon \hat{B}_w(w) \tag{B.67a}$$

$$\text{s.t.}\ \ Cw = b(x). \tag{B.67b}$$

Considering again the formulation with a purely quadratic terminal cost, the respective vectors and matrices can easily be derived as

$$H = 2\begin{bmatrix} R & & \cdots & & 0 \\ & Q & & & \\ \vdots & & \ddots & & \vdots \\ & & & R & \\ 0 & & \cdots & & P \end{bmatrix}, \qquad F = 0 \in \mathbb{R}^{n \times n_w}, \qquad Y = 2Q, \tag{B.68a}$$

$$C = \begin{bmatrix} -B & I & 0 & 0 & 0 & 0 & \cdots & 0 & 0 & 0 \\ 0 & -A & -B & I & 0 & 0 & \cdots & 0 & 0 & 0 \\ 0 & 0 & 0 & -A & -B & I & \cdots & 0 & 0 & 0 \\ \vdots & \vdots & \vdots & \vdots & \vdots & \vdots & \ddots & \vdots & \vdots & \vdots \\ 0 & 0 & 0 & 0 & 0 & 0 & \cdots & I & 0 & 0 \\ 0 & 0 & 0 & 0 & 0 & 0 & \cdots & -A & -B & I \end{bmatrix}, \quad b(x) = \begin{bmatrix} Ax \\ 0 \\ 0 \\ \vdots \\ 0 \\ 0 \end{bmatrix}, \tag{B.68b}$$

while the relaxed logarithmic barrier function for the stacked input and state constraints is given by

$$\hat{B}_{xw}(w,x) = \sum_{i=1}^{q} w_r^i \left(\hat{B}(-G^i w + E^i x + d^i) + \ln(d^i) \right) \tag{B.69}$$

with $q = N(q_x + q_u) + q_x$, $w_r \in \mathbb{R}^q_{++}$ denoting the recentering vector, and

$$G = \begin{bmatrix} 0 & \cdots & & 0 \\ C_u & & & \\ \vdots & C_x & & \vdots \\ & & \ddots & \\ & & C_u & \\ 0 & & \cdots & C_x \end{bmatrix}, \quad E = \begin{bmatrix} -C_x \\ 0 \\ \\ \vdots \\ \\ 0 \end{bmatrix}, \quad d = \begin{bmatrix} d_x \\ d_u \\ d_x \\ \vdots \\ d_u \\ d_x \end{bmatrix}. \tag{B.70}$$

Note that, as in the condensed formulation, the first block row of the inequality constraints is exclusively related to the given current system state x and will therefore, in general, not affect the solution of problem (B.67). However, we include it here in order to be consistent with the original formulation (3.16), which also imposes a constraint and barrier function on the initial state. The recentering vector $w_r \in \mathbb{R}^q_{++}$ can again be constructed by stacking the vectors w_x and w_u in a suitable way. Using the above definition of the constraint matrices, this for example leads to $w_r = [w_x^\top, w_u^\top, w_x^\top, \ldots, w_u^\top, w_x^\top]^\top$.

B.10 A step size bound for Newton's method with backtracking

For the Newton method search direction given in (4.38c), the Armijo condition (4.37a) can with the corresponding squared Newton decrement

$$\lambda(U,x)^2 := \nabla_U \hat{J}_N(U,x)^\top \left(\nabla_U^2 \hat{J}_N(U,x)\right)^{-1} \nabla_U \hat{J}_N(U,x) \tag{B.71}$$

be rewritten as

$$\hat{J}_N(U+sp,x) \leq \hat{J}_N(U,x) - c_1 s\, \lambda(U,x)^2 . \tag{B.72}$$

On the other hand, applying Taylor's Theorem to the left-hand side of (B.72), we get

$$\hat{J}_N(U+sp,x) = \hat{J}_N(U,x) + s\nabla_U \hat{J}_N(U,x)^\top p + \frac{1}{2} s^2 p^\top \nabla_U^2 \hat{J}_N(U+\tau sp,x) p \tag{B.73}$$

for some $\tau \in (0,1)$, see for example Theorem 2.1 in (Nocedal and Wright, 1999). Exploiting now the upper and lower bounds from (4.8a) and using $\nabla_U \hat{J}_N(U,x)^\top p = -\lambda(U,x)^2$ as well as $p^\top \nabla_U^2 \hat{J}_N(U,x) p = \lambda(U,x)^2$, we obtain

$$\hat{J}_N(U+sp,x) \leq \hat{J}_N(U,x) - s\Big(1 - \frac{L}{2\mu} s\Big) \lambda(U,x)^2. \tag{B.74}$$

This directly implies that (B.72), and hence (4.37a), will be satisfied for any step size satisfying $s \leq \bar{s} := 2\mu(1-c_1)/L$. The corresponding backtracking line search will therefore always terminate with $s = \rho^j \geq \rho\bar{s}$, which reveals that $j \leq \lceil 1 + \log_\rho(2\mu(1-c_1)/L) \rceil$.

Appendix C

Technical Proofs

C.1 Proof of Theorem 2.3

The proof uses standard arguments and is very similar to the one for conventional MPC, cf. (Mayne et al., 2000; Rawlings and Mayne, 2009; Grüne and Pannek, 2011). We first show recursive feasibility of problem (2.23). For any $x_0 \in \mathcal{X}_N^\circ$, there exists by definition of the set $\mathcal{X}_N$ an optimal input sequence $\tilde{\boldsymbol{u}}^*(x_0) = \{\tilde{u}_0^*, \ldots, \tilde{u}_{N-1}^*\}$ that guarantees strict satisfaction of all input, state, and terminal set constraints and results in a feasible open-loop state sequence $\boldsymbol{x}^*(x_0) = \{x_0, x_1^*, \ldots, x_N^*\}$ with $x_N^* \in \mathcal{X}_f^\circ$. The successor state that results by applying the first element of this input sequence to the underlying system dynamics is given by $x_0^+ = f(x_0, \tilde{u}_0^*) = x_1^*$. Due to the properties of the terminal set $\mathcal{X}_f$ and the corresponding auxiliary control law $k_f(\cdot)$, the (suboptimal) input sequence $\tilde{\boldsymbol{u}}^+(x_0) = \{\tilde{u}_1^*, \ldots, \tilde{u}_{N-1}^*, k_f(x_N^*)\}$ is feasible for the initial state x_0^+ and results in the strictly feasible open-loop state sequence $\boldsymbol{x}(\tilde{\boldsymbol{u}}^+(x_0), x_0^+) = \{x_0^+, x_2^*, \ldots, x_N^*, f(x_N^*, k_f(x_N^*))\}$ with $f(x_N^*, k_f(x_N^*)) \in \mathcal{X}_f^\circ$, see properties *i)* and *ii)* of Assumption 2.5. Thus, for any $x_0 \in \mathcal{X}_N^\circ$, there exists a strictly feasible input sequence that ensures that the successor state $x_0^+ = f(x_0, \tilde{u}_0^*)$ will again lie in the interior of the feasible set $\mathcal{X}_N$, which implies recursive feasibility of the barrier function based open-loop optimal control problem (2.23) as well as strict satisfaction of the underlying input and state constraints for all $k \in \mathbb{N}$.
In a second step, we are now going to show that the associated value function will decrease along trajectories of the closed-loop system (2.25) and can, therefore, be used as a Lyapunov function. In particular, we verify in the following that the value function $\tilde{J}_N^* : \mathbb{R}^n \to \mathbb{R}_+$ in (2.23) satisfies

$$\tilde{J}_N^*(x_0^+) - \tilde{J}_N^*(x_0) \leq -\tilde{\ell}(x_0, \tilde{u}_0^*(x_0)) \;\; \forall\, x_0 \in \mathcal{X}_N^\circ\,. \tag{C.1}$$

First, note that $\tilde{J}_N^*(x_0^+) - \tilde{J}_N^*(x_0) \leq \tilde{J}_N(\tilde{\boldsymbol{u}}^+(x_0), x_0^+) - \tilde{J}_N^*(x_0)$, where $\tilde{J}_N(\tilde{\boldsymbol{u}}^+(x_0), x_0^+)$ denotes the value of the cost function evaluated for the suboptimal input sequence $\tilde{\boldsymbol{u}}^+(x_0)$. Moreover, for any $x_0 \in \mathcal{X}_N^\circ$ it holds that

$$\begin{aligned} \tilde{J}_N(\tilde{\boldsymbol{u}}^+(x_0), x_0^+) - \tilde{J}_N^*(x_0) &= \tilde{F}(f(x_N^*, k_f(x_N^*)) - \tilde{F}(x_N^*) + \tilde{\ell}(x_N^*, k_f(x_N^*)) - \tilde{\ell}(x_0, \tilde{u}_0^*) & \text{(C.2a)} \\ &\leq -\tilde{\ell}(x_0, \tilde{u}_0^*)\,, & \text{(C.2b)} \end{aligned}$$

where we first used a telescoping sum to arrive at (C.2a) and then exploited the fact that $x_N^* \in \mathcal{X}_f^\circ$, which implies $\tilde{F}(f(x_N^*, k_f(x_N^*))) - \tilde{F}(x_N^*) + \tilde{\ell}(x_N^*, k_f(x_N^*)) \leq 0$ due to property *iii*) of Assumption 2.5. Thus, in combination with the arguments above, we have $\tilde{J}_N^*(x_0^+) - \tilde{J}_N^*(x_0) \leq -\tilde{\ell}(x_0, \tilde{u}_0^*)$ for any $x_0 \in \mathcal{X}_N^\circ$, and (C.1) holds. Finally, Assumptions 2.1, 2.2 and 2.4 ensure that $\tilde{J}_N^* : \mathcal{X}_N^\circ \to \mathbb{R}_+$ is a well-defined and positive definite function with $\tilde{J}_N^*(x) \to \infty$ whenever $x \to \partial\mathcal{X}_N$. Thus, in combination with (C.1) and the fact that $\tilde{\ell}(x,u) \geq \ell(x,u) \geq c\|(x,u)\|^2$, see Assumption 2.2, it can be used as a Lyapunov function that proves asymptotic stability of the origin of system (2.25) with region of attraction $\mathcal{X}_N^\circ$, cf. Appendix A.1. This completes the proof.

C.2 Proof of Lemma 3.1

The result for the terminal set $\mathcal{X}_f$ follows directly from the fact that $\hat{B}_\mathrm{f}(\cdot)$ is convex and strictly monotone in the argument $z = 1 - \varphi_\mathrm{f}(x)$. Thus, whenever $\hat{B}_\mathrm{f}(x) \leq \bar{\beta}_\mathrm{f}(\delta)$ for some $x \in \mathbb{R}^n$, this implies that $1 - \varphi_\mathrm{f}(x) \geq 0$, which is by definition equivalent to $x \in \mathcal{X}_f$. Concerning the state constraints, we are going to demonstrate in the following based on a proof by contradiction that every state that lies in the sublevel set $\mathcal{S}_{\hat{B}_\mathrm{x}}$ also has to lie in the constraint set $\mathcal{X}$. To this end, let us assume that there exists an $\bar{x} \in \mathcal{S}_{\hat{B}_\mathrm{x}}$ with $\bar{x} \notin \mathcal{X}$. Note that the latter assumption implies that there exists at least one constraint $i \in \{1, \ldots, q_\mathrm{x}\}$ such that $C_\mathrm{x}^i\bar{x} > d_\mathrm{x}^i$, while all other constraints may be satisfied. Let now $\lambda \in (0,1)$ be a suitable scaling factor such that $C_\mathrm{x}\lambda\bar{x} = \lambda C_\mathrm{x}\bar{x} \leq d_\mathrm{x}$ while, in particular, $C_\mathrm{x}^i\lambda\bar{x} = \lambda C_\mathrm{x}^i\bar{x} = d_\mathrm{x}^i$ with i denoting the index of the previously violated constraint. Note that such a scaling factor always exists and that $\bar{x}' = \lambda\bar{x}$ can be interpreted as the projection of the infeasible state $\bar{x}$ onto the boundary of the polytope $\mathcal{X}$. However, due to convexity and positive definiteness of the relaxed barrier function $\hat{B}_\mathrm{x}(\cdot)$ and the assumption that $\bar{x} \in \mathcal{S}_{\hat{B}_\mathrm{x}}$, this would imply that $\hat{B}_\mathrm{x}(\lambda\bar{x}) \leq \lambda\hat{B}_\mathrm{x}(\bar{x}) \leq \lambda\bar{\beta}_\mathrm{x}(\delta) < \bar{\beta}_\mathrm{x}(\delta)$. Thus, there would exist a point $\bar{x}' = \lambda\bar{x}$ with $\hat{B}_\mathrm{x}(\bar{x}') < \bar{\beta}_\mathrm{x}(\delta)$ on the boundary of the set $\mathcal{X}$, which obviously is a contradiction to the definition of $\bar{\beta}_\mathrm{x}(\delta)$ given in Definition 3.4. As a consequence, $\bar{x} \in \mathcal{S}_{\hat{B}_\mathrm{x}}$ with $\bar{x} \notin \mathcal{X}$ cannot exist, revealing that $x \in \mathcal{X}$ whenever $x \in \mathcal{S}_{\hat{B}_\mathrm{x}}$, i.e., $\mathcal{S}_{\hat{B}_\mathrm{x}} \subseteq \mathcal{X}$. Similar arguments can be used for the input constraint case, leading to $\mathcal{S}_{\hat{B}_\mathrm{u}} \subseteq \mathcal{U}$.

C.3 Proof of Theorem 3.1

The proof consists of three parts and is closely related to that of Theorem B.1 for the nonrelaxed case, see Appendix B.4. First, we show that the underlying input, state, and terminal set constraints are not violated for any $x_0 \in \hat{\mathcal{X}}_N(\delta)$; then we use standard MPC arguments to show that the value function $\hat{J}_N^*(x(k))$ will decrease along trajectories of the closed-loop system when applying the feedback $u(k) = \hat{u}_0^*(x(k))$; finally, we use this result to conclude that the resulting input and state sequences will also be strictly feasible at all later time steps and that the origin of the closed-loop system is asymptotically stable with region of attraction $\hat{\mathcal{X}}_N(\delta)$. Rigorous definitions of the applied Lyapunov stability concepts can be found in Appendix A.1.

Part 1) Consider an arbitrary but fixed $x_0 \in \hat{\mathcal{X}}_N(\delta)$ and let $\hat{\boldsymbol{u}}^*(x_0) = \{\hat{u}_0^*, \ldots, \hat{u}_{N-1}^*\}$ and $\boldsymbol{x}^*(x_0) = \{x_0^* = x_0, x_1^*, \ldots, x_N^*\}$ denote the associated optimal open-loop input and state sequences. Since the cost function in (3.16) is a sum of positive definite terms, it holds that $\varepsilon\hat{B}_\mathrm{x}(x_k^*) < \hat{J}_N^*(x_0)$, $\varepsilon\hat{B}_\mathrm{u}(\hat{u}_k^*) < \hat{J}_N^*(x_0)$ for any $k = 0, \ldots, N-1$ as well as $\varepsilon\hat{B}_\mathrm{f}(x_N^*) < \hat{J}_N^*(x_0)$ for any $x_0 \in \hat{\mathcal{X}}_N(\delta) \setminus \{0\}$. By definition of the set $\hat{\mathcal{X}}_N(\delta)$, this immediately implies that $\hat{B}_\mathrm{x}(x_k^*) < \bar{\beta}(\delta)$, $\hat{B}_\mathrm{u}(\hat{u}_k^*) < \bar{\beta}(\delta)$, and $\hat{B}_\mathrm{f}(x_N^*) < \bar{\beta}(\delta)$. Recalling the definition of $\bar{\beta}(\delta)$ and exploiting Lemma 3.1, this reveals that $x_k^* \in \mathcal{X}^\circ$, $\hat{u}_k^* \in \mathcal{U}^\circ$ for any $k = 0, \ldots, N-1$ as well as $x_N^* \in \mathcal{X}_f^\circ$ whenever $x_0 \in \hat{\mathcal{X}}_N(\delta) \setminus \{0\}$. The case $x_0 = 0$ is trivial.

Part 2) The arguments of Part 1 essentially reveal that $x_0 \in \mathcal{X}_N^\circ$ whenever $x_0 \in \hat{\mathcal{X}}_N(\delta)$. Consider again an arbitrary but fixed $x_0 \in \hat{\mathcal{X}}_N(\delta)$ and let the associated successor state be given by $x_0^+ = Ax_0 + B\hat{u}_0^* = x_1^*$. Due to the linear dynamics and the properties of the terminal set $\mathcal{X}_f$, see A3 of Assumption 3.2, the candidate solution $\hat{\boldsymbol{u}}^+(x_0) = \{\hat{u}_1^*, \ldots, \hat{u}_{N-1}^*, Kx_N^*\}$ provides a suboptimal input sequence for the initial state x_0^+, that results in the open-loop state sequence $\boldsymbol{x}(\hat{\boldsymbol{u}}^+(x_0), x_0^+) = \{x_0^+, x_2^*, \ldots, x_N^*, A_K x_N^*\}$ with $A_K x_N^* \in \mathcal{X}_f^\circ$. Now we can use basically the same arguments as in the proof of Theorem B.1 to show that

$$\hat{J}_N^*(x_0^+) - \hat{J}_N^*(x_0) \leq -\hat{\ell}(x_0, \hat{u}_0^*(x_0)) \quad \forall\, x_0 \in \hat{\mathcal{X}}_N(\delta). \tag{C.3}$$

In particular, due to suboptimality of the candidate solution $\hat{\boldsymbol{u}}^+(x_0)$, it holds that $\hat{J}_N^*(x_0^+) - \hat{J}_N^*(x_0) \leq \hat{J}_N(\hat{\boldsymbol{u}}^+(x_0), x_0^+) - \hat{J}_N^*(x_0)$, where $\hat{J}_N(\hat{\boldsymbol{u}}^+(x_0), x_0^+)$ denotes the value of the cost function evaluated for the suboptimal input sequence $\hat{\boldsymbol{u}}^+(x_0)$. On the other hand, the assumed properties of terminal set and terminal weight ensure that $\hat{J}_N(\hat{\boldsymbol{u}}^+(x_0), x_0^+) - \hat{J}_N^*(x_0) = \hat{F}(A_K x_N^*) - \hat{F}(x_N^*) + \hat{\ell}(x_N^*, Kx_N^*) - \hat{\ell}(x_0, \hat{u}_0^*) \leq -\hat{\ell}(x_0, \hat{u}_0^*)$ for any $x_0 \in \hat{\mathcal{X}}_N(\delta)$. More precisely, we know from Part 1 that $x_N^* = x_N(\hat{\boldsymbol{u}}^*(x_0), x_0) \in \mathcal{X}_f^\circ$ and, therefore,

$$\hat{F}(A_K x_N^*) - \hat{F}(x_N^*) + \hat{\ell}(x_N^*, Kx_N^*) \tag{C.4a}$$
$$= \|A_K x_N^*\|_P^2 - \|x_N^*\|_P^2 + \|x_N^*\|_Q^2 + \|Kx_N^*\|_R^2 + \varepsilon\hat{B}_K(x_N^*) + \varepsilon\left(\hat{B}_\mathrm{f}(A_K x_N^*) - \hat{B}_\mathrm{f}(x_N^*)\right) \tag{C.4b}$$
$$\leq \varepsilon\big(\hat{B}_\mathrm{f}(A_K x_N^*) - \hat{B}_\mathrm{f}(x_N^*)\big) = \varepsilon\big(\hat{B}(1 - \varphi_\mathrm{f}(A_K x_N^*)) - \hat{B}(1 - \varphi_\mathrm{f}(x_N^*))\big) \leq 0\,. \tag{C.4c}$$

Here, the first inequality follows from the choice of $\mathcal{X}_f$ and P according to Assumption 3.2 and the fact that $\hat{B}_K(x) \leq x^\top M x\ \forall\, x \in \mathcal{X}_f \subseteq \mathcal{N}_K$. The second inequality follows from the monotonicity of the relaxed logarithmic barrier function $\hat{B}_\mathrm{f}(\cdot)$ and the assumption that $\varphi_\mathrm{f}(A_K x) \leq \varphi_\mathrm{f}(x)$ for any $x \in \mathcal{X}_f$, see A3 of Assumption 3.2. Thus, in total, $\hat{J}_N^*(x_0^+) - \hat{J}_N^*(x_0) \leq -\hat{\ell}(x_0, \hat{u}_0^*(x_0))$ for any $x_0 \in \hat{\mathcal{X}}_N(\delta)$, and (C.3) holds.

Part 3) The fact that the value function decreases implies that $\hat{J}_N^*(x_0^+) \leq \hat{J}_N^*(x_0) \leq \varepsilon\bar{\beta}(\delta)$ and hence $x_0^+ \in \hat{\mathcal{X}}_N(\delta)$ for any $x_0 \in \hat{\mathcal{X}}_N(\delta)$. By repeating this argument, the resulting closed-loop system state satisfies $x(k) \in \hat{\mathcal{X}}_N(\delta)\ \forall\, k \geq 0$ for any $x(0) \in \hat{\mathcal{X}}_N(\delta)$, which shows that (3.2) will be strictly satisfied by all future states and inputs. Moreover, in combination with (C.3) this ensures that for any $x(0) \in \hat{\mathcal{X}}_N(\delta)$

$$\hat{J}_N^*(x(k+1)) - \hat{J}_N^*(x(k)) \leq -\hat{\ell}(x(k), \hat{u}_0^*(x(k)))\ \ \forall\, k \in \mathbb{N}\,. \tag{C.5}$$

Due to the design of the relaxed barrier functions, $\hat{\ell}(x,u) \geq \ell(x,u) \geq \|x\|_Q^2$, while $\hat{J}_N^* : \mathbb{R}^n \to \mathbb{R}_+$ is a well-defined, positive definite, and radially unbounded function. Hence, it can be used as a Lyapunov function for the closed-loop system (3.19), proving asymptotic stability of the origin with a guaranteed region of attraction of at least $\hat{\mathcal{X}}_N(\delta)$.

C.4 Proof of Lemma 3.2

For any $\delta \in \mathbb{R}_{++}$ satisfying Assumptions 3.1 and 3.3, the value function $\hat{J}_N^* : \mathbb{R}^n \to \mathbb{R}_+$ associated to problem (3.16) is a positive definite, radially unbounded, and convex function. In particular, convexity follows from the facts that all cost function terms in (3.16a) are convex and that convexity is preserved under affine transformations as well as when minimizing over a subset of variables (Boyd and Vandenberghe, 2004). Thus, as both ε and $\bar{\beta}(\delta)$ are strictly positive, $\hat{\mathcal{X}}_N(\delta) := \{x \in \mathbb{R}^n \,|\, \hat{J}_N^*(x) \leq \varepsilon\bar{\beta}(\delta)\}$ is a nonempty, compact, and convex set. Furthermore, as shown in the proof of Theorem 3.1, any initial condition $x_0 \in \hat{\mathcal{X}}_N(\delta)$ results in strictly feasible input and state sequences, which implies that $\hat{\mathcal{X}}_N(\delta) \subseteq \mathcal{X}_N^\circ$. This completes the proof of the first claim.

We now prove the second claim, i.e., that $\hat{\mathcal{X}}_N(\delta)$ will contain any compact subset $\mathcal{X}_0$ of $\mathcal{X}_N^\circ$ if we choose the relaxation parameter sufficiently small. We proceed as follows: first we show in Part 1 that for any *fixed* $x_0 \in \mathcal{X}_N^\circ$, there exists a relaxation parameter $\delta_0'(x_0) \in \mathbb{R}_{++}$ such that $x_0 \in \hat{\mathcal{X}}_N(\delta)$ for any $\delta \leq \delta_0'(x_0)$. In part 2, we then show the claimed result by uniformly choosing $\bar{\delta}_0 = \min_{x_0 \in \mathcal{X}_0} \delta_0'(x_0)$. In order to make the dependency on the relaxation parameter more explicit, we write in the following $\hat{J}_N^*(x;\delta)$ instead of $\hat{J}_N^*(x)$.

Part 1) Assume $x_0 \in \mathcal{X}_N^\circ$ and let $\tilde{\boldsymbol{u}}^*(x_0) = \{\tilde{u}_0^*(x_0), \ldots, \tilde{u}_{N-1}^*(x_0)\}$, $\tilde{\boldsymbol{x}}^*(x_0) = \{\tilde{x}_0^*(x_0) = x_0, \tilde{x}_1^*(x_0), \ldots, \tilde{x}_N^*(x_0)\}$, and $\tilde{J}_N^*(x_0)$ denote the solution of the corresponding *nonrelaxed* problem formulation (B.10). Note that, due to the definition of the set $\mathcal{X}_N$, this solution is guaranteed to exists, and that the resulting open-loop input and state sequences will strictly satisfy the polytopic constraints (3.2). As a consequence, there always exists $\delta_0(x_0) \in \mathbb{R}_{++}$, defined as $\delta_0(x_0) := \min\{-C_\mathrm{x}^i \tilde{x}_k(x_0) + d_\mathrm{x}^i, -C_\mathrm{u}^j \tilde{u}_k(x_0) + d_\mathrm{u}^j, 1 - \varphi_\mathrm{f}(\tilde{x}_N^*(x_0)), i = 1, \ldots, q_\mathrm{x}, j = 1, \ldots, q_\mathrm{u}, k = 0, \ldots, N-1\}$, with the property that the solutions of relaxed and nonrelaxed formulation will be equivalent for all $\delta \leq \delta_0(x_0)$, i.e., $\hat{\boldsymbol{u}}^*(x_0) = \tilde{\boldsymbol{u}}^*(x_0)$, $\hat{\boldsymbol{x}}^*(x_0) = \tilde{\boldsymbol{x}}^*(x_0)$, $\hat{J}_N^*(x_0;\delta) = \tilde{J}_N^*(x_0)$. In principle, $\delta_0(x_0)$ is nothing else than the minimal distance of the optimal open-loop state and input sequences of the nonrelaxed formulation to the boundaries of the respective constraint sets (which is guaranteed to be strictly positive). Thus, by choosing the relaxation parameter δ smaller than this value, we can ensure that the optimal solution of the relaxed problem will be the same as that of the nonrelaxed problem, cf. (Hauser and Saccon, 2006). Note that $\delta_0(x_0)$ is a continuous function of x_0 since both $\tilde{\boldsymbol{u}}^*(x_0)$ and $\tilde{\boldsymbol{x}}^*(x_0)$ are continuous due to the smooth and strictly convex problem formulation (B.10)[1]. Let us further define

$$\delta_0'(x_0) := \max\{\delta \in \mathbb{R}_{++} \,|\, \delta \leq \delta_0(x_0),\, \tilde{J}_N^*(x_0) \leq \varepsilon\bar{\beta}(\delta)\}. \tag{C.6}$$

[1]For a detailed discussion on continuity and regularity properties of solutions to parametric optimization problems see for example Fiacco (1983) as well as Section 4.1 in the book of Bonnans and Shapiro (2000).

On the one hand, this definition ensures that $\delta_0'(x_0) \leq \delta_0(x_0)$ and, hence, $\tilde{J}_N^*(x_0) = \hat{J}_N^*(x_0;\delta) = \hat{J}_N^*(x_0;\delta_0(x_0))$ for all $\delta \leq \delta_0'(x_0)$. On the other hand, the second condition in (C.6) implies that $\tilde{J}_N^*(x_0) = \hat{J}_N^*(x_0;\delta) \leq \varepsilon\bar{\beta}(\delta)$, such that $x_0 \in \hat{\mathcal{X}}_N(\delta)$ for any $\delta \leq \delta_0'(x_0)$, cf. Definition 3.5. We may picture the construction of $\delta_0'(x_0)$ as decreasing the relaxation parameter δ from $\delta(x_0)$ until also the second condition in the foregoing definition of $\delta_0'(x_0)$ is satisfied. As $\hat{J}_N^*(x_0;\delta) = \hat{J}_N^*(x_0;\delta_0(x_0)) = \tilde{J}_N^*(x_0)$ for all $\delta \leq \delta_0(x_0)$ and, on the other hand,$\bar{\beta}(\delta)$ grows without bound for decreasing δ, this can always be achieved. Thus, $\delta'(x_0)$ exists for any $x_0 \in \mathcal{X}_N^\circ$. Moreover, since $\bar{\beta}(\delta)$ is continuous in δ and both $\delta_0(x_0)$ and $\tilde{J}_N^*(x_0)$ are continuous in x_0, $\delta_0'(x_0)$ is a continuous function of x_0. Note that continuity of $\bar{\beta}(\delta)$ follows from noting that the constraints in (3.27a) and (3.27b) are independent from δ and that $\hat{B}_\mathrm{x}(x)$ and $\hat{B}_\mathrm{u}(u)$ are continuous in both x (respectively u) and δ and possess, in addition, compact sublevel sets for any fixed value of δ, see (Fiacco, 1983, Theorem 2.2) as well as (Bonnans and Shapiro, 2000, Proposition 4.4).

Part 2) Consider now an arbitrary compact set $\mathcal{X}_0 \subseteq \mathcal{X}_N^\circ$ and define $\bar{\delta}_0 \in \mathbb{R}_{++}$ as

$$\bar{\delta}_0 = \min_{x_0 \in \mathcal{X}_0} \delta_0'(x_0)\,. \tag{C.7}$$

Due to the continuity of $\delta_0'(x_0)$ (see the arguments above) and the compactness of $\mathcal{X}_0$, this value always exists and ensures that $\mathcal{X}_0 \subseteq \hat{\mathcal{X}}_N(\delta)$ for any $\delta \leq \bar{\delta}_0$.

C.5 Proof of Theorem 3.2

Due to Assumption 3.4 and the fact that the terminal set $\mathcal{X}_\mathrm{f}$ is nonempty, there exists for any $x_0 \in \mathbb{R}^n$ well-defined open-loop input and state sequences $\hat{\boldsymbol{u}}^*(x_0) = \{\hat{u}_0^*, \dots, \hat{u}_{N-1}^*\}$ and $\boldsymbol{x}^*(x_0) = \{x_0^* = x_0, x_1^*, \dots, x_N^*\}$ with $x_N^* = x_N(\hat{\boldsymbol{u}}^*(x_0), x_0) \in \mathcal{X}_f^\circ$. Note that the elements of $\hat{\boldsymbol{u}}^*(x_0)$ and $\boldsymbol{x}^*(x_0)$ will, in general, not necessarily satisfy the input and state constraints (3.2). However, the fact that $x_N^* \in \mathcal{X}_f^\circ$ suffices to apply basically the same arguments as in part ii) of the proof of Theorem 3.1. In particular, due to the properties of the terminal set $\mathcal{X}_f$, see A3 of Assumption B.1, the candidate solution $\hat{\boldsymbol{u}}^+(x_0) = \{\hat{u}_1^*, \dots, \hat{u}_{N-1}^*, Kx_N^*\}$ provides a suboptimal input sequence for the successor state $x_0^+ = Ax_0 + B\hat{u}_0^*(x_0)$ that results in the open-loop state sequence $\boldsymbol{x}(\hat{\boldsymbol{u}}^+(x_0), x_0^+) = \{x_0^+, x_2^*, \dots, x_N^*, A_K x_N^*\}$ with $A_K x_N^* \in \mathcal{X}_f^\circ$. Then, exploiting the suboptimality of the candidate solution as well as the assumed properties of the terminal set and the terminal weight P, it is rather straightforward to show that for any $x(0) \in \mathbb{R}^n$

$$\hat{J}_N^*(x(k+1)) - \hat{J}_N^*(x(k)) \leq -\hat{\ell}(x(k), \hat{u}_0^*(x(k))) \;\; \forall\, k \in \mathbb{N}\,, \tag{C.8}$$

see Part ii) of the proof of Theorem 3.1. Thus, the associated value function will in this case decrease along trajectories of the closed-loop system for any initial condition $x(0) \in \mathbb{R}$. Again, $\hat{\ell}(x,u) \geq \ell(x,u) \geq \|x\|_Q^2$, while $\hat{J}_N^* : \mathbb{R}^n \to \mathbb{R}_+$ is a well-defined, positive definite, and radially unbounded function due to the design of the relaxed barrier functions. Hence, it can be used as a Lyapunov function for the closed-loop system (3.19), proving in this case global asymptotic stability of the origin.

C.6 Proof of Lemma 3.3

Based on Taylor's Theorem (Nocedal and Wright, 1999, Theorem 2.1), we know that $\hat{B}_z(z) = \hat{B}_z(0) + [\nabla \hat{B}_z(0)]^\top z + \frac{1}{2} z^\top \nabla^2 \hat{B}_z(sz) z$ for some $s \in (0,1)$. Due the assumed recentering, the first two terms vanish, revealing that $\hat{B}_z(z) = \frac{1}{2} z^\top \nabla^2 \hat{B}_z(sz) z$ for some $s \in (0,1)$. Thus, it obviously holds that $\hat{B}_z(z) \leq \frac{1}{2} z^\top M' z \; \forall z \in \mathbb{R}^{n_z}$ for any $M' \in \mathbb{S}_{++}^{n_z}$ that is a global upper bound on the Hessian $\nabla^2 \hat{B}_z(z)$. When considering the quadratic relaxing function $\beta(\cdot;\delta)$ from (3.8), this Hessian is given by

$$\nabla^2 \hat{B}_z(z) = C^\top \mathrm{diag}(D_1(z), \ldots, D_q(z))\, C, \tag{C.9a}$$

$$\text{where } D_i(z) = \begin{cases} \frac{w_z^i}{(-C^i z + d^i)^2} & -C^i z + d^i > \delta \\ \frac{w_z^i}{\delta^2} & -C^i z + d^i \leq \delta\,. \end{cases} \tag{C.9b}$$

As $D_i(z) \leq \frac{w_z^i}{\delta^2} \; \forall z \in \mathbb{R}^{n_z}$, it follows immediately that $\nabla^2 \hat{B}_z(z) \preceq \frac{1}{\delta^2} C^\top \mathrm{diag}\,(w_z)\, C$ for any $z \in \mathbb{R}^{n_z}$. Combining this upper bound with the previous arguments, it follows that $\hat{B}_z(z) \leq z^\top M z \; \forall z \in \mathbb{R}^{n_z}$ with $M := \frac{1}{2\delta^2} C^\top \mathrm{diag}\,(w_z)\, C$. Note that the matrix M is guaranteed to be positive definite as the existence of $\bar{z} \neq 0$ with the property $\bar{z}^\top M \bar{z} = 0$ would directly imply $C\bar{z} = 0$ and therefore $\kappa \bar{z} \in \mathcal{P}$ for any $\kappa \in \mathbb{R}_+$, which contradicts the fact that $\mathcal{P}$ as a polytope is bounded.

C.7 Proof of Theorem 3.3

The proof is very similar to the one of Theorem 3.2, with the exception that the restriction of the terminal state to a suitable set around the origin (and hence the assumption of controllability) is not required anymore.

First note that the cost function $\hat{J}_N : \mathbb{R}^{Nm} \times \mathbb{R}^n \to \mathbb{R}_+$ is globally defined due to the relaxed state and input constraints and the choice of a purely quadratic cost function. In combination with the strict convexity properties implied by the recentering and the positive definiteness of R, this shows that a unique and well-defined solution $\hat{\boldsymbol{u}}^*(x)$ exists for any $x \in \mathbb{R}^n$. As discussed in the following, it is furthermore straightforward to show that the value function $\hat{J}_N^*(x(k))$ decreases under the applied MPC feedback for all $x(k) \in \mathbb{R}^n$. To this end, consider an arbitrary $x_0 \in \mathbb{R}^n$ and let $\hat{\boldsymbol{u}}^*(x_0) = \{\hat{u}_0^*, \ldots, \hat{u}_{N-1}^*\}$ and $\boldsymbol{x}^*(x_0) = \{x_0, x_1^*, \ldots, x_N^*\}$ denote the associated optimal open-loop input and state sequences. Furthermore, define the corresponding suboptimal input and state sequences $\hat{\boldsymbol{u}}^+(x_0) = \{\hat{u}_1^*, \ldots, \hat{u}_{N-1}^*, Kx_N^*\}$ and $\boldsymbol{x}^+(x_0) = \{x_1^*, \ldots, x_N^*, A_K x_N^*\}$ for the successor state $x_0^+ = Ax_0 + B\hat{u}_0^*$, where $A_K := A + BK$ and K is chosen according to (3.34). Then, the choice of the terminal cost matrix P ensures that

$$\hat{J}_N(\hat{\boldsymbol{u}}^+(x_0), x_0^+) - \hat{J}_N^*(x_0) \tag{C.10a}$$

$$= -\hat{\ell}(x_0, \hat{u}_0^*(x_0)) + \|A_K x_N^*\|_P^2 - \|x_N^*\|_P^2 + \|x_N^*\|_Q^2 + \|Kx_N^*\|_R^2 + \varepsilon \hat{B}_K(x_N^*) \tag{C.10b}$$

$$\leq -\hat{\ell}(x_0, \hat{u}_0^*(x_0))\,, \tag{C.10c}$$

where we used the global quadratic bound on $\hat{B}_K(\cdot)$ from Lemma 3.4 and the choice of P according to (3.34b). As obviously $\hat{J}_N^*(x_0^+) - \hat{J}_N^*(x_0) \leq \hat{J}_N(\hat{\boldsymbol{u}}^+(x_0), x_0^+) - \hat{J}_N^*(x_0)$ and due to the fact that $x_0 \in \mathbb{R}^n$ was arbitrary, this shows that

$$\hat{J}_N^*(x(k+1)) - \hat{J}_N^*(x(k)) \leq -\hat{\ell}(x(k), \hat{u}_0^*(x(k))) \ \ \forall\, k \in \mathbb{N}, \tag{C.11}$$

for any $x(0) \in \mathbb{R}^n$. Again, $\hat{J}_N^* : \mathbb{R}^n \to \mathbb{R}_+$ is a positive definite and radially unbounded function, while $\hat{\ell}(x, u) \geq \ell(x, u) \geq \|x\|_Q^2$ with $Q \in \mathbb{S}_{++}^n$. Thus, the value function can be used as a Lyapunov function that proves global asymptotic stability of the origin of the closed-loop system (3.19).

C.8 Proof of Theorem 3.4

Due to Assumption 3.5, there exist for any $x_0 \in \mathbb{R}^n$ and any input sequence $\hat{\boldsymbol{u}}(x_0)$ with resulting terminal state $x_N = x_N(\hat{\boldsymbol{u}}(x_0), x_0)$ suitable tail sequences $\boldsymbol{v}(x_N)$ and $\boldsymbol{z}(x_N)$ that will steer the predicted system state from x_N to the origin in T steps. Moreover, as the input and state constraint barrier functions are relaxed, see Assumption 3.1, both the stage and terminal cost are always well defined. Hence, also $\hat{\boldsymbol{u}}^*(x_0)$ and $\hat{J}_N^*(x_0)$ are defined for any $x_0 \in \mathbb{R}^n$, which shows that problem (3.16) always admits a feasible solution.
As before, we now show that the value function $\hat{J}_N^*(x(k))$ will continuously decrease under the applied feedback for any initial condition $x(0) \in \mathbb{R}^n$. To this end, consider an arbitrary $x_0 \in \mathbb{R}^n$ and let $\hat{\boldsymbol{u}}^*(x_0) = \{\hat{u}_0^*, \ldots, \hat{u}_{N-1}^*\}$ and $\boldsymbol{x}^*(x_0) = \{x_0, x_1^*, \ldots, x_N^*\}$ denote the associated optimal open-loop input and state sequences. Furthermore, let $\boldsymbol{v}^*(x_0) := \boldsymbol{v}(x_N^*(x_0))$ and $\boldsymbol{z}^*(x_0) := \boldsymbol{z}(x_N^*(x_0))$ be the corresponding tail sequences for the resulting predicted terminal state $x_N^*(x_0) = x_N(\hat{\boldsymbol{u}}^*(x_0), x_0)$, cf. Assumption 3.5. Consider now the successor state $x_0^+ = Ax_0 + B\hat{u}_0^*$ and note that $\hat{\boldsymbol{u}}^+(x_0) = \{\hat{u}_1^*, \ldots, \hat{u}_{N-1}^*, v_0^*\}$ and $\boldsymbol{v}^+(x_0) = \{v_1^*, \ldots, v_{T-1}^*, 0\}$ are suboptimal input sequences which steer the state from x_0^+ to the origin in a finite number of $N+T-1$ steps. The resulting state sequences are given by $\boldsymbol{x}^+(x_0) = \{x_1^*, \ldots, x_N^*, z_1^*\}$ and $\boldsymbol{z}^+(x_0) = \{z_1^*, \ldots, z_{T-1}^*, 0\}$. Using the above suboptimal input and state sequences as well as the fact that the tail sequences within the terminal cost are simply appended by zero values, it is straightforward to show that

$$\hat{J}_N(\hat{\boldsymbol{u}}^+(x_0), x_0^+) - \hat{J}_N^*(x_0) = -\hat{\ell}(x_0, \hat{u}_0^*(x_0)). \tag{C.12}$$

As it obviously holds that $\hat{J}_N^*(x_0^+) - \hat{J}_N^*(x_0) \leq \hat{J}_N(\hat{\boldsymbol{u}}^+(x_0), x_0^+) - \hat{J}_N^*(x_0)$, and due to the fact that $x_0 \in \mathbb{R}^n$ was arbitrary, this shows that for any $x(0) \in \mathbb{R}^n$

$$\hat{J}_N^*(x(k+1)) - \hat{J}_N^*(x(k)) \leq -\hat{\ell}(x(k), \hat{u}_0^*(x(k))) \ \ \forall\, k \in \mathbb{N}. \tag{C.13}$$

Again, $\hat{J}_N^* : \mathbb{R}^n \to \mathbb{R}_+$ is positive definite and radially unbounded, while $\hat{\ell}(x, u) \geq \ell(x, u) \geq \|x\|_Q^2$ with $Q \in \mathbb{S}_{++}^n$. Thus, by (C.13), the value function represents a Lyapunov function for the closed-loop system, proving global asymptotic stability of the origin.

C.9 Proof of Lemma 3.5

The claim $\hat{J}_N^*(x) \leq \hat{J}_\infty^*(x)$ follows trivially from optimality and the fact that the stage cost $\hat{\ell} : \mathbb{R}^n \times \mathbb{R}^m \to \mathbb{R}_+$ is positive definite in both of its arguments.
Furthermore, the quadratic upper bounds stated in Lemma 3.3 ensure that

$$\hat{\ell}(x,u) \leq \bar{\ell}(x,u) := \|x\|^2_{Q+\varepsilon M_\mathrm{x}} + \|u\|^2_{R+\varepsilon M_\mathrm{u}} \;\forall\; (x,u) \in \mathbb{R}^n \times \mathbb{R}^m. \tag{C.14}$$

If we now consider the infinite-horizon LQR problem related to the quadratic upper bound on the right-hand side of(C.14), i.e.,

$$\begin{aligned} \bar{J}_\infty^*(x) = \min_u \sum_{k=0}^{\infty} \bar{\ell}(x_k, u_k) && \text{(C.15a)} \\ \text{s.t. } x_{k+1} = Ax_k + Bu_k,\; x_0 = x\,, && \text{(C.15b)} \end{aligned}$$

we know that the associated optimal solution can be directly characterized by the pair (K,P) obtained from the discrete-time algebraic Riccati equation

$$\begin{aligned} K &= -\left(R + B^\top PB + \varepsilon M_\mathrm{u}\right)^{-1} B^\top PA && \text{(C.16a)} \\ P &= (A+BK)^\top P(A+BK) + K^\top(R+\varepsilon M_\mathrm{u})K + Q + \varepsilon M_\mathrm{x}\,, && \text{(C.16b)} \end{aligned}$$

which is nothing else than (3.34). In particular, it holds that $\bar{J}_\infty^*(x) = x^\top Px$ for any $x \in \mathbb{R}^n$. On the other hand, letting $\bar{\boldsymbol{u}}(x) = \{\bar{u}_0, \bar{u}_1, \ldots,\}$ and $\bar{\boldsymbol{x}}(x) = \{\bar{x}_0, \bar{x}_1, \ldots,\}$ denote the corresponding optimal input and state sequences, it follows immediately from (C.14) that for any $x \in \mathbb{R}^n$,

$$\hat{J}_\infty^*(x) \leq \sum_{k=0}^{\infty} \hat{\ell}(\bar{x}_k, \bar{u}_k) \leq \sum_{k=0}^{\infty} \bar{\ell}(\bar{x}_k, \bar{u}_k) = x^\top Px\,, \tag{C.17}$$

which proves our second claim and thus completes the proof. Note that a positive definite solution $P \in \mathbb{S}^n_{++}$ to (C.16), respectively (3.34), always exists due to positive definiteness of Q and R and the assumption that (A,B) is stabilizable.

C.10 Proof of Theorem 3.5

While large parts of the proof follow similar arguments as in (Grüne, 2012), we nevertheless present it here in a self-contained form for the sake of completeness. First we show the claimed stability result and derive the bound for the critical horizon $\bar{N} \in \mathbb{N}_+$ given in (3.44). In the second part, we then use the key result from the first part in order to prove (3.45) with α_N as defined in the theorem.

Part 1) For given $x \in \mathbb{R}^n$, let $\hat{\boldsymbol{u}}(x) = \{\hat{u}_0^*, \hat{u}_1^*, \ldots, \hat{u}_{N-1}^*\}$ and $\boldsymbol{x}^*(x) = \{x_0^* = x, x_1^*, \ldots, x_{N-1}^*\}$ denote the optimal open-loop input and state sequences associated to problem (3.40) for a given prediction horizon $N \geq 2$. Furthermore, let us introduce $\hat{\ell}_k := \hat{\ell}(x_k^*, \hat{u}_k^*)$ as well as $V_p(x) := \hat{J}_p^*(x)$ for $p = 1, \dots, N$, which in particular implies that $V_N(x) = \sum_{k=0}^{N-1} \hat{\ell}_k$. Moreover, define

$$\hat{\ell}^*(x) := \min_{u \in \mathbb{R}^m} \hat{\ell}(x,u) = \|x\|_Q^2 + \varepsilon \hat{B}_\mathrm{x}(x), \tag{C.18}$$

which by design has the property that $\hat{\ell}_k \geq \hat{\ell}^*(x_k^*)$. In addition, $Q \in \mathbb{S}_{++}^n$ and the upper bound on the relaxed barrier function $\hat{B}_\mathrm{x}(\cdot)$ given in Lemma 3.4 ensure that

$$\|x\|_Q^2 \leq \hat{\ell}^*(x) \leq \|x\|_{Q+\varepsilon M_\mathrm{x}}^2 \tag{C.19}$$

for any $x \in \mathbb{R}^n$. In combination with (3.43) from Lemma 3.5, this reveals that for any $N \in \mathbb{N}_+$ and any $x \in \mathbb{R}^n$

$$V_N(x) \leq \gamma \hat{\ell}^*(x), \tag{C.20}$$

for any $\gamma \geq \lambda_{\max}(PQ^{-1})$. Note that it follows directly from the characterization of P in (3.34) that $\gamma \in (1, \infty)$. A direct consequence of this result is that

$$V_{N-p+1}(x_p^*) \leq \gamma \hat{\ell}^*(x_p^*) \leq \gamma \hat{\ell}_p \tag{C.21}$$

for any $p = 1, \dots, N-1$. Let us now denote the optimal input sequence related to $V_2(x_{N-1}^*)$ by $\tilde{\boldsymbol{u}}^* = \{\tilde{u}_0^*, \tilde{u}_1^*\}$ and consider $\bar{\boldsymbol{u}}(x) = \{\hat{u}_1^*, \dots, \hat{u}_{N-2}^*, \tilde{u}_0^*, \tilde{u}_1^*\}$ as a candidate solution for the successor state $x^+ = Ax + B\hat{u}_0^* = x_1^*$. Then, the suboptimality of $\bar{\boldsymbol{u}}(x)$ in combination with the above arguments ensures that

$$V_N(x^+) \leq \hat{J}_N(\bar{\boldsymbol{u}}(x), x) = \sum_{k=1}^{N-2} \hat{\ell}_k + V_2(x_{N-1}^*) \tag{C.22a}$$

$$= \sum_{k=0}^{N-1} \hat{\ell}_k - \hat{\ell}_0 - \hat{\ell}_{N-1} + V_2(x_{N-1}^*) \tag{C.22b}$$

$$\leq V_N(x) - \hat{\ell}_0 + (\gamma - 1)\hat{\ell}_{N-1}, \tag{C.22c}$$

where we inserted $V_N(x) = \sum_{k=0}^{N-1} \hat{\ell}_k$ and exploited the fact that $V_2(x_{N-1}^*) \leq \gamma \hat{\ell}_{N-1}$ due to (C.21). Following the arguments of Grüne (2012), we now derive an upper bound for $\hat{\ell}_{N-1}$ in terms of $\hat{\ell}_0$. To this end, note that the dynamic programming principle and again (C.20) guarantee that for any $p = 0, \dots, N-2$ the following holds

$$\sum_{k=p}^{N-1} \hat{\ell}_k = V_{N-p}(x_p^*) \leq \gamma \hat{\ell}^*(x_p^*) \leq \gamma \hat{\ell}_p. \tag{C.23}$$

By subtracting $\hat{\ell}_p$ on both sides and dividing by $(\gamma - 1)$, this implies that

$$\hat{\ell}_p \geq \frac{1}{\gamma - 1} \sum_{k=p+1}^{N-1} \hat{\ell}_k, \tag{C.24}$$

by which we obtain

$$\hat{\ell}_p + \sum_{k=p+1}^{N-1} \hat{\ell}_k \geq \frac{1}{\gamma - 1} \sum_{k=p+1}^{N-1} \hat{\ell}_k + \sum_{k=p+1}^{N-1} \hat{\ell}_k = \frac{\gamma}{\gamma - 1} \sum_{k=p+1}^{N-1} \hat{\ell}_k. \tag{C.25}$$

Writing down this inequality in an inductive fashion reveals that

$$p = N-2: \qquad \hat{\ell}_{N-2} + \hat{\ell}_{N-1} \geq \frac{\gamma}{\gamma - 1} \hat{\ell}_{N-1} \tag{C.26a}$$

$$p = N-3: \qquad \hat{\ell}_{N-3} + \hat{\ell}_{N-2} + \hat{\ell}_{N-1} \geq \frac{\gamma}{\gamma - 1} \left(\hat{\ell}_{N-2} + \hat{\ell}_{N-1} \right) \geq \left(\frac{\gamma}{\gamma - 1} \right)^2 \hat{\ell}_{N-1} \tag{C.26b}$$

$$\vdots$$

$$p = 1: \qquad \hat{\ell}_1 + \hat{\ell}_2 + \ldots + \hat{\ell}_{N-1} \geq \frac{\gamma}{\gamma - 1} \left(\hat{\ell}_2 + \ldots + \hat{\ell}_{N-1} \right) \geq \left(\frac{\gamma}{\gamma - 1} \right)^{N-2} \hat{\ell}_{N-1} \tag{C.26c}$$

and thus

$$\hat{\ell}_{N-1} \leq \left(\frac{\gamma - 1}{\gamma} \right)^{N-2} \sum_{k=1}^{N-1} \hat{\ell}_k \,. \tag{C.27}$$

On the other hand, (C.24) implies for $p = 0$ that $\sum_{k=1}^{N-1} \hat{\ell}_k \leq (\gamma - 1)\hat{\ell}_0$. Inserting this relation into (C.27) then finally results in

$$\hat{\ell}_{N-1} \leq \gamma \left(\frac{\gamma - 1}{\gamma} \right)^{N-1} \hat{\ell}_0 \,, \tag{C.28}$$

which provides the upper bound for $\hat{\ell}_{N-1}$ in terms of $\hat{\ell}_0$ that we were looking for. By combining now (C.22) and (C.28), we obtain

$$V_N(x^+) \leq V_N(x) - \alpha_N \hat{\ell}_0 \,, \qquad \alpha_N := 1 - \frac{(\gamma - 1)^N}{\gamma^{N-2}} \,. \tag{C.29}$$

As $\gamma \in (1, \infty)$, it holds by definition that $\alpha_N < 1$. Furthermore, it is straightforward to show that $\alpha_N > 0$ whenever

$$N > \frac{2 \ln(\gamma)}{\ln(\gamma) - \ln(\gamma - 1)} \,. \tag{C.30}$$

Obviously, this is satisfied for any N that is chosen according to (3.44). Assuming that this is the case, i.e. $N \geq \bar{N}$, and returning to the original notation, (C.29) allows to conclude that for any $x(0) \in \mathbb{R}^n$, the closed-loop value function will decrease according to

$$\hat{J}_N^*(x(k+1)) - \hat{J}_N^*(x(k)) \leq -\alpha_N \hat{\ell}(x(k), \hat{u}_0^*(x(k))) \tag{C.31}$$

with $\alpha_N \in (0, 1)$. As before $\hat{J}_N^* : \mathbb{R}^n \to \mathbb{R}_+$ is a positive definite and radially unbounded function, and $\hat{\ell}(x, u) \geq \ell(x, u) \geq \|x\|_Q^2$ with $Q \in \mathbb{S}_{++}^n$. Thus, whenever $N \geq \bar{N}$, then the value function related to (3.40) represents a suitable Lyapunov function, proving global asymptotic stability of the origin of the associated closed-loop system (3.19).

Part 2) We now want to prove our second claim, i.e., that (3.45) holds with α_N as defined above and that α_N is monotonically increasing with $\lim_{N \to \infty} \alpha_N = 1$. To this end, assume

$N \geq \bar{N}$, by which we know that (C.31) holds. Summing up over both sides and exploiting that $\lim_{k\to\infty} \hat{J}_N^*(x(k)) = 0$, we obtain

$$-\hat{J}_N^*(x(k)) \leq -\alpha_N \sum_{i=k}^{\infty} \hat{\ell}(x(k), \hat{u}_0^*(x(k))) = -\alpha_N \hat{J}_\infty^{\mathrm{cl}}(x(k)) \tag{C.32}$$

and thus, by virtue of Lemma 3.5,

$$\hat{J}_\infty^{\mathrm{cl}}(x) \leq \hat{J}_N^*(x)/\alpha_N \leq \hat{J}_\infty^*(x)/\alpha_N \tag{C.33}$$

for any $x = x(k) \in \mathbb{R}^n$, which is nothing else than (3.45). In order to prove the claimed monotonicity of the performance index α_N, note that

$$\alpha_{N+1} = 1 - \frac{(\gamma-1)^{N+1}}{\gamma^{N-1}} = 1 - \left(\frac{\gamma-1}{\gamma}\right)\frac{(\gamma-1)^N}{\gamma^{N-2}} = \alpha_N + \frac{(\gamma-1)^N}{\gamma^{N-1}} > \alpha_N\,. \tag{C.34}$$

In addition, it is also straightforward to verify that for $\gamma \in (1, \infty)$

$$\lim_{N\to\infty} \alpha_N = \lim_{N\to\infty}\left(1 - \frac{(\gamma-1)^N}{\gamma^{N-2}}\right) = \lim_{N\to\infty}\left(1 - \gamma^2\left(1-\frac{1}{\gamma}\right)^N\right) = 1\,. \tag{C.35}$$

Thus, also the second claim is proven, which concludes the proof.

C.11 Proof of Theorem 3.6

The proof consists of three parts. First we show that the relaxed barrier functions for the state and input constraints satisfy for any $k \in \mathbb{N}$ the inequalities

$$\varepsilon\hat{B}_{\mathrm{x}}(x(k)) \leq \hat{J}_N^*(x(k)) - x(k)^\top P_{\mathrm{uc}}^* x(k)\,, \tag{C.36a}$$
$$\varepsilon\hat{B}_{\mathrm{u}}(u(k)) \leq \hat{J}_N^*(x(k)) - x(k)^\top P_{\mathrm{uc}}^* x(k)\,, \tag{C.36b}$$

where $P_{\mathrm{uc}}^* \in \mathbb{S}_{++}^n$ is the solution to the discrete-time algebraic Riccati equation (DARE) associated to the unconstrained infinite-horizon LQR problem for the setup (A, B, Q, R), i.e., to (3.58) with $M_{\mathrm{x}} = 0$, $M_{\mathrm{u}} = 0$. In a second step, we use (C.36) for deriving a time-varying upper bound on the maximal constraint violations. Finally, we exploit the decrease of the value function in order to show the claimed monotonicity result.

Part 1. Based on Theorem 3.3 we know that property (3.60) holds. In particular, this ensures that the closed-loop system is asymptotically stable and that $\lim_{k\to\infty} \hat{J}_N^*(x(k)) = 0$. Summing up over all future sampling instants and using a telescoping sum on the left-hand side, we get that $\hat{J}_N^*(x(k)) \geq \sum_{i=k}^{\infty} \hat{\ell}(x(i), u(i))$, where $u(i) = \hat{u}_0^*(x(i))$. Furthermore, $\sum_{i=k}^{\infty} \hat{\ell}(x(i), u(i)) = \sum_{i=k}^{\infty} \|x(i)\|_Q^2 + \|u(i)\|_R^2 + \varepsilon \sum_{i=k}^{\infty} \hat{B}_{\mathrm{x}}(x(i)) + \hat{B}_{\mathrm{u}}(u(i))$, whereas it holds on the other hand that $\sum_{i=k}^{\infty} \|x(i)\|_Q^2 + \|u(i)\|_R^2 \geq x(k)^\top P_{\mathrm{uc}}^* x(k)$ due to optimality of the respective unconstrained infinite-horizon LQR solution. In combination this yields

$$\hat{J}_N^*(x(k)) \geq x(k)^\top P_{\mathrm{uc}}^* x(k) + \varepsilon \sum_{i=k}^{\infty} \hat{B}_{\mathrm{x}}(x(i)) + \hat{B}_{\mathrm{u}}(u(i))\,. \tag{C.37}$$

Since due to the recentering of the underlying barrier functions all terms in the sum on the right-hand side are guaranteed to be positive definite, we finally obtain (C.36).

Part 2. Based on these observations, let us now define the time-varying upper bound on the right-hand side of (C.36) for ease of notation as

$$\hat{\alpha}(k) := \hat{J}_N^*(x(k)) - x(k)^\top P_{\mathrm{uc}}^* x(k) \,. \tag{C.38}$$

Note that $\hat{\alpha}(k) \in \mathbb{R}_+$ by definition, see (C.37). As the relaxed barrier functions are positive definite and radially unbounded, property (C.36) directly implies that also the maximally possible violations of the associated input and state constraints are bounded. In particular, for any $\varepsilon \in \mathbb{R}_{++}$, $\delta \in \mathbb{R}_{++}$ and any realization of the system state $x(k) \in \mathbb{R}^n$, upper bounds for the maximal violations of state and input constraints are given by

$$\hat{z}_{\mathrm{x}}^i(k) = \max_{\xi} \left\{ C_{\mathrm{x}}^i \xi - d_{\mathrm{x}}^i \mid \varepsilon \hat{B}_{\mathrm{x}}(\xi) \leq \hat{\alpha}(k) \right\}, \quad i = 1, \ldots, q_{\mathrm{x}}, \tag{C.39a}$$

$$\hat{z}_{\mathrm{u}}^j(k) = \max_{v} \left\{ C_{\mathrm{u}}^j v - d_{\mathrm{u}}^j \mid \varepsilon \hat{B}_{\mathrm{u}}(v) \leq \hat{\alpha}(k) \right\}, \quad j = 1, \ldots, q_{\mathrm{u}}, \tag{C.39b}$$

where $\hat{\alpha}(k)$ is defined according to (C.38). Thus, (3.61) holds with the elements of the vectors $\hat{z}_{\mathrm{x}}(k) \in \mathbb{R}^{q_{\mathrm{x}}}$ and $\hat{z}_{\mathrm{u}}(k) \in \mathbb{R}^{q_{\mathrm{u}}}$ being given by (C.39). Note that the optimization problems in (C.39) are convex due to the fact that the involved barrier functions are convex. Moreover, they always admit a feasible and well-defined solution as the barrier functions are positive definite and continuous and $\alpha(k) \in \mathbb{R}_+$ for all $k \in \mathbb{N}$.

Part 3. In the following, we show that the upper bound $\hat{\alpha}(k)$ as defined in (C.38) is monotonically decreasing and that $\lim_{k\to\infty} \hat{\alpha}(k) = 0$. First, note that the value function decrease stated in (3.60) implies that

$$\hat{\alpha}(k+1) - \hat{\alpha}(k) \leq -\hat{\ell}(x(k), u(k)) + x(k)^\top P_{\mathrm{uc}}^* x(k) - x(k+1)^\top P_{\mathrm{uc}}^* x(k+1) \,. \tag{C.40}$$

Now recall $\hat{\ell}(x(k), u(k)) = \|x(k)\|_Q^2 + \|u(k)\|_R^2 + \varepsilon \hat{B}_{\mathrm{x}}(x(k)) + \varepsilon \hat{B}_{\mathrm{u}}(u(k))$ and note that $\|x(k)\|_Q^2 + \|u(k)\|_R^2 \geq x(k)^\top P_{\mathrm{uc}}^* x(k) - x(k+1)^\top P_{\mathrm{uc}}^* x(k+1)$ due to the principle of optimality. Inserting these relations into (C.40) finally reveals that

$$\hat{\alpha}(k+1) - \hat{\alpha}(k) \leq -\varepsilon(\hat{B}_{\mathrm{x}}(x(k)) + \hat{B}_{\mathrm{u}}(u(k))) \,. \tag{C.41}$$

Since $\hat{B}_{\mathrm{x}}(\cdot)$ and $\hat{B}_{\mathrm{u}}(\cdot)$ are positive definite, this ensures that $\alpha(k)$ decreases strictly monotonically over time. Moreover, it follows directly from (C.38) and the convergence of the closed-loop system state to the origin that $\lim_{k\to\infty} \hat{\alpha}(k) = 0$. Now, as $\hat{\alpha}(k)$ decreases monotonically, the same holds for the upper bounds $\hat{z}_{\mathrm{x}}^i(k)$ and $\hat{z}_{\mathrm{u}}^j(k)$ given in (C.39), which proves the claimed strict monotonicity result. Finally, as both $x(k)$ and $u(k)$ asymptotically converge to the origin (see Theorem 3.3), there always exists a finite $k_0 \in \mathbb{N}$ such that $\hat{z}_{\mathrm{x}}(k) \leq 0$, $\hat{z}_{\mathrm{u}}(k) \leq 0$ for any $k \geq k_0$. Note that this last fact may also be deduced from inserting $\lim_{k\to\infty} \hat{\alpha}(k) = 0$ into (C.39), noting thereby that $\hat{z}_{\mathrm{x}}(k)$ and $\hat{z}_{\mathrm{u}}(k)$ will asymptotically converge to the strictly negative vectors $-d_{\mathrm{x}}$ and $-d_{\mathrm{u}}$.

C.12 Proof of Theorem 3.7

We prove the claimed result for the case of zero constraint violations, i.e. $\hat{z}_{\mathrm{x,tol}} = 0$, $\hat{z}_{\mathrm{u,tol}} = 0$, which then automatically comprises also all cases with positive tolerances. The proof consists of two parts. First, we show that there exists for any $\delta \in \mathbb{R}_{++}$ a compact and nonempty set $\hat{\mathcal{X}}_N(\delta) \subseteq \mathcal{X}$ such that for any $x(0) \in \hat{\mathcal{X}}_N(\delta)$, the state and input constraints will be satisfied for all future sampling instants, that is, (3.65) will hold with $\hat{z}_{\mathrm{x,tol}} = 0$, $\hat{z}_{\mathrm{u,tol}} = 0$. In fact, this set will turn out to be nothing else than the set $\hat{\mathcal{X}}_N(\delta)$ from Corollary 3.2. In a second step, we then show that for any compact set $\mathcal{X}_0 \subseteq \mathcal{X}_N^\circ$ there exists $\bar{\delta}_0 \in \mathbb{R}_{++}$ such that the set inclusion $\mathcal{X}_0 \subseteq \hat{\mathcal{X}}_N(\delta)$ whenever $\delta \leq \bar{\delta}_0$. In order to make the influence of the relaxation parameter on the value function more explicit, we write in the following $\hat{J}_N^*(x;\delta)$ instead of $\hat{J}_N^*(x)$.

Part 1. Inspired by the above arguments and the key idea behind Corollary 3.2, let us for a given relaxation parameter $\delta \in \mathbb{R}_{++}$ define the scalars $\bar{\beta}_{\mathrm{x}}(\delta), \bar{\beta}_{\mathrm{u}}(\delta), \bar{\beta}(\delta) \in \mathbb{R}_{++}$ as

$$\bar{\beta}_{\mathrm{x}}(\delta) := \min_{i,\xi}\{\hat{B}_{\mathrm{x}}(\xi) \,|\, C_{\mathrm{x}}\xi = d_{\mathrm{x}},\; C_{\mathrm{x}}^i\xi = d_{\mathrm{x}}^i\},\; i = 1,\ldots,q_{\mathrm{x}}\,, \tag{C.42a}$$

$$\bar{\beta}_{\mathrm{u}}(\delta) := \min_{j,v}\{\hat{B}_{\mathrm{u}}(v) \,|\, C_{\mathrm{u}}v = d_{\mathrm{u}},\; C_{\mathrm{u}}^j v = d_{\mathrm{u}}^j\},\; j = 1,\ldots,q_{\mathrm{u}}\,, \tag{C.42b}$$

$$\bar{\beta}(\delta) := \min\{\bar{\beta}_{\mathrm{x}}(\delta), \bar{\beta}_{\mathrm{u}}(\delta)\}\,. \tag{C.42c}$$

Note that $\bar{\beta}_{\mathrm{x}}(\delta)$ and $\bar{\beta}_{\mathrm{u}}(\delta)$ can be interpreted as lower bounds for the values that are attained by $\hat{B}_{\mathrm{x}} : \mathbb{R}^n \to \mathbb{R}_+$ and $\hat{B}_{\mathrm{u}} : \mathbb{R}^m \to \mathbb{R}_+$ on the boundaries of the constraint sets $\mathcal{X}$ and $\mathcal{U}$, respectively. As a consequence, the $\bar{\beta}(\delta)$-sublevel sets of $\hat{B}_{\mathrm{x}}(\cdot)$ and $\hat{B}_{\mathrm{u}}(\cdot)$ will always be contained within the sets $\mathcal{X}$ and $\mathcal{U}$, see Lemma 3.1 on page 44. Based on this observation and inspired by the proof of Theorem 3.6, we introduce

$$\hat{\alpha}(x;\delta) := \hat{J}_N^*(x;\delta) - x^\top P_{\mathrm{uc}}^*\, x\,, \tag{C.43}$$

and define the set $\hat{\mathcal{X}}_N(\delta) \subseteq \mathbb{R}^n$ as

$$\hat{\mathcal{X}}_N(\delta) := \left\{x \in \mathbb{R}^n \,|\, \hat{\alpha}(x;\delta) \leq \varepsilon\bar{\beta}(\delta)\right\}\,. \tag{C.44}$$

Here, $\bar{\beta}(\delta)$ is given by (C.42) and $P_{\mathrm{uc}}^* \in \mathbb{S}_{++}^n$ refers again to the DARE solution related to the infinite-horizon LQR problem for (A, B, Q, R). Suppose now that $x(0) \in \hat{\mathcal{X}}_N(\delta)$. Based on the proof of Theorem 3.6, we know that $\varepsilon\hat{B}_{\mathrm{x}}(x(k)) \leq \hat{\alpha}(x(k);\delta)$, $\varepsilon\hat{B}_{\mathrm{u}}(u(k)) \leq \hat{\alpha}(x(k);\delta)$, see (C.36) as well as the definition of $\hat{\alpha}(\cdot;\delta)$ in (C.43). In addition, $\hat{\alpha}(x(k);\delta) \leq \hat{\alpha}(x(0);\delta)$ for all $k \in \mathbb{N}$, see (C.41). So if $x(0) \in \hat{\mathcal{X}}_N(\delta)$, then $\hat{B}_{\mathrm{x}}(x(k)) \leq \bar{\beta}(\delta)$, $\hat{B}_{\mathrm{u}}(u(k)) \leq \bar{\beta}(\delta)$, which by the definition of $\bar{\beta}(\delta)$ and Lemma 3.1 implies that $x(k) \in \mathcal{X}$, $u(k) \in \mathcal{U}$ for all $k \in \mathbb{N}$. Thus, for given $\delta \in \mathbb{R}_{++}$, (3.65) holds with $\hat{z}_{\mathrm{x,tol}} = 0$, $\hat{z}_{\mathrm{u,tol}} = 0$ whenever $x(0) \in \hat{\mathcal{X}}_N(\delta)$.

Part 2. In the following, we are going to prove our second claim, namely the existence of $\bar{\delta}_0 \in \mathbb{R}_{++}$ with the property that $\mathcal{X}_0 \subset \hat{\mathcal{X}}_N(\delta)$ for any $\delta \leq \bar{\delta}_0$. Based on the introduced notation, it obviously holds that for arbitrary $x_0 \in \mathbb{R}^n$

$$x_0 \in \hat{\mathcal{X}}_N(\delta) \;\Leftrightarrow\; \hat{\alpha}(x_0;\delta) \leq \varepsilon\bar{\beta}(\delta)\,. \tag{C.45}$$

We now proceed as follows: first we show in *a)* that for any *fixed* $x_0 \in \mathcal{X}_0 \subseteq \mathcal{X}_N^\circ$, there exists a $\delta_0'(x_0)$ such that the right-hand side of (C.45) is satisfied for all $\delta \leq \delta_0'(x_0)$; in part *b)*, we then show the claimed result for $\mathcal{X}_0$ by *uniformly* choosing $\bar{\delta}_0 = \min_{x_0 \in \mathcal{X}_0} \delta_0'(x_0)$.
a) The key idea underlying the following arguments is to show that for any $x_0 \in \mathcal{X}_0$, the term $\hat{\alpha}(x_0;\delta)$ will stay bounded when δ is decreased below a certain threshold. As, on the other hand, $\bar{\beta}(\delta)$ grows without bound for $\delta \to 0$, this reveals that the right-hand side of (C.45) can always be satisfied by choosing δ sufficiently small.
Consider a fixed $x_0 \in \mathcal{X}_0 \subseteq \mathcal{X}_N^\circ$ and let $\tilde{U}^*(x_0)$ and $\tilde{J}_N^*(x_0) = \tilde{J}_N(\tilde{U}^*(x_0), x_0)$ denote the optimal solution and value function associated to the related *nonrelaxed* barrier function based problem formulation with additional zero terminal state constraint, i.e.,

$$\tilde{J}_N^*(x) = \min_U \sum_{k=0}^{N-1} \tilde{\ell}(x_k, u_k) \tag{C.46a}$$

$$\text{s.t. } x_{k+1} = Ax_k + Bu_k,\ x_N = 0,\ x_0 = x\,. \tag{C.46b}$$

Here, $\tilde{\ell}(x,u) := \|x\|_Q^2 + \|u\|_R^2 + \varepsilon B_\mathrm{u}(u) + \varepsilon B_\mathrm{x}(x)$, where $B_\mathrm{x} : \mathcal{X}^\circ \to \mathbb{R}_+$ and $B_\mathrm{u} : \mathcal{U}^\circ \to \mathbb{R}_+$ refer to the nonrelaxed versions of the relaxed logarithmic barrier functions $\hat{B}_\mathrm{u} : \mathbb{R}^m \to \mathbb{R}_+$ and $B_\mathrm{x} : \mathbb{R}^n \to \mathbb{R}_+$. It can be shown that under the assumption that $\begin{bmatrix} A^{N-1}B & \cdots & AB & B \end{bmatrix}$ has full row rank n, the optimal solution $\tilde{U}^*(x_0)$ associated to (C.46) is unique for any $x_0 \in \mathcal{X}_N^\circ$ and, moreover, that $\tilde{U}^* : \mathcal{X}_N^\circ \to \mathbb{R}^{Nm}$ is continuous in the system state x_0, see Lemma B.5 in Appendix B.7. Note that the additional constraint $x_N = 0$ is required in order to ensure that $\tilde{J}_N^*(x_0)$ does not depend on the relaxation parameter δ via the quadratic terminal cost in (3.56) and (3.58). By definition, $\tilde{U}^*(x_0)$ exists for any $x_0 \in \mathcal{X}_0 \subseteq \mathcal{X}_N^\circ$ and results in strictly feasible state and input sequences $\{\tilde{x}_0, \tilde{x}_1, \ldots \tilde{x}_N\}$ and $\{\tilde{u}_0, \tilde{u}_1, \ldots \tilde{u}_{N-1}\}$, respectively. Now define

$$\delta_0(x_0) := \min_{i,j,k} \left\{ -C_\mathrm{x}^i \tilde{x}_k + d_\mathrm{x}^i, -C_\mathrm{u}^j \tilde{u}_k + d_\mathrm{u}^j \right\} \tag{C.47}$$

for $i = 1, \ldots, q_\mathrm{x}$, $j = 1, \ldots, q_\mathrm{u}$, $k = 0, \ldots, N-1$, which characterizes the minimal distance of the nonrelaxed open-loop trajectories to the boundaries of the respective constraint sets. Due to the definition of the relaxed barrier functions, the resulting cost function values of relaxed and nonrelaxed formulation will be identical, i.e., $\hat{J}_N(\tilde{U}^*(x_0), x_0; \delta) = \tilde{J}_N^*(x_0)$, whenever we choose $\delta \leq \delta_0(x_0)$, see also (Hauser and Saccon, 2006). In addition, it always holds due to optimality that $\hat{J}_N^*(x_0;\delta) \leq \hat{J}_N(\tilde{U}^*(x_0), x_0; \delta)$ and, thus, $\hat{J}_N^*(x_0;\delta) \leq \tilde{J}_N^*(x_0)$ for any $\delta \leq \delta_0(x_0)$. This shows that the optimal value function, and hence also $\hat{\alpha}(x_0;\delta)$ in (C.45), will stay bounded when we decrease the relaxation parameter δ below $\delta_0(x_0)$. In particular, it holds that

$$\hat{\alpha}(x_0;\delta) = \hat{J}_N^*(x_0;\delta) - x_0^\top P_\mathrm{uc}^* x_0 \leq \tilde{J}_N^*(x_0) - x_0^\top P_\mathrm{uc}^* x_0 \tag{C.48}$$

whenever $\delta \leq \delta_0(x_0)$. As the smooth and convex problem formulation ensures that $\tilde{U}^*(x_0)$ will be continuous, see Lemma B.5 in the Appendix, $\delta_0(x_0)$ is also continuous.

Let us now choose $\delta_0'(x_0) \leq \delta_0(x_0)$ in such a way that the right-hand side of (C.45) is satisfied for all $\delta \leq \delta_0'(x_0)$, which then implies that $x_0 \in \hat{\mathcal{X}}_N(\delta)$. In particular, we define

$$\delta_0'(x_0) := \max\left\{\delta \in \mathbb{R}_{++} \,\middle|\, \delta \leq \delta_0(x_0),\ \tilde{J}_N^*(x_0) - x_0^\top P_{\text{uc}}^* x_0 \leq \varepsilon\bar{\beta}(\delta)\right\}. \tag{C.49}$$

In combination with (C.48), this definition ensures that the right-hand side of (C.45) will be satisfied for any $\delta \leq \delta_0'(x_0)$. Hence, $x_0 \in \hat{\mathcal{X}}_N(\delta)$ for any $\delta \leq \delta_0'(x_0)$. Note that we may picture the construction of $\delta_0'(x_0)$ as decreasing δ from $\delta(x_0)$ until also the second condition in the foregoing definition of $\delta_0'(x_0)$ is satisfied. As $\bar{\beta}(\delta)$ grows without bound for decreasing δ, this can always be achieved, which shows that $\delta'(x_0)$ exists for any $x_0 \in \mathcal{X}_N^\circ$. Moreover, since $\bar{\beta}(\delta)$ is continuous in δ and both $\delta_0(x_0)$ and $\tilde{J}_N^*(x_0)$ are continuous in x_0, $\delta_0'(x_0)$ is also continuous. Note that continuity of $\bar{\beta}(\delta)$ follows from noting that the constraints in (C.42a) and (C.42b) are independent from δ and that the relaxed barrier functions $\hat{B}_\text{x}(x)$ and $\hat{B}_\text{u}(u)$ are continuous in both x (respectively u) and δ and possess, in addition, compact sublevel sets for any fixed value of δ, see (Fiacco, 1983, Theorem 2.2) as well as (Bonnans and Shapiro, 2000, Proposition 4.4).

b) Consider now the compact set $\mathcal{X}_0 \subseteq \mathcal{X}_N^\circ$ and define $\bar{\delta}_0 \in \mathbb{R}_{++}$ as $\bar{\delta}_0 = \min_{x_0 \in \mathcal{X}_0} \delta_0'(x_0)$, where $\delta_0'(x_0)$ is given by (C.49). Due to the continuity of $\delta_0'(x_0)$ and the compactness of $\mathcal{X}_0$, this value always exists by virtue of the Weierstraß extreme value theorem. Furthermore, the above arguments ensure that for any $\delta \leq \bar{\delta}_0$ it is guaranteed that $x_0 \in \hat{\mathcal{X}}_N(\delta)$ whenever $x_0 \in \mathcal{X}_0$. Thus, $\mathcal{X}_0 \subseteq \hat{\mathcal{X}}_N(\delta)$ for any $\delta \leq \bar{\delta}_0$, which proves our second claim. As mentioned above, the discussed zero tolerance case comprises all other cases with nonzero tolerances $\hat{z}_{\text{x,tol}} \in \mathbb{R}_+^{q_x}$, $\hat{z}_{\text{u,tol}} \in \mathbb{R}_+^{q_u}$, and the proof is complete.

C.13 Proof of Lemma 3.6

Let $U := \begin{bmatrix} u_0^\top & \cdots & u_{N-1}^\top \end{bmatrix}^\top \in \mathbb{R}^{Nm}$ and consider the vectorized formulation of problem (3.71), cf. (3.46). Due to the discussed design of the barrier functions and $Q \in \mathbb{S}_{++}^n$, $R \in \mathbb{S}_{++}^m$, the overall cost function $\hat{J}_N : \mathbb{R}^{Nm} \times \mathbb{R}^n \to \mathbb{R}_+$ is a globally defined, positive definite, twice continuously differentiable, and strongly convex function. Consequently, the associated optimizer $\hat{U}^* : \mathbb{R}^n \to \mathbb{R}^{Nm}$ is for any $x \in \mathbb{R}^n$ uniquely characterized by the condition that $\nabla_U \hat{J}_N(\hat{U}^*(x), x) = 0$. By applying the implicit function theorem (see for example (Polyak, 1987, Theorem 2)) to this vector equation and exploiting strong convexity, i.e., the fact that the Hessian $\nabla_U^2 \hat{J}_N(U, x)$ is positive definite and therefore invertible for any $(U, x) \in \mathbb{R}^{Nm} \times \mathbb{R}^n$, it follows immediately that there always exists a unique and continuously differentiable optimizer $\hat{U}^* : \mathbb{R}^n \to \mathbb{R}^{Nm}$ that leads to a twice continuously differentiable value function $\hat{J}_N^*(x) = \hat{J}_N(\hat{U}^*(x), x)$. Moreover, since minimizing over the first argument preserves convexity, the value function will always be convex in x.

In a second step, we now show that $\hat{J}_N^* : \mathbb{R}^n \to \mathbb{R}_+$ satisfies in addition the quadratic upper and lower bounds stated in (3.75). To this end, note that optimality arguments yield that for any $x \in \mathbb{R}^n$

$$x^\top P_{\text{uc}}^* x \leq \hat{J}_N^*(x) \leq \hat{J}_N(\bar{U}, x), \tag{C.50}$$

where $P_{uc^*} \in \mathbb{S}^n_{++}$ denotes the cost matrix related to the unconstrained infinite-horizon LQR problem and $\bar{U} \in \mathbb{R}^{Nm}$ refers to an arbitrary input vector. A detailed derivation of the lower bound can be found in the proof of Theorem 3.6, see in particular (C.37) on page 209. The upper bound follows directly from suboptimality of $\bar{U} \in \mathbb{R}^{Nm}$. Choosing for example $\bar{u}_k = K(A+BK)^k x$ for some stabilizing $K \in \mathbb{R}^{m\times n}$ and using the quadratic upper bounds from Lemma 3.4, the right-hand side can be bounded as $\hat{J}_N(\bar{U},x) \le x^\top \bar{P}_K x$ with

$$\bar{P}_K := \sum_{k=0}^{N-1} A_K^{k\top}(Q+\varepsilon M_x + K^\top(R+\varepsilon M_u)K)A_K^k + A_K^{N\top} P A_K^N \tag{C.51}$$

for $A_K = A+BK$. In combination with (C.50), this finally reveals that (3.75) holds with $a = \lambda_{\min}(P^*_{uc})$ and $b = \lambda_{\max}(\bar{P}_K)$, which completes our proof.

C.14 Proof of Theorem 3.8

The proof consist of two parts. In the first part, we use the properties stated in Lemma 3.6 to show that the gradient of the value function satisfies a linear growth condition. In the second part, we then use this result to show that the value function satisfies in addition to (3.75) also

$$\hat{J}^*_N(Ax+B\hat{u}^*_0(x)+w) - \hat{J}^*_N(x) \le -\gamma_1\|x\|^2 + \gamma_2\|w\|^2 \quad \forall\, x \in \mathbb{R}^n, w \in \mathbb{R}^n \tag{C.52}$$

for suitably chosen γ_1, $\gamma_2 \in \mathbb{R}_{++}$, which allows to employ it as an ISS Lyapunov function for the closed-loop system (3.72).

Part 1. In the following, we show that any convex C^1 function $V : \mathbb{R}^n \to \mathbb{R}_+$ that satisfies $a\|x\|^2 \le V(x) \le b\|x\|^2\ \forall\, x \in \mathbb{R}^n$ also satisfies $\|\nabla V(x)\| \le d\|x\|\ \forall\, x \in \mathbb{R}^n$, where $d \in \mathbb{R}_{++} < \infty$ depends on a, b. To this end, consider $x = x_0 + tv$ for $t \in \mathbb{R}_{++}$ and $v := \nabla V(x_0)/\|\nabla V(x_0)\|$, $\nabla V(x_0) \neq 0$, which describes a ray that goes from $x_0 \in \mathbb{R}^n$ in the direction of $\nabla V(x_0)$. Then, it holds due to convexity of the function $V(\cdot)$ that

$$V(x) \ge V(x_0) + t\nabla V(x_0)^\top v = V(x_0) + t\|\nabla V(x_0)\| \tag{C.53}$$

for any $x_0 \in \mathbb{R}^n$ and any $t \in \mathbb{R}_{++}$. On the other hand, the assumed quadratic upper and lower bounds ensure that $V(x) \le b\|x\|^2 = b\|x_0+tv\|^2$ and $V(x_0) \ge a\|x_0\|^2$. Inserting these relations into (C.53), using the triangle inequality as well as $\|v\| = 1$, and dividing by t yields

$$\|\nabla V(x_0)\| \le \frac{b-a}{t}\|x_0\|^2 + 2b\,\|x_0\| + bt \quad \forall\, t \in \mathbb{R}_{++}, \tag{C.54}$$

As (C.54) holds for all $t \in \mathbb{R}_{++}$, we may minimize over the right hand side, which yields $t^* = \sqrt{1-\frac{a}{b}}\|x_0\|$. Inserting $t = t^*$ into (C.54) then finally reveals that

$$\|\nabla V(x_0)\| \le 2(\sqrt{b(b-a)} + b)\,\|x_0\| \tag{C.55}$$

for all $x_0 \in \mathbb{R}^n$ satisfying $\nabla V(x_0) \neq 0$. As this inequality is also trivially satisfied whenever $\nabla V(x_0) = 0$, this proves the above claim with $d = 2(\sqrt{b(b-a)} + b) \in \mathbb{R}_{++}$. Furthermore, it follows immediately that $d < \infty$ whenever a and b are bounded.

Part 2. In combination with Lemma 3.6, the result of Part 1 implies that the value function satisfies $a\|x\|^2 \leq \hat{J}_N^*(x) \leq b\|x\|^2$ as well as $\|\nabla \hat{J}_N^*(x)\| \leq d\|x\|$ for all $x \in \mathbb{R}^n$. Furthermore, we know from our nominal stability result, and in particular from Equation (3.73), that $\hat{J}_N^*(x^+) - \hat{J}_N^*(x) \leq -c\|x\|^2$, where $x^+ = Ax + B\hat{u}_0^*(x)$ denotes the next nominal system state and $c \in \mathbb{R}_{++}$ may be chosen as $c = \lambda_{\min}(Q)$. Note that this directly implies that $c < b$ for b as defined in Lemma 3.6. Moreover, $a\|x^+\|^2 \leq \hat{J}_N^*(x^+) \leq \hat{J}_N^*(x) - c\|x\|^2 \leq (b-c)\|x\|^2$, and hence $\|x^+\|^2 \leq \frac{b-c}{a}\|x\|^2$ for any $x \in \mathbb{R}^n$. In the following, we use these results to prove that $V(x) = \hat{J}_N^*(x)$ also satisfies (C.52) and therefore represents an ISS Lyapunov function for system (3.72), see Definition A.8 in Appendix A.2. To this end, note that the differentiability and convexity of $\hat{J}_N^*(\cdot)$ imply that for any $x^+ \in \mathbb{R}^n$ and any $w \in \mathbb{R}^n$

$$\hat{J}_N^*(x^+ + w) \leq \hat{J}_N^*(x^+) + \nabla \hat{J}_N^*(x^+ + w)^\top w \,. \tag{C.56}$$

Moreover, using the linear growth condition for the norm of the gradient as well as Young's inequality, we obtain

$$\nabla \hat{J}_N^*(x^+ + w)^\top w \leq \|\nabla \hat{J}_N^*(x^+ + w)\|\|w\| \leq d\|x^+ + w\|\|w\| \tag{C.57a}$$

$$\leq d\left(\frac{\|x^+\|^2}{2\kappa} + \frac{\kappa\|w\|^2}{2}\right) + d\|w\|^2 \tag{C.57b}$$

which is guaranteed to hold for any $\kappa \in \mathbb{R}_{++}$. Using now $\hat{J}_N^*(x^+) \leq \hat{J}_N^*(x) - c\|x\|^2$ as well as $\|x^+\|^2 \leq \frac{b-c}{a}\|x\|^2$, inserting into (C.56) finally reveals that

$$\hat{J}_N^*(x^+ + w) - \hat{J}_N^*(x) \leq -\gamma_1\|x\|^2 + \gamma_2\|w\|^2 \quad \forall\, x \in \mathbb{R}^n, w \in \mathbb{R}^n \,, \tag{C.58}$$

with $\gamma_1 := c - \frac{d(b-c)}{2a\kappa}$ and $\gamma_2 := d\left(1 + \frac{\kappa}{2}\right)$. Note that $\gamma_1 > 0$, $\gamma_2 > 0$ can always be ensured by choosing the auxiliary parameter κ large enough. Thus, $V(x) = \hat{J}_N^*(x)$ satisfies (A.6) with $\alpha_1(\|x\|) = a\|x\|^2$, $\alpha_2(\|x\|) = b\|x\|^2$, $\alpha_3(\|x\|) = \gamma_1\|x\|^2$, $\sigma(\|w\|) = \gamma_2\|w\|^2$, which shows that it represents an ISS Lyapunov function for the closed-loop system. In combination with Theorem A.2, this completes the proof.

C.15 Proof of Lemma 3.8

The proof consists of two parts. First, we derive upper and lower bounds for the second derivatives of the relaxed barrier function based cost function (3.84). In the second part, we then use these bounds in combination with the implicit function theorem to show the claimed Lipschitz continutiy result.

Part 1. First note that the barrier function $\hat{B}_{\mathrm{xu}} : \mathbb{R}^{Nm} \times \mathbb{R}^n \to \mathbb{R}_+$ in (3.84) may in general be written as

$$\hat{B}_{\mathrm{xu}}(U, x) = \sum_{i=1}^{q} w^i \left(\hat{B}(z_i(U, x)) - \ln\left(d^i\right)\right) , \tag{C.59}$$

where $z_i(U,x) = -G^iU + E^ix + d^i$ and $\hat{B} : \mathbb{R} \to \mathbb{R}$ refers to the primordial relaxed logarithmic barrier function defined according to (3.4). Furthermore, $w \in \mathbb{R}^q_{++}$ denotes a suitable recentering vector that may for example be constructed by stacking the vectors w_x and w_u in a suitable way. Based on this formulation, the Hessian of the cost function with respect to U can easily be computed as

$$\nabla^2_U \hat{J}_N(U,x) = H + \varepsilon G^\top \mathrm{diag}(w^1\hat{D}(z_1(U,x)),\ldots,w^q\hat{D}(z_q(U,x)))G\,, \tag{C.60}$$

where $\hat{D} : \mathbb{R} \to \mathbb{R}_+$ refers to the second derivative of the primordial relaxed logarithmic barrier function, that is

$$\hat{D}(z) = \begin{cases} \frac{1}{z^2} & z > \delta \\ \frac{1}{\delta^2} & z \leq \delta \end{cases}. \tag{C.61}$$

As $0 \leq \hat{D}(z) \leq 1/\delta^2$ for any $z \in \mathbb{R}$, we can conclude that the Hessian in (C.60) satisfies the upper and lower bounds

$$\mu I \preceq \nabla^2_U \hat{J}_N(U,x) \preceq LI\,, \quad \forall\,(U,x) \in \mathbb{R}^{Nm} \times \mathbb{R}^n \tag{C.62}$$

with $\mu := \lambda_{\min}(H) \in \mathbb{R}_{++}$ and $L := \lambda_{\max}(H + \varepsilon/\delta^2\, G^\top \mathrm{diag}(w)G) \in \mathbb{R}_{++} < \infty$. A direct consequence of this result is that

$$1/L \leq \|(\nabla^2_U \hat{J}_N(U,x))^{-1}\| \leq 1/\mu\,, \quad \forall\,(U,x) \in \mathbb{R}^{Nm} \times \mathbb{R}^n\,. \tag{C.63}$$

Similar arguments can be used to show that $\|\nabla^2_{Ux}\hat{J}_N(U,x)\| \leq L'$ with $L' := \|F\| + \varepsilon/\delta^2\,\|G^\top \mathrm{diag}(w)E\|$. Note that we made in the above derivations use of the fact that for a symmetric positive definite matrix A, it holds that $\|A\| = \lambda_{\max}(A)$, $\|A^{-1}\| = (\lambda_{\min}(A))^{-1}$.

Part 2. Now note that, due to the strong convexity of $\hat{J}_N(U,x)$ with respect to U, the optimal solution $\hat{U}^*(x)$ is for any $x \in \mathbb{R}^n$ uniquely characterized by the condition that $\nabla_U \hat{J}_N(\hat{U}^*(x),x) = 0$. Applying the implicit function theorem (Polyak, 1987, Theorem 2) to this parametric equation yields that $\hat{U}^* : \mathbb{R}^n \to \mathbb{R}^{Nm}$ is uniquely defined as well as continuously differentiable with

$$\nabla_x \hat{U}^*(x) = -\left(\nabla^2_U \hat{J}_N(\hat{U}^*(x),x)\right)^{-1} \nabla^2_{Ux}\hat{J}_N(\hat{U}^*(x),x)\,. \tag{C.64}$$

Using now the submultiplicativity of the matrix norm together with the bounds derived above, we obtain that for any $x \in \mathbb{R}^n$

$$\|\nabla_x \hat{U}^*(x)\| \leq \|(\nabla^2_U \hat{J}_N(U,x))^{-1}\|\|\nabla^2_{Ux}\hat{J}_N(U,x)\| \leq L'/\mu\,. \tag{C.65}$$

Furthermore, for any $x_1, x_2 \in \mathbb{R}^n$, the mean value theorem for multivariable functions asserts that $\hat{U}^*(x_2)$ in (3.87) may be expressed as

$$\hat{U}^*(x_2) = \hat{U}^*(x_1) + \int_0^1 \nabla_x \hat{U}^*(x_1 + s(x_2 - x_1))\mathrm{d}s\,(x_2 - x_1) \tag{C.66}$$

By exploiting (C.65), this finally reveals that

$$\|\hat{U}^*(x_1) - \hat{U}^*(x_2)\| \leq L'/\mu\,\|x_1 - x_2\|\,, \tag{C.67}$$

by which (3.87) holds with $L_U = L'/\mu \in \mathbb{R}_{++}$. This completes the proof.

C.16 Proof of Theorem 3.9

The proof follows in a quite straightforward way from Theorem 3.8 and the Lipschitz property that is ensured by Lemma 3.8. In particular, note that the system state dynamics of the overall closed-loop system (3.83) can be rewritten as

$$x(k+1) = Ax(k) + B\hat{u}_0^*(x(k)) + \tilde{w}(k), \tag{C.68}$$

where we introduced the artificial disturbance

$$\tilde{w}(k) := B\Pi_0\left(\hat{U}_0^*(x(k)+e(k)) - \hat{U}_0^*(x(k))\right) + w(k). \tag{C.69}$$

Applying the ISS result of Theorem 3.8 to (C.68) ensures that there exist $\beta_{\mathrm{x}} \in \mathcal{KL}$ and $\gamma_{\mathrm{x}} \in \mathcal{K}$ such that for all $\xi = x(0) \in \mathbb{R}^n$ and all $k \in \mathbb{N}_+$

$$\|x(k,\xi,\tilde{\boldsymbol{w}})\| \leq \beta_{\mathrm{x}}(\|\xi\|,k) + \gamma_{\mathrm{x}}(\|\tilde{\boldsymbol{w}}_{[k-1]}\|). \tag{C.70}$$

Furthermore, applying Lemma 3.8 to (C.69) reveals $\|\tilde{w}(k)\| \leq L_U\|B\|\|e(k)\| + \|w(k)\|$, which implies that the term $\|\tilde{\boldsymbol{w}}_{[k-1]}\|$ in the above inequality satisfies

$$\|\tilde{\boldsymbol{w}}_{[k-1]}\| \leq L_U\|B\|\|\boldsymbol{e}_{[k-1]}\| + \|\boldsymbol{w}_{[k-1]}\|. \tag{C.71}$$

Now note in addition that Assumption 3.6 directly implies that there exist $\beta_{\mathrm{e}} \in \mathcal{KL}$ and $\gamma_{\mathrm{e}} \in \mathcal{K}$ such that for all $\epsilon = e(0) \in \mathbb{R}^n$ and all $k \in \mathbb{N}_+$

$$\|e(k,\epsilon,\boldsymbol{w},\boldsymbol{v})\| \leq \beta_{\mathrm{e}}(\|\epsilon\|,k) + \gamma_{\mathrm{e}}(\|(\boldsymbol{w},\boldsymbol{v})_{[k-1]}\|). \tag{C.72}$$

By finally inserting (C.71) into (C.70) and exploiting the upper bound for $\|\boldsymbol{e}_{[k-1]}\|$ that is implied by (C.72), it then follows that there exist $\beta \in \mathcal{KL}$ and $\gamma_{\mathrm{w}}, \gamma_{\mathrm{v}} \in \mathcal{K}$ such that for all $\zeta = (\xi,\epsilon) = (x(0),e(0)) \in \mathbb{R}^n \times \mathbb{R}^n$ and all $k \in \mathbb{N}_+$

$$\|z(k,\epsilon,\boldsymbol{w},\boldsymbol{v})\| \leq \beta(\|\zeta\|,k) + \gamma_{\mathrm{w}}(\|\boldsymbol{w}_{[k-1]}\|) + \gamma_{\mathrm{v}}(\|\boldsymbol{v}_{[k-1]}\|), \tag{C.73}$$

where $z(k) = (x(k),e(k)) \in \mathbb{R}^n \times \mathbb{R}^n$ denotes the overall system state. Explicit expressions for suitable $\beta, \gamma_{\mathrm{w}}$, and γ_{v} in dependence of $\beta_{\mathrm{x}}, \beta_{\mathrm{e}}, \gamma_{\mathrm{x}}$, and γ_{e} can be easily derived from the above arguments and are not stated here for the sake of simplicity.

C.17 Proof of Lemma 3.9

The proof consists of two parts. First we show that the a priori error dynamics are ISS with respect to $\boldsymbol{w},\boldsymbol{v}$. In the second part, we show the claimed result for the a posteriori error dynamics by exploiting the relation between a priori and a posteriori error. While the proof is conceptually similar to the results presented by Reif and Unbehauen (1999) and Huang et al. (2012) for the extended Kalman filter, the obtained ISS results are global instead of local and can be shown based on easily verifiable standard assumptions.

Part 1. Define $\bar{a} := \|A\|$, $\bar{c} := \|C\|$, $\bar{q} := \|Q_\mathrm{w}\|$, $\bar{r} := \|R_\mathrm{v}\|$, $\underline{q} := \lambda_{\min}(Q_\mathrm{w})$, $\underline{r} := \lambda_{\min}(R_\mathrm{v})$ as well as $\Pi_k^+ := (P^+)^{-1}$, $\Pi_k^- := (P_k^-)^{-1}$. Detectability of (A, C) ensures that all elements of the sequence $\{P_k^+\}$ are uniformly bounded and satisfy $0 \preceq P_k^+ \preceq \bar{p}I$ with finite $\bar{p} \in \mathbb{R}_{++}$ for all $k \in \mathbb{N}$ and any initialization $P_0^+ \in \mathbb{S}_+^n$, see Theorem 4.1 in (Caines and Mayne, 1970). Using (3.93b) and $Q_\mathrm{w} \in \mathbb{S}_{++}^n$, this implies that $\underline{p}'I \preceq P_k^- \preceq \bar{p}'I$ for all $k \in \mathbb{N}_+$, where $\underline{p}' := \underline{q}$ and $\bar{p}' := \bar{a}^2\bar{p} + \bar{q}$. Moreover, inserting the definition of the Kalman gain from (3.93a) into (3.93b) directly yields that

$$P_k^+ = P_k^- - P_k^- C^\top (C P_k^- C^\top + R_\mathrm{v})^{-1} C P_k^- \,, \tag{C.74}$$

which, by applying the matrix inversion lemma, gives

$$P_k^+ = \left((P_k^-)^{-1} + C^\top R_\mathrm{v}^{-1} C \right)^{-1} . \tag{C.75}$$

Due to the upper and lower bounds on P_k^-, the inverse on the right hand side exists for all $k \in \mathbb{N}_+$ and is positive definite. In particular, $P_k^+ \geq \underline{p}I$ with *strictly* positive $\underline{p} = \underline{p}'\underline{r}/(\underline{r} + \bar{c}^2\underline{p}') \in \mathbb{R}_{++}$. Thus, P_k^+ and P_k^- are both upper and lower bounded for all $k \in \mathbb{N}_+$ and, hence, the same holds for Π_k^+ and Π_k^- viz. $1/\bar{p}\, I \preceq \Pi_k^+ \preceq 1/\underline{p}\, I$ and $1/\bar{p}'\, I \preceq \Pi_k^- \preceq 1/\underline{p}'\, I$. For a more general derivation of similar upper and lower bounds based on (uniform) detectability and controllability assumptions, see for example (Jazwinski, 1970, Chapter 7.6).

We are now in a position to start our considerations for the a priori estimation error $\zeta_k = \hat{x}_k^- - x_k$, which evolves according to the dynamics

$$\zeta_{k+1} = A(I - K_k C)\zeta_k + d_k \,, \tag{C.76}$$

where $d_k := A K_k v_k - w_k$, cf. (3.79) and (3.92). Note that we make use of subindices for the sake of a simplified representation. Following the approach of Reif and Unbehauen (1999) and Huang et al. (2012), we define the time-varying ISS Lyapunov function candidate $V(\zeta_k) = \zeta_k^\top \Pi_k^- \zeta_k$, which satisfies due to the bounds derived above

$$\frac{1}{\bar{p}'}\|\zeta_k\|^2 \leq V(\zeta_k) \leq \frac{1}{\underline{p}'}\|\zeta_k\|^2 \quad \forall \zeta_k \in \mathbb{R}^n \,. \tag{C.77}$$

Furthermore, Lemma 6 in (Reif and Unbehauen, 1999) implies that for any $k \in \mathbb{N}_+$

$$\Pi_{k+1}^- \preceq \tilde{A}_k^{-\top} \left[\Pi_k^- - \Pi_k^- \left(\Pi_k^+ + A^T Q_w^{-1} A \right)^{-1} \Pi_k^- \right] \tilde{A}_k^{-1} \,, \tag{C.78}$$

where $\tilde{A}_k := A(I - K_k C)$. Note that the inverse of $\tilde{A}_k$ exists as both A and $(I - K_k C) = P_k^+ \Pi_k^-$ have full rank. We also have that $\|\tilde{A}_k\| \leq \bar{a}(1 + \bar{k}\bar{c})$ with $\bar{k} := \bar{p}'\bar{c}/\underline{r}$, cf. (3.93a). Consider now

$$V(\zeta_{k+1}) = \zeta_{k+1}^\top \Pi_{k+1}^- \zeta_{k+1} = \zeta_k^\top \tilde{A}_k^\top \Pi_{k+1}^- \tilde{A}_k \zeta_k + 2 d_k^\top \Pi_{k+1}^- \tilde{A} \zeta_k + d_k^\top \Pi_{k+1}^- d_k \tag{C.79}$$

and let us treat each of the terms on the right-hand side separately. First, we note that, using (C.78), the first term satisfies

$$\zeta_k^\top \tilde{A}_k^\top \Pi_{k+1}^- \tilde{A}_k \zeta_k \leq \zeta_k^\top \Pi_k^- \zeta_k - \zeta_k^\top \Pi_k^- \left(\Pi_k^+ + A^T Q_w^{-1} A\right)^{-1} \Pi_k^- \zeta_k \tag{C.80a}$$

$$\leq V(\zeta_k) - \kappa \|\zeta_k\|^2, \quad \kappa := 1/\bar{p}'^2 \left(1/\underline{p}' + \bar{a}^2/\underline{q}\right)^{-1} \tag{C.80b}$$

where we made use of the upper and lower bounds defined above. Furthermore, by using Young's inequality, the second term in (C.79) can be upper bounded as

$$2d_k^\top \Pi_{k+1}^- \tilde{A}_k \zeta_k \leq s\|\zeta_k\|^2 + \frac{\bar{a}^2(1+\bar{k}\bar{c})^2}{s\underline{p}'^2}\|d_k\|^2 \tag{C.81}$$

where $s \in \mathbb{R}_{++}$ is a free parameter. Finally, the third term in (C.79) obviously satisfies

$$d_k^\top \Pi_{k+1}^- d_k \leq 1/\underline{p}' \, \|d_k\|^2 . \tag{C.82}$$

By combining (C.80)–(C.82) with (C.79), we therefore get that

$$V(\zeta_{k+1}) \leq V(\zeta_k) - \gamma_1 \|\zeta_k\|^2 + \gamma_2 \|d_k\|^2 \tag{C.83}$$

where $\gamma_1 := \kappa - s \in \mathbb{R}_{++}$ and $\gamma_2 := \bar{a}^2(1+\bar{k}\bar{c})^2/(s\underline{p}'^2) + 1/\underline{p}' \in \mathbb{R}_{++} < \infty$ for all $0 < s < \kappa$. Together with (C.77) this shows that $V(\zeta_k)$ represents a (time-varying) ISS Lyapunov function for the a priori estimation error dynamics (C.76). In particular, exploiting (C.77) yields that

$$V(\zeta_{k+1}) \leq \beta V(\zeta_k) + \gamma_2 \|d_k\|^2 \tag{C.84}$$

with $\beta := 1 - \gamma_1 \underline{p}'$, for which based on the definition of κ in (C.80b) it can be shown easily that $\beta \in (0, 1)$. Hence, by using a geometric series and recalling that the a priori error is only defined for strictly positive k, we obtain

$$V(\zeta_k) \leq \beta^k V(\zeta_1) + \frac{\gamma_2}{1-\beta}\|\boldsymbol{d}_{[k-1]}\|^2 \;\; \forall k \in \mathbb{N}_+. \tag{C.85}$$

Using again (C.77), this finally implies that the a priori estimation error satisfies

$$\|\zeta_k\| \leq c_\zeta \beta_\zeta^k \|\zeta_1\| + \gamma_\zeta \|\boldsymbol{d}_{[k-1]}\| \quad \forall k \in \mathbb{N}_+, \tag{C.86}$$

where we introduced $c_\zeta := \sqrt{\bar{p}'/\underline{p}'}$, $\beta_\zeta := \sqrt{\beta} \in (0,1)$, and $\gamma_\zeta := \sqrt{\bar{p}'\gamma_2/(1-\beta)}$.

Part 2. We now derive a similar result for the a posteriori error. To this end, recall that $e_k = (I - K_k C)\zeta_k + K_k v_k$ and, hence,

$$\|e_k\| \leq (1+\bar{k}\bar{c})\|\zeta_k\| + \bar{k}\|v_k\| . \tag{C.87}$$

On the other hand, we know from above that $\zeta_{k+1} = Ae_k - w_k$, which implies that

$$\|\zeta_1\| \leq \bar{a}\|e_0\| + \|w_0\| . \tag{C.88}$$

Moreover, recalling the definition of $d_k = AK_k v_k - w_k$, we get

$$\|\boldsymbol{d}_{[k-1]}\| \leq \bar{a}\bar{k}\|\boldsymbol{v}_{[k-1]}\| + \|\boldsymbol{w}_{[k-1]}\| \,. \tag{C.89}$$

Now, by inserting (C.86) into (C.87) and exploiting (C.88)–(C.89), we finally obtain after some algebraic manipulations that

$$\|e_k\| \leq c_e \beta_e^k \|e_0\| + \gamma_w \|\boldsymbol{w}_{[k-1]}\| + \gamma_v \|\boldsymbol{v}_{[k]}\| \tag{C.90}$$

with $\beta_e \in (0, 1)$ and $c_e, \gamma_w, \gamma_v \in \mathbb{R}_{++}$ given by

$$\beta_e := \beta_\zeta \,, \quad c_e := (1 + \bar{k}\bar{c})\bar{a}\, c_\zeta \tag{C.91a}$$

$$\gamma_w := (1 + \bar{k}\bar{c})(c_\zeta + \gamma_\zeta), \; \gamma_v := \bar{k} + (1 + \bar{k}\bar{c})\bar{a}\bar{k}\gamma_\zeta \,. \tag{C.91b}$$

Noting the equivalence of (3.96) and (C.90) for $\beta = \beta_e$, $c = c_e$, and $e(0) = e_0$, this completes the proof.

C.18 Proof of Theorem 4.1

The proof consists of three parts. First, we show that the relaxed barrier function based cost function is twice continuously differentiable and satisfies the (strong) convexity and Lipschitz conditions stated in (4.8). In a second step, we then show that also the associated Hessian is globally Lipschitz continuous, i.e., that (4.9) holds. In the third step, we finally prove that for any arbitrary $x \in \mathbb{R}^n$, the resulting cost function is self-concordant in the optimization variable U.

Part 1. As discussed in Appendix B.9, the overall barrier function $\hat{B}_{xu} : \mathbb{R}^{Nm} \times \mathbb{R}^n \to \mathbb{R}_+$ can be formulated as

$$\hat{B}_{xu}(U, x) = \sum_{i=1}^{q} w^i \left(\hat{B}\big(- G^i U + E^i x + d^i \big) - \ln\left(d^i\right) \right) , \tag{C.92}$$

where $G \in \mathbb{R}^{q \times Nm}$, $d \in \mathbb{R}^q$, and $E \in \mathbb{R}^{q \times n}$ are the corresponding constraint matrices, $w \in \mathbb{R}^q_{++}$ denotes a suitably constructed recentering vector, and $\hat{B} : \mathbb{R} \to \mathbb{R}$ refers to the primordial relaxed logarithmic barrier function defined according to (3.4). As both the underlying relaxing function $\beta(\cdot; \delta)$ and the resulting relaxed logarithmic barrier function $\hat{B}(\cdot)$ are twice continuously differentiable, the same holds for the overall, combined barrier function in (C.92). Consequently, the relaxed barrier function based cost function $\hat{J}_N : \mathbb{R}^{Nm} \times \mathbb{R}^n \to \mathbb{R}_+$ given in (4.6) is twice continuously differentiable in both of its arguments. In particular, using the definition of the primordial relaxed logarithmic barrier function, the second derivative with respect to U can be computed easily as

$$\nabla_U^2 \hat{J}_N(U, x) = H + \varepsilon G^\top \mathrm{diag}\left(w^1 \hat{D}(z_1(U, x)), \ldots, w^q \hat{D}(z_q(U, x)) \right) G \,, \tag{C.93}$$

where $z_i(U,x) := -G^iU + E^ix + d^i$. Furthermore, the function $\hat{D} : \mathbb{R} \to \mathbb{R}_+$ characterizing the diagonal elements on the right-hand side is defined as

$$\hat{D}(z) = \begin{cases} \frac{1}{z^2} & z > \delta \\ \frac{1}{\delta^2} & z \leq \delta, \end{cases} \tag{C.94}$$

which is obviously nothing else than the second derivative of the primordial relaxed barrier function $\hat{B}(\cdot)$. Note now that $0 \leq \hat{D}(z) \leq \frac{1}{\delta^2}$ for any $z \in \mathbb{R}$, which immediately yields that

$$\mu I \preceq \nabla^2_U \hat{J}_N(U,x) \preceq LI \quad \forall (U,x) \in \mathbb{R}^{Nm} \times \mathbb{R}^n, \tag{C.95}$$

where $\mu := \lambda_{\min}(H) \in \mathbb{R}_{++}$ and $L := \lambda_{\max}(H + \frac{\varepsilon}{\delta^2} G^\top \mathrm{diag}(w) G) \in \mathbb{R}_{++}$. Thus, the cost function is not only strongly convex but we can also give a global upper bound on the Hessian, and (4.8a) holds. Furthermore, by means of the mean value theorem for multivariable functions (see also Nocedal and Wright, 1999, Theorem 2.1) we obtain

$$\nabla_U \hat{J}_N(U,x) - \nabla_U \hat{J}_N(U',x) = \int_0^1 \nabla^2_U \hat{J}_N(U + s(U-U'),x)\,\mathrm{d}s\,(U-U'), \tag{C.96}$$

and hence, by virtue of (C.95), $\|\nabla_U \hat{J}_N(U,x) - \nabla_U \hat{J}_N(U',x)\| \leq L\|U-U'\|$ for all $U,U' \in \mathbb{R}^{Nm}$ and any $x \in \mathbb{R}^n$, which proves (4.8b).

Part 2. We now show that in addition also (4.9) holds, i.e., the global Lipschitz condition on the Hessian. To this end, note that (C.93) directly implies that

$$\|\nabla^2_U \hat{J}_N(U,x) - \nabla^2_U \hat{J}_N(U',x)\| \leq \varepsilon \bar{w} q \, \|G\|^2 \max_{i \in \{1,\dots,q\}} |\hat{D}(z_i(U,x)) - \hat{D}(z_i(U',x))|, \tag{C.97}$$

where $\bar{w} := \max(w)$. This reveals that (4.9) will hold if the Hessian of the primordial relaxed barrier function is globally Lipschitz continuous, that is,

$$|\hat{D}(z) - \hat{D}(z')| \leq L_D |z - z'| \quad \forall z, z' \in \mathbb{R} \tag{C.98}$$

for some $L_D \in \mathbb{R}_+$. In order to show this result and derive an explicit expression for L_D, we now distinguish the following cases: $z \leq \delta \wedge z' \leq \delta$, $z \geq \delta \wedge z' \geq \delta$, and $z \geq \delta \wedge z' \leq \delta$ (which also includes the case $z \leq \delta \wedge z' \geq \delta$ by interchangeability of z and z'). For $z \leq \delta \wedge z' \leq \delta$, it obviously holds that $\hat{D}(z) = \hat{D}(z') = \frac{1}{\delta^2}$ and hence $\hat{D}(z) = \hat{D}(z') = 0$. As a consequence, (C.98) holds with $L_D = 0$. For the second case, we get that

$$|\hat{D}(z) - \hat{D}(z')| = \left|\frac{1}{z^2} - \frac{1}{z'^2}\right| = \left|\frac{z'^2 - z^2}{z^2 z'^2}\right| \leq \left|\frac{z' + z}{z^2 z'^2}\right| |z - z'| \leq \frac{2}{\delta^3}|z - z'|, \tag{C.99}$$

where we exploited in the last step that $z \geq \delta$, $z' \geq \delta$. Furthermore, for the third case with $z \geq \delta$ and $z' \leq \delta$, we obtain the following

$$\begin{aligned} |\hat{D}(z) - \hat{D}(z')| &= |\hat{D}(z) - \hat{D}(\delta) + \hat{D}(\delta) - \hat{D}(z')| && \text{(C.100a)} \\ &\leq |\hat{D}(z) - \hat{D}(\delta)| + \underbrace{|\hat{D}(\delta) - \hat{D}(z')|}_{=0} && \text{(C.100b)} \\ &\leq \left|\frac{1}{z^2} - \frac{1}{\delta^2}\right| = \left|\frac{\delta^2 - z^2}{z^2\delta^2}\right| \leq \left|\frac{\delta + z}{z^2\delta^2}\right| |z - \delta| \leq \frac{2}{\delta^3}|z - \delta|. && \text{(C.100c)} \end{aligned}$$

However, as $z' \leq \delta \leq z$, it holds that $|z - \delta| \leq |z - z'|$, which together with (C.100) implies that also in this case

$$|\hat{D}(z) - \hat{D}(z')| \leq \frac{2}{\delta^3}|z - z'|. \tag{C.101}$$

Thus, putting the results for the three separate cases together, we can conclude that (C.98) holds with $L_D = \frac{2}{\delta^3}$. Inserting this result into (C.97) and exploiting that $|z_i(U,x) - z_i(U',x)| = |G^i(U - U')| \leq \|G^i\|\|U - U'\|$ then finally yields that

$$\|\nabla^2_U \hat{J}_N(U,x) - \nabla^2_U \hat{J}_N(U',x)\| \leq L'\|U - U'\| \tag{C.102}$$

with the Lipschitz constant $L' \in \mathbb{R}_{++}$ being defined as $L' := 2\varepsilon\bar{w}q/\delta^3\|G\|^2 \max_i \|G^i\|$. Note that $\max_i \|G^i\| \leq \sqrt{q}\|G\|$, which implies that a more compact (but potentially also more conservative) expression for L' is given by $L' := 2\varepsilon\bar{w}q^{3/2}/\delta^3\|G\|^3$. In both cases, the resulting Lipschitz constant L' does only depend on the constraints and the known barrier function parameters ε, δ, and $\bar{w}$.

Part 3. We conclude the proof by showing that the relaxed barrier function based cost function $\hat{J}_N(\cdot, x)$ is for any $x \in \mathbb{R}^n$ a self-concordant function according to Definition 2.3. We make use of the following auxiliary result.

Lemma C.1. *Let $\hat{B} : \mathbb{R} \to \mathbb{R}$ denote the primordial relaxed logarithmic barrier function introduced in Definition 3.1 and consider the associated relaxed barrier function for the positive orthant of dimension n, given by $f(x) = \sum_{i_1}^{n} \hat{B}(x_i)$. Then, $f : \mathbb{R}^n \to \mathbb{R}$ is a self-concordant function.*

Proof. The proof uses similar arguments as applied by Renegar (2001) in the context of nonrelaxed logarithmic barrier functions, see also Appendix B.1. Note that $\mathcal{D}_f = \mathbb{R}^n$ and that

$$\nabla^2 f(x) = \operatorname{diag}\big(\hat{D}(x_1), \ldots, \hat{D}(x_n)\big) \in \mathbb{S}^n_{++} \tag{C.103}$$

where $\hat{D} : \mathbb{R} \to \mathbb{R}_+$ is defined according to (C.94). Consider now $\mathcal{B}(x,1) := \{y \in \mathbb{R}^n : \|y - x\|_{\nabla^2 f(x)} < 1\}$, cf. Definition 2.3. Obviously, $\mathcal{B}(x,1) \subseteq \mathcal{D}_f$ for any $x \in \mathcal{D}_f = \mathbb{R}^n$, which shows that condition *i*) of Definition 2.3 is trivially satisfied. Furthermore, concerning condition *ii*) of Definition 2.3, we can state the following for any $v, x, y \in \mathbb{R}^n$:

$$\|v\|^2_{\nabla^2 f(y)} = \sum_{i=1}^{n} v_i^2 \hat{D}(y_i) = \sum_{i=1}^{n} v_i^2 \hat{D}(x_i)\frac{\hat{D}(y_i)}{\hat{D}(x_i)} \leq \|v\|^2_{\nabla^2 f(x)} \max_i \frac{\hat{D}(y_i)}{\hat{D}(x_i)}. \tag{C.104}$$

Thus, in order to show that (2.19) holds, we need to show that the quotient on the right-hand side satisfies for any $x \in \mathbb{R}^n$, any $y \in \mathcal{B}(x,1)$, and any $i \in \{1, \ldots, n\}$

$$\frac{\hat{D}(y_i)}{\hat{D}(x_i)} \leq \frac{1}{(1 - \|y - x\|_{\nabla^2 f(x)})^2}. \tag{C.105}$$

In order to prove that this is indeed true, we need to distinguish the following four cases, depending on whether the relaxation is active or inactive for x_i and y_i, respectively.

i) $x_i > \delta, y_i > \delta$. In this case, the relaxation is inactive and $\hat{D}(y_i)/\hat{D}(x_i) = x_i^2/y_i^2$. However, as we already observed in the nonrelaxed case in Appendix B.1,

$$1 - \|y - x\|_{\nabla^2 f(x)} = 1 - \sqrt{\sum_{i=1}^{n} \left(\frac{y_i - x_i}{x_i} \right)^2} \leq 1 - \left| \frac{y_i}{x_i} - 1 \right| \leq \frac{y_i}{x_i} \quad \forall i \in \{1, \dots, n\}, \tag{C.106}$$

which shows that $x_i/y_i \leq 1/(1 - \|y - x\|_{\nabla^2 f(x)})$ for all $i \in \{1, \dots, n\}$. Inserting this relation into the quotient above, then reveals that (C.105) holds.
ii) $x_i \leq \delta$, $y_i \leq \delta$. In this case, the relaxation is active for both x_i and y_i. As a result, it follows immediately that

$$\frac{\hat{D}(y_i)}{\hat{D}(x_i)} = \frac{\delta^2}{\delta^2} = 1 \leq \frac{1}{(1 - \|y - x\|_{\nabla^2 f(x)})^2}. \tag{C.107}$$

Note that the inequality follows in this case from the fact that $y \in \mathcal{B}(x, 1)$ implies per definition that $\|y - x\|_{\nabla^2 f(x)} < 1$, which directly shows that $1 - \|y - x\|_{\nabla^2 f(x)} \in (0, 1]$.
iii) $x_i \geq \delta, y_i \leq \delta$. In this case

$$\frac{\hat{D}(y_i)}{\hat{D}(x_i)} = \frac{x_i^2}{\delta^2} \leq \frac{x_i^2}{y_i^2} \leq \frac{1}{(1 - \|y - x\|_{\nabla^2 f(x)})^2}, \tag{C.108}$$

where we exploited again (C.106).
iv) $x_i \leq \delta, y_i \geq \delta$. In this case,

$$\frac{\hat{D}(y_i)}{\hat{D}(x_i)} = \frac{\delta^2}{y_i^2} \leq \frac{y_i^2}{y_i^2} = 1 \leq \frac{1}{(1 - \|y - x\|_{\nabla^2 f(x)})^2}, \tag{C.109}$$

where we exploited again that $y \in \mathcal{B}(x, 1)$ and hence $\|y - x\|_{\nabla^2 f(x)} < 1$. Thus, (C.105) holds for any possible combination of x_i and y_i. In combination with (C.104) this reveals that the right-hand side inequality in (2.19) holds for any $x \in \mathbb{R}^n$ whenever $y \in \mathcal{B}(x, 1)$. As shown by Renegar (2001), the inequality on the left-hand side is redundant. Thus, condition *ii)* from Definition 2.3 is satisfied and f is a self-concordant function. ■

Assuming that the weighting vector $w \in \mathbb{R}^q_{++}$ has been chosen according to Appendix B.3, satisfying therefore $w = [w_x^\top, w_u^\top, w_x^\top, \dots, w_u^\top, w_x^\top]^\top \geq \mathbb{1}$, and exploiting the fact that self-concordance is preserved under addition, affine transformation, and multiplication with a value greater than one (Renegar, 2001; Boyd and Vandenberghe, 2004), we can based on Lemma C.1 directly conclude that the relaxed barrier function (C.92) is a self-concordant function in $\mathcal{U}$. As a consequence, the cost function given in (4.6) is also self-concordant since it is the sum of a quadratic (and hence self-concordant) function and the self-concordant relaxed barrier function $\hat{B}_{xu}(\cdot, x)$. This completes the proof.

C.19 Proof of Theorem 4.2

We rewrite the closed-loop system (4.18) as

$$x(k+1) = Ax(k) + B\Pi_0 \hat{U}^*(x(k)) + w(k), \quad w(k) := B\Pi_0 \left(\hat{U}(k) - \hat{U}^*(x(k))\right). \quad \text{(C.110)}$$

From Theorem 3.8 in Section 3.6.2 we know that the system on the left-hand side is input-to-state stable with respect to the additive disturbance $w(k)$. In particular, there exist $\beta_\mathrm{x} \in \mathcal{KL}$ and $\gamma_\mathrm{x} \in \mathcal{K}$ such that

$$\|x(k)\| \leq \beta_\mathrm{x}(\|x(0)\|, k) + \gamma_\mathrm{x}(\|\boldsymbol{w}_{[k-1]}\|) \quad \text{(C.111)}$$

for any initial condition $x(0) \in \mathbb{R}^n$ and any $k \in \mathbb{N}$. Furthermore, the suboptimality bound in (4.19) and the fact that $\|\Pi_0\| = 1$ ensure that

$$\|w(k)\| = \|B\Pi_0 \left(\hat{U}(k) - \hat{U}^*(x(k))\right)\| \leq \|B\| \|\hat{U}(k) - \hat{U}^*(x(k))\| \leq \|B\| \epsilon(k), \quad \text{(C.112)}$$

which obviously implies $\|\boldsymbol{w}_{[k-1]}\| \leq \|B\| \|\boldsymbol{\epsilon}_{[k-1]}\|$. Inserting this relation into (C.111) reveals that (4.20) holds with $\beta(r,s) = \beta_\mathrm{x}(r,s)$ and $\gamma(r) = \gamma_\mathrm{x}(\|B\| r)$.

C.20 Proof of Theorem 4.3

Due to the design of the underlying primordial relaxed logarithmic barrier function $\hat{B}(\cdot)$, the applied recentering, and $Q \in \mathbb{S}^n_{++}$, $R \in \mathbb{S}^n_{++}$, $P \in \mathbb{S}^n_{++}$, both the barrier function based stage cost $\hat{\ell} : \mathbb{R}^n \times \mathbb{R}^m \to \mathbb{R}_+$ and the quadratic terminal cost $\hat{F} : \mathbb{R}^n \to \mathbb{R}_+$ are guaranteed to be continuous, positive definite, and radially unbounded. Moreover, the eliminated predicted system states $x_k(U, x)$, $k = 1, \ldots, N$, are affine in (U, x). As a consequence, the resulting overall cost function $\hat{J}_N : \mathbb{R}^{Nm} \times \mathbb{R}^n \to \mathbb{R}_+$ given in (4.6) is a continuous, positive definite, and radially unbounded function. This directly implies that there exist $\underline{\alpha}, \bar{\alpha} \in \mathcal{K}_\infty$ such that for all $(U, x) \in \mathbb{R}^{Nm} \times \mathbb{R}^n$ it holds that

$$\underline{\alpha}(\|(U,x)\|) \leq \hat{J}_N(U,x) \leq \bar{\alpha}(\|(U,x)\|), \quad \text{(C.113)}$$

see for example Lemma 4.3 in (Khalil, 2002). In the following, we show that the cost function will in addition always decrease to zero along trajectories of the closed-loop system. To this end, assume given arbitrary initial conditions $(U_0, x_0) \in \mathbb{R}^{Nm} \times \mathbb{R}^n$. Consider now the next system state $x_0^+ = Ax_0 + B\Pi_0 U_0$ as well as the corresponding warm-start solution $U_0^+ = \Psi_\mathrm{s}(U_0, x_0)$, with the shift operator $\Psi_\mathrm{s}(\cdot,\cdot)$ being defined according to (4.23)–(4.24). Obviously U_0^+ is nothing else than the vectorized (and suboptimal) version of the shifted candidate solution $\hat{\boldsymbol{u}}^+(x_0)$ that we applied in the stability proof of Theorem 3.3. As a consequence, we can observe that for $x_N = x_N(U_0, x_0)$

$$\begin{aligned}
&\hat{J}_N(U_0^+, x_0^+) - \hat{J}_N(U_0, x_0) && \text{(C.114a)}\\
&\quad = -\hat{\ell}(x_0, \Pi_0 U_0) + \|A_K x_N\|_P^2 - \|x_N\|_P^2 + \|x_N\|_Q^2 + \|Kx_N\|_R^2 + \varepsilon \hat{B}_K(x_N) && \text{(C.114b)}\\
&\quad \leq -\hat{\ell}(x_0, \Pi_0 U_0), && \text{(C.114c)}
\end{aligned}$$

where we made use of a telescoping sum and exploited the global quadratic bound on $\hat{B}_K(\cdot)$ from Lemma 3.4 as well as the choice of P according to (4.5). Thus, U_0^+ represents a globally valid warm-start solution for the next system state that will always lead to a decrease in the cost function. Furthermore, Assumption 4.1 ensures that the optimization algorithm update, i.e., applying the operator $\Psi_o(\cdot,\cdot)$, will lead to an even further decrease. In particular, for any number $i \in \mathbb{N}_+$ of optimization algorithm iterations,

$$\hat{J}_N(\Phi^i(U_0,x_0),x_0^+) \leq \hat{J}_N(U_0^+,x_0^+) - \gamma(U_0^+,x_0^+) \tag{C.115a}$$
$$\leq \hat{J}_N(U_0,x_0) - \hat{\ell}(x_0,\Pi_0\, U_0) - \gamma(U_0^+,x_0^+)\,. \tag{C.115b}$$

As the above holds for any arbitrary $(U_0,x_0) \in \mathbb{R}^{Nm} \times \mathbb{R}^n$, we can conclude that the extended state of system (4.21) satisfies for any $k \in \mathbb{N}$ the decrease condition

$$\hat{J}_N(U(k+1),x(k+1)) - \hat{J}_N(U(k),x(k)) \leq -\hat{\ell}(x(k),\Pi_0\, U(k)) - \gamma(U^+(k),x^+(k))\,, \tag{C.116}$$

where $U^+(k) := \Psi_s(U(k),x(k))$ and $x^+(k) = Ax(k) + B\Pi_0\, U(k)$. As both $\hat{\ell}(\cdot,\cdot)$ and $\gamma(\cdot,\cdot)$ are nonnegative, this implies that $(U,x) = (0,0)$ is stable. In particular the combined plant and optimizer state satisfies $\|(U(k),x(k))\| \leq \underline{\alpha}^{-1}(\bar{\alpha}(\|(U(0),x(0))\|))$ for any $(U(0),x(0)) \in \mathbb{R}^{Nm} \times \mathbb{R}^n$. Moreover, the only solution that can stay identically in the set

$$\Omega = \left\{(U,x) \in \mathbb{R}^{Nm} \times \mathbb{R}^n : \hat{\ell}(x,\Pi_0\, U) + \gamma(\Psi_s(U,x), Ax + B\Pi_0\, U) = 0\right\}, \tag{C.117}$$

i.e., in the set where the right-hand side of (C.116) vanishes, is given by $(U,x) = (0,0)$. This can be shown as follows. Positive definiteness of $\hat{\ell}(\cdot,\cdot)$ and $\gamma(\cdot,\cdot) \geq 0$ imply $x = 0$ as well as $\Pi_0\, U = 0$ and, hence, $x^+ = Ax + B\Pi_0\, U = 0$. However, together with strong convexity of the cost function and the property that $\gamma(U,x) = 0 \Leftrightarrow \nabla_U \hat{J}_N(U,x) = 0 \Leftrightarrow U = \hat{U}^*(x)$ (see Assumption 4.1), the additional condition $\gamma(U^+,x^+) = 0$ then implies that $U^+ = \hat{U}^*(x^+) = \hat{U}^*(0) = 0$. As $U^+ = \Psi_s(U,x)$ is defined according to the shift operator given in (4.23), it follows immediately that $u_k = 0$ for $i = k,\ldots,N-1$. Together with $\Pi_0\, U = u_0 = 0$ and $x = 0$, this finally shows that $(U,x) = (0,0)$. By combining this result with (C.113) and (C.116), global asymptotic stability of the origin then follows from the Barbashin-Krasovskii Theorem, see for example Corollary 2.31 in (Halanay and Rasvan, 2000) or (for a continuous-time version) Corollary 4.2 in (Khalil, 2002).

C.21 Proof of Theorem 4.4

The proof is very similar to that of Theorem 3.6 and consists of three parts. First we show that the relaxed barrier functions for the input and state constraints satisfy for any $k \in \mathbb{N}$

$$\varepsilon\hat{B}_x(x(k)) \leq \hat{J}_N(U(k),x(k)) - x(k)^\top P_{uc}^*\, x(k) \tag{C.118a}$$
$$\varepsilon\hat{B}_u(u(k)) \leq \hat{J}_N(U(k),x(k)) - x(k)^\top P_{uc}^*\, x(k)\,, \tag{C.118b}$$

where $P_{uc}^* \in \mathbb{S}_{++}^n$ is the solution to the discrete-time algebraic Riccati equation related to the infinite-horizon LQR problem for the setup (A,B,Q,R). In a second step, we use (C.118) for deriving a time-varying upper bound on the maximally possible constraint violations. Finally, we exploit the decrease of the cost function in order to show the claimed monotonicity result.

Part 1. Based on the proof of Theorem 4.3 and the nonnegativity of $\gamma(\cdot,\cdot)$, we know that

$$\hat{J}_N(U(i+1),x(i+1)) - \hat{J}_N(U(i),x(i)) \leq -\hat{\ell}(x(i),u(i)), \tag{C.119}$$

for any $i \in \mathbb{N}$. Furthermore, we know that the closed-loop system is asymptotically stable and that $\lim_{i\to\infty} \hat{J}_N(U(i),x(i)) = 0$, see Theorem 4.3. Summing up over all future sampling instants beginning at $i = k$ and using a telescoping sum on the left-hand side, we get from (C.119) that $\hat{J}_N(U(k),x(k)) \geq \sum_{i=k}^{\infty} \hat{\ell}(x(i),u(i))$ for any $k \in \mathbb{N}$. Furthermore, it holds that $\sum_{i=k}^{\infty} \hat{\ell}(x(i),u(i)) \geq x(k)^\top P_{\mathrm{uc}}^* x(k) + \varepsilon \sum_{i=k}^{\infty} \hat{B}_{\mathrm{x}}(x(i)) + \hat{B}_{\mathrm{u}}(u(i))$ due to the definition of $\hat{\ell}(\cdot,\cdot)$ and the optimality of the unconstrained infinite-horizon LQR solution. In combination, this yields

$$\hat{J}_N(U(k),x(k)) \geq x(k)^\top P_{\mathrm{uc}}^* x(k) + \varepsilon \sum_{i=k}^{\infty} \hat{B}_{\mathrm{x}}(x(i)) + \hat{B}_{\mathrm{u}}(u(i)). \tag{C.120}$$

Since due to the recentering of the underlying barrier functions all terms in the sum on the right-hand side are guaranteed to be positive, we finally obtain (C.118).

Part 2. Based on these observations, let us now define the time-varying upper bound on the right-hand side of (C.118) for the ease of notation as

$$\hat{\alpha}(k) := \hat{J}_N(U(k),x(k)) - x(k)^\top P_{\mathrm{uc}}^* x(k). \tag{C.121}$$

Note that $\hat{\alpha}(k) \in \mathbb{R}_+$ by definition, see (C.120). It then follows immediately from (C.118) and (C.121) that upper bounds for the maximal violations of state and input constraints are given by

$$\hat{z}_{\mathrm{x}}^i(k) = \max_{\xi} \left\{ C_{\mathrm{x}}^i \xi - d_{\mathrm{x}}^i \mid \varepsilon \hat{B}_{\mathrm{x}}(\xi) \leq \hat{\alpha}(k) \right\}, \quad i = 1,\dots,q_{\mathrm{x}} \tag{C.122a}$$

$$\hat{z}_{\mathrm{u}}^j(k) = \max_{v} \left\{ C_{\mathrm{u}}^j v - d_{\mathrm{u}}^j \mid \varepsilon \hat{B}_{\mathrm{u}}(v) \leq \hat{\alpha}(k) \right\}, \quad j = 1,\dots,q_{\mathrm{u}}. \tag{C.122b}$$

Thus, (4.27) holds with the elements of $\hat{z}_{\mathrm{x}}(k) \in \mathbb{R}^{q_{\mathrm{x}}}$ and $\hat{z}_{\mathrm{u}}(k) \in \mathbb{R}^{q_{\mathrm{u}}}$ being given by (C.122). Note that the optimization problems in (C.122) are convex as the involved barrier functions are convex and the objective functions are linear in the respective variables.

Part 3. Let us now show that $\hat{\alpha}(k)$ given in (C.121) is strictly monotonically decreasing and that $\lim_{k\to\infty} \hat{\alpha}(k) = 0$. First, note that the cost function decrease implies that

$$\hat{\alpha}(k+1) - \hat{\alpha}(k) \leq -\hat{\ell}(x(k),u(k)) - x(k+1)^\top P_{\mathrm{uc}}^* x(k+1) + x(k)^\top P_{\mathrm{uc}}^* x(k). \tag{C.123}$$

Now recall $\hat{\ell}(x(k),u(k)) = \|x(k)\|_Q^2 + \|u(k)\|_R^2 + \varepsilon \hat{B}_{\mathrm{x}}(x(k)) + \varepsilon \hat{B}_{\mathrm{u}}(u(k))$ and note that $\|x(k)\|_Q^2 + \|u(k)\|_R^2 \geq x(k)^\top P_{\mathrm{uc}}^* x(k) - x(k+1)^\top P_{\mathrm{uc}}^* x(k+1)$ due to the principle of optimality. Inserting these relations into (C.123) finally reveals that

$$\hat{\alpha}(k+1) - \hat{\alpha}(k) \leq -\varepsilon(\hat{B}_{\mathrm{x}}(x(k)) + \hat{B}_{\mathrm{u}}(u(k))). \tag{C.124}$$

Since $\hat{B}_x(\cdot)$ and $\hat{B}_u(\cdot)$ are positive definite, this ensures that $\alpha(k)$ decreases strictly monotonically over time. Moreover, it follows directly from (C.124) and the convergence of the closed-loop system state to the origin that $\lim_{k\to\infty}\hat{\alpha}(k) = 0$. As $\hat{\alpha}(k)$ decreases monotonically, the same holds for the upper bounds $\hat{z}_x^i(k)$ and $\hat{z}_u^j(k)$ given in (C.122), which proves the claimed strict monotonicity result. Finally, as both $x(k)$ and $u(k)$ asymptotically converge to the origin (see Theorem 4.3), there always exists a finite $k_0 \in \mathbb{N}$ such that $\hat{z}_x(k) \leq 0$, $\hat{z}_u(k) \leq 0$ for any $k \geq k_0$.

C.22 Proof of Theorem 4.5

We prove the result for the case of zero constraint violations, i.e. $\hat{z}_{x,\text{tol}} = 0$, $\hat{z}_{u,\text{tol}} = 0$, which then also comprises all cases with positive tolerances. The proof consists of two parts. First, we show that for any $\delta \in \mathbb{R}_{++}$ there exists a compact and nonempty set $\hat{\mathcal{Z}}_N(\delta) \subseteq \mathbb{R}^{Nm} \times \mathbb{R}^n$ such that for any $(U_0, x_0) \in \hat{\mathcal{Z}}_N(\delta)$, the state and input constraints will be satisfied also for all future steps, i.e., (4.29) will hold with $\hat{z}_{x,\text{tol}} = 0$, $\hat{z}_{u,\text{tol}} = 0$. Then, we show in the second part that for any compact set $\mathcal{X}_0 \subseteq \mathcal{X}_N^\circ$ there exists $\bar{\delta}_0 \in \mathbb{R}_{++}$ and $\gamma \in \mathcal{K}_\infty$ such that if $x_0 \in \mathcal{X}_0$, $\delta \leq \bar{\delta}_0$, and $U_0 \in \mathcal{B}^{Nm}_{\gamma(\delta)}(\hat{U}^*(x_0))$, then $(U_0, x_0) \in \hat{\mathcal{Z}}_N(\delta)$.

Part 1. Following the key idea behind the proofs of Theorems 3.6 and 3.7, let us for a given relaxation parameter $\delta \in \mathbb{R}_{++}$ define the scalars $\bar{\beta}_x(\delta), \bar{\beta}_u(\delta), \bar{\beta}(\delta) \in \mathbb{R}_{++}$ as

$$\bar{\beta}_x(\delta) := \min_{i,\xi}\{\hat{B}_x(\xi) \,|\, C_x\xi \leq d_x,\ C_x^i\xi = d_x^i\},\ i = 1,\dots,q_x, \tag{C.125a}$$

$$\bar{\beta}_u(\delta) := \min_{j,v}\{\hat{B}_u(v) \,|\, C_u v \leq d_u,\ C_u^j v = d_u^j\},\ j = 1,\dots,q_u, \tag{C.125b}$$

$$\bar{\beta}(\delta) = \min\{\bar{\beta}_x(\delta), \bar{\beta}_u(\delta)\}, \tag{C.125c}$$

which can be interpreted as lower bounds for the values that are attained by the relaxed barrier functions on the boundaries of the constraint sets $\mathcal{X}$ and $\mathcal{U}$, respectively. As shown by Lemma 3.1 in Section 3.5, the $\bar{\beta}(\delta)$-sublevel sets of $\hat{B}_x(\cdot)$ and $\hat{B}_u(\cdot)$ will always be contained within the sets $\mathcal{X}, \mathcal{U}$. Based on this observation and inspired by the proof of Theorem 4.4, we introduce

$$\hat{\alpha}(U, x) := \hat{J}_N(U, x) - x^\top P_{\text{uc}}^* x \tag{C.126}$$

and define the set $\hat{\mathcal{Z}}_N(\delta) \subset \mathbb{R}^{Nm} \times \mathbb{R}^n$ as

$$\hat{\mathcal{Z}}_N(\delta) := \left\{(U, x) \in \mathbb{R}^{Nm} \times \mathbb{R}^n \,|\, \hat{\alpha}(U, x) \leq \varepsilon\bar{\beta}(\delta)\right\}, \tag{C.127}$$

where $\bar{\beta}(\delta)$ is given by (C.125) and $P_{\text{uc}}^* \in \mathbb{S}_{++}^n$ is again the solution to the DARE related to the infinite-horizon LQR problem for (A, B, Q, R). Suppose now that $(U_0, x_0) \in \hat{\mathcal{Z}}_N(\delta)$. Based on the proof of Theorem 4.4, we know that $\varepsilon\hat{B}_x(x(k)) \leq \hat{\alpha}(U(k), x(k))$, $\varepsilon\hat{B}_u(u(k)) \leq \hat{\alpha}(U(k), x(k))$, see (C.118). In addition, $\hat{\alpha}(U(k), x(k)) \leq \hat{\alpha}(U_0, x_0)$ for all $k \in \mathbb{N}$, cf. (C.124). So if $(U_0, x_0) \in \hat{\mathcal{Z}}_N(\delta)$, then $\hat{B}_x(x(k)) \leq \bar{\beta}(\delta)$, $\hat{B}_u(u(k)) \leq \bar{\beta}(\delta)$, which by the definition of $\bar{\beta}(\delta)$ and Lemma 3.1 implies that $x(k) \in \mathcal{X}$, $u(k) \in \mathcal{U}$ for all $k \in \mathbb{N}$. Thus, for given $\delta \in \mathbb{R}_{++}$, (4.29) holds with $\hat{z}_{\mathrm{x,tol}} = 0$, $\hat{z}_{\mathrm{u,tol}} = 0$ whenever $(U_0, x_0) \in \hat{\mathcal{Z}}_N(\delta)$.

Part 2. In the following, we are going to prove our second claim, namely the existence of $\bar{\delta}_0 \in \mathbb{R}_{++}$ and $\gamma \in \mathcal{K}_\infty$ with the properties specified above. As we need to vary the relaxation parameter, we make the influence of δ on the cost function more explicit. In particular, we will use $\hat{J}_N(U, x; \delta)$ to denote the value of the cost function for $(U, x) \in \mathbb{R}^{Nm} \times \mathbb{R}^n$ and a given relaxation parameter $\delta \in \mathbb{R}_{++}$. In accordance with (C.126), we further define $\hat{\alpha}(U, x; \delta) := \hat{J}_N(U, x; \delta) - x^\top P^*_{\mathrm{uc}} x$. Based on this notation, it obviously holds that

$$(U_0, x_0) \in \hat{\mathcal{Z}}_N(\delta) \Leftrightarrow \hat{\alpha}(U_0, x_0; \delta) \leq \varepsilon \bar{\beta}(\delta) \,. \tag{C.128}$$

We now proceed as follows: first we show in *a)* that for any *fixed* $x_0 \in \mathcal{X}_0 \subseteq \mathcal{X}_N^\circ$ and a corresponding suitably chosen U_0 there exists a $\delta_0'(x_0)$ such that (C.128) holds for all $\delta \leq \delta_0'(x_0)$; in part *b)*, we then show the claimed result for $\mathcal{X}_0$ by *uniformly* choosing $\bar{\delta}_0 = \min_{x_0 \in \mathcal{X}_0} \delta_0'(x_0)$.

a) The key idea behind the following arguments is to show that for any $x_0 \in \mathcal{X}_0 \subset \mathcal{X}_N^\circ$ and suitably chosen U_0, the term $\hat{\alpha}(U_0, x_0; \delta)$ will stay bounded when δ is decreased below a certain threshold. As, on the other hand, $\bar{\beta}(\delta)$ grows without bound for $\delta \to 0$, the right-hand side of (C.128) can then always be satisfied by choosing δ sufficiently small.

Consider a fixed $x_0 \in \mathcal{X}_0 \subseteq \mathcal{X}_N^\circ$ and let $\tilde{U}^*(x_0)$ and $\tilde{J}_N^*(x_0) = \tilde{J}_N(\tilde{U}^*(x_0), x_0)$ denote the optimal solution and value function associated to the related *nonrelaxed* barrier function based problem formulation with additional zero terminal state constraint, see (C.46) in the proof of Theorem 3.7. As shown by Lemma B.5 in Appendix B.7, the optimal solution $\tilde{U}^*(x_0)$ associated to this problem is unique for any $x_0 \in \mathcal{X}_N^\circ$ and $\tilde{U}^* : \mathcal{X}_N^\circ \to \mathbb{R}^{Nm}$ is a continuous map. As before, the additional constraint $x_N = 0$ is required in order to ensure that $\tilde{J}_N^*(x_0)$ does not depend on δ via the quadratic terminal cost that is used in the relaxed formulation. By definition, $\tilde{U}^*(x_0)$ exists for any $x_0 \in \mathcal{X}_0$ and results in strictly feasible state and input sequences $\{\tilde{x}_0, \tilde{x}_1, \ldots \tilde{x}_N\}$ and $\{\tilde{u}_0, \tilde{u}_1, \ldots \tilde{u}_{N-1}\}$, respectively. Now define

$$\delta_0(x_0) := \min_{i,j,k} \left\{ -C_{\mathrm{x}}^i \tilde{x}_k + d_{\mathrm{x}}^i,\ -C_{\mathrm{u}}^j \tilde{u}_k + d_{\mathrm{u}}^j \right\} \tag{C.129}$$

for $i = 1, \ldots, q_{\mathrm{x}}$, $j = 1, \ldots, q_{\mathrm{u}}$, $k = 0, \ldots, N-1$, which characterizes the minimal distance of the nonrelaxed open-loop trajectories to the boundaries of the respective constraint sets. Now note that $\hat{J}_N(\tilde{U}^*(x_0), x_0; \delta) = \tilde{J}_N^*(x_0)$, whenever we choose $\delta \leq \delta_0(x_0)$, since in this case no constraint relaxation is active. In addition, it always holds due to optimality that $\hat{J}_N^*(x_0; \delta) \leq \hat{J}_N(\tilde{U}^*(x_0), x_0; \delta)$ and, thus, $\hat{J}_N^*(x_0; \delta) \leq \tilde{J}_N^*(x_0)$ for all $\delta \leq \delta_0(x_0)$. As $\tilde{U}^*(x_0)$ is a continuous map, $\delta_0(x_0)$ in (C.129) is also continuous in x_0.

Next, we derive an upper bound for the cost function $\hat{J}_N(U_0, x_0; \delta)$ for U_0 in the neighborhood of $\hat{U}^*(x_0)$. Assume $\delta \leq \delta_0(x_0)$ and consider $U_0 \in \mathcal{B}_{\gamma_0}^{Nm}(\hat{U}^*(x_0))$, i.e. $U_0 = \hat{U}^*(x_0) + \gamma_0 v$, $\|v\| \leq 1$, for $\gamma_0 \in \mathbb{R}_{++}$. Due to Taylor's Theorem, see (Nocedal and Wright, 1999, Theorem 2.1), and the fact that the recentering ensures $\nabla_U \hat{J}_N(\hat{U}^*(x_0), x_0; \delta) = 0$, it holds that

$$\hat{J}_N(U_0, x_0; \delta) = \hat{J}_N(\hat{U}^*(x_0), x_0; \delta) + \frac{1}{2}\gamma_0^2 v^\top \nabla_U^2 \hat{J}_N(\hat{U}^*(x_0) + sv, x_0; \delta) v \tag{C.130}$$

for some $s \in (0,1)$. However, together with $\hat{J}_N(\hat{U}^*(x_0), x_0; \delta) = \hat{J}_N^*(x_0; \delta) \leq \tilde{J}_N^*(x_0)$ from above, this implies

$$\hat{J}_N(U_0, x_0; \delta) \leq \tilde{J}_N^*(x_0) + \frac{\gamma_0^2}{2} \left\| H + \frac{\varepsilon}{\delta^2} G^\top \text{diag}(w) G \right\|, \tag{C.131}$$

where we used $\|v\| \leq 1$ and the fact that $\nabla_U^2 \hat{J}_N(U, x; \delta) \preceq H + \frac{\varepsilon}{\delta^2} G^\top \text{diag}(w) G$ for any $(U, x) \in \mathbb{R}^{Nm} \times \mathbb{R}^n$, see Theorem 4.1 and the associated proof in Section C.18. Let us now choose $\gamma_0 = \gamma(\delta) = \delta / \|H\|^{1/2}$. Then, it follows from (C.131) that for any $x_0 \in \mathcal{X}_0$, any $\delta \leq \delta_0(x_0)$, and any $U_0 \in \mathcal{B}^{Nm}_{\gamma(\delta)}(\hat{U}^*(x_0))$

$$\hat{J}_N(U_0, x_0; \delta) \leq \tilde{J}_N^*(x_0) + \frac{1}{2} \left(\delta^2 + \frac{\varepsilon}{\|H\|} \left\| G^\top \text{diag}(w) G \right\| \right). \tag{C.132}$$

Note that due to the fact that the underlying global quadratic upper bounds depend on the choice of the relaxation parameter, the matrix H, and hence also $\|H\|$, depend via the terminal cost matrix P given in (4.5) directly on δ. That is, $P = P(\delta)$ and $H = H(\delta)$. It is therefore not immediately clear how the right-hand side of (C.132) will behave for decreasing values of δ. However, we can state the following. As the (strong) solution of the DARE (4.5) is continuous and monotonic in the δ-dependent matrices M_x and M_u (Lancaster et al., 1987), it follows that the solution $P \in \mathbb{S}^n_{++}$ is continuous and monotonic in δ and that, in particular, $P(\delta) \succeq P(\delta_0)$ for all $\delta \leq \delta_0$. Likewise, the matrix $H = H(\delta) \in \mathbb{S}^n_{++}$ as given in Appendix B.9 is continuous and monotonic in δ and $H(\delta) \succeq H(\delta_0)$ for all $\delta \leq \delta_0$. This observation hast two important implications. First, the derived monotonicity relations imply together with (C.132) that we can state for any $U_0 \in \mathcal{B}^{Nm}_{\gamma(\delta)}(\hat{U}^*(x_0))$ with $\delta \leq \delta_0(x_0)$, the following upper bound on the cost function

$$\hat{J}_N(U_0, x_0; \delta) \leq \tilde{J}_N^*(x_0) + \frac{1}{2} \left(\delta_0^2 + \frac{\varepsilon}{\|H(\delta_0)\|} \left\| G^\top \text{diag}(w) G \right\| \right). \tag{C.133}$$

Here, we used that if $H(\delta) \succeq H(\delta_0)$ for all $\delta \leq \delta_0$, then $\|H(\delta)\| \geq \|H(\delta_0)\|$ due to monotonicity of the spectral matrix norm. Thus, we conclude that the cost function value, and hence also $\hat{\alpha}(U_0, x_0; \delta)$ in (C.128), will stay bounded for any $\delta \leq \delta_0(x_0)$ if we choose the optimizer initialization U_0 sufficiently close to $\hat{U}^*(x_0)$. Second, the fact that $H(\delta)$ is continuous and monotonically increasing for decreasing δ reveals that $\gamma : \delta \mapsto \delta / \|H(\delta)\|^{1/2}$ is continuous and strictly monotonically increasing. In addition, it is straightforward to show that $\lim_{\delta \to 0} \gamma(\delta) = 0$ and that $\gamma(\delta) \to \infty$ for $\delta \to \infty$. Thus, $\gamma \in \mathcal{K}_\infty$.
Let us now choose $\delta_0'(x_0) \leq \delta_0(x_0)$ in such a way that the right-hand side of (C.128) is satisfied for all $\delta \leq \delta_0'(x_0)$, which then implies that $(U_0, x_0) \in \hat{\mathcal{Z}}_N(\delta)$. In particular, let

$$\delta_0'(x_0) := \max \left\{ \delta \in \mathbb{R}_{++} \,\middle|\, \delta \leq \delta_0(x_0), \right. \tag{C.134}$$

$$\left. \tilde{J}_N^*(x_0) + \frac{1}{2} \left(\delta_0(x_0)^2 + \frac{\varepsilon}{\|H(\delta_0(x_0))\|} \|G^\top \text{diag}(w) G\| \right) - x_0^\top P_{\text{uc}}^* x_0 \leq \varepsilon \bar{\beta}(\delta) \right\}.$$

In combination with the definition of $\hat{\alpha}(U,x;\delta)$, it follows from (C.133) and (C.134) that the right-hand side of (C.128) will be satisfied for any $\delta \leq \delta_0'(x_0)$ whenever $U_0 \in \mathcal{B}^{Nm}_{\gamma(\delta)}(\hat{U}^*(x_0))$. Hence, $(U_0, x_0) \in \hat{\mathcal{Z}}_N(\delta)$ for any $\delta \leq \delta_0'(x_0)$ whenever $U_0 \in \mathcal{B}^{Nm}_{\gamma(\delta)}(\hat{U}^*(x_0))$. Note that we may picture the construction of $\delta_0'(x_0)$ as decreasing δ from $\delta(x_0)$ until also the second condition in the foregoing definition of $\delta_0'(x_0)$ is satisfied. As $\bar{\beta}(\delta)$ grows without bound for decreasing δ, this can always be achieved, which shows that $\delta'(x_0)$ exists for any $x_0 \in \mathcal{X}_N^\circ$. Moreover, since $\bar{\beta}(\delta)$ is continuous in δ and both $\delta_0(x_0)$ and $\tilde{J}_N^*(x_0)$ are continuous in x_0, see Lemma B.5, $\delta_0'(x_0)$ in (C.134) is also continuous. As already discussed in the proof of Theorem 3.7, continuity of $\bar{\beta}(\delta)$ follows from noting that the constraints in (C.125a) and (C.125b) are independent from δ and that the relaxed barrier functions $\hat{B}_\mathrm{x}(x)$ and $\hat{B}_\mathrm{u}(u)$ are continuous in both x (respectively u) and δ and possess, in addition, compact sublevel sets for any fixed value of δ, see (Fiacco, 1983, Theorem 2.2) as well as (Bonnans and Shapiro, 2000, Proposition 4.4).
b) Consider now the compact set $\mathcal{X}_0 \subseteq \mathcal{X}_N^\circ$ and define $\bar{\delta}_0 \in \mathbb{R}_{++}$ as $\bar{\delta}_0 = \min_{x_0 \in \mathcal{X}_0} \delta_0'(x_0)$. Due to the continuity of $\delta_0'(x_0)$ and the compactness of $\mathcal{X}_0$, this value always exists by virtue of the Weierstraß extreme value theorem. Then, by the above arguments, $(U_0, x_0) \in \hat{\mathcal{Z}}_N(\delta)$ for any $x_0 \in \mathcal{X}_0$ and any $\delta \leq \bar{\delta}_0$ if we choose U_0 as $U_0 \in \mathcal{B}^{Nm}_{\gamma(\delta)}(\hat{U}^*(x_0))$ with $\gamma(\delta) = \delta / \|H(\delta)\|^{1/2} \in \mathcal{K}_\infty$, which proves our second claim.
As mentioned above, the zero tolerance case comprises all other cases with nonzero tolerances $\hat{z}_{\mathrm{x,tol}} \in \mathbb{R}_+^{q_x}$, $\hat{z}_{\mathrm{u,tol}} \in \mathbb{R}_+^{q_u}$, and the proof is complete.

C.23 Proof of Theorem 4.6

Due to the assumption that the optimization algorithm operator $\Phi^{i_\mathrm{T}(k)} : \mathbb{R}^{Nm} \times \mathbb{R}^n \to \mathbb{R}^{Nm}$ is continuous, the right hand side of (4.30) is continuous in all of its arguments for any given sequence $\boldsymbol{i}_\mathrm{T} = \{i_\mathrm{T}(0), i_\mathrm{T}(1), \ldots\}$. Thus, the claimed IISS result follows directly from the fact that global asymptotic stability and integral input-to-state stability are in the discrete-time case equivalent, see the main result of Angeli (1999), which is for the sake of convenience recalled in Theorem A.4 in Appendix A.2.

C.24 Proof of Theorem 4.7

Let us first reformulate the closed-loop (4.30) in such a way that the influence of the optimization error becomes more explicit. Let $\hat{x}(k) = x(k) + e(k)$ denote the inexact state estimate and let $\hat{x}^+(k) = A\hat{x}(k) + B\Pi_0 U(k)$ be the corresponding predicted next nominal system state. As can be seen from (4.30) and the definition of the optimization algorithm operator in (4.22)–(4.25), the iterative optimization will for each $k \in \mathbb{N}$ try to find the minimum of $\hat{J}_N(U, \hat{x}^+(k))$. Consequently, at each sampling step $k \in \mathbb{N}$, the optimization error may be written as

$$\epsilon(k) = \Phi^{i_\mathrm{T}(k)}\left(U(k), \hat{x}(k)\right) - \hat{U}^*(\hat{x}^+(k)) = U(k+1) - \hat{U}^*(\hat{x}^+(k)) \,, \tag{C.135}$$

by which system (4.30) can be reformulated as

$$x(k+1) = A\,x(k) + B\Pi_0\,U(k) + w(k)\,, \tag{C.136a}$$

$$U(k+1) = \hat{U}^*\left(A\big(x(k)+e(k)\big) + B\Pi_0\,U(k)\right) + \epsilon(k)\,. \tag{C.136b}$$

The proof is then based on the following two auxiliary results from Chapter 3: *i*) when applying the exact optimal feedback $u(k) = \Pi_0\,\hat{U}^*(x(k))$, the associated closed-loop state dynamics (4.7) are ISS with with respect to arbitrary additive disturbances, see Theorem 3.8, and *ii*) the optimal solution $\hat{U}^*(x)$ related to (4.6) is Lipschitz continuous in x, see Lemma 3.8. Note now that, based on (C.136b), the system state dynamics (C.136a) can be rewritten as

$$x(k+1) = Ax(k) + B\Pi_0\,\hat{U}^*(x(k)) + \tilde{w}(k) \tag{C.137}$$

with $\tilde{w}(k)$ being defined as

$$\tilde{w}(k) = B\Pi_0\left(\hat{U}^*(\hat{x}^+(k-1)) - \hat{U}^*(x(k)) + \epsilon(k-1)\right) + w(k)\,, \tag{C.138}$$

where $\hat{x}^+(k-1) = A(x(k-1)+e(k-1)) + B\Pi_0 U(k-1)$. Due to the aforementioned ISS result provided by Theorem 3.8, there exist $\beta_{\mathrm{x}} \in \mathcal{KL}$ and $\gamma_{\mathrm{x}} \in \mathcal{K}$ such that for $\xi = x(0)$,

$$\|x(k,\xi,\tilde{\boldsymbol{w}})\| \le \beta_{\mathrm{x}}(\|\xi\|,k) + \gamma_{\mathrm{x}}(\|\tilde{\boldsymbol{w}}_{[k-1]}\|)\,. \tag{C.139}$$

Furthermore, by noting that $\hat{x}^+(k-1)$ can via (C.136a) be expressed as $\hat{x}^+(k-1) = x(k) + Ae(k-1) - w(k-1)$ and by exploiting the Lipschitz continuity of $\hat{U}^*(\cdot)$, we get

$$\|\tilde{w}(k)\| \le L_U\|A\|\|B\|\|e(k-1)\| + \|B\|\|\epsilon(k-1)\| + L_U\|B\|\|w(k-1)\| + \|w(k)\|\,, \tag{C.140}$$

which reveals that for any $k \in \mathbb{N}_+$

$$\|x(k,\xi,\boldsymbol{w},\boldsymbol{e},\boldsymbol{\epsilon})\| \le \beta_{\mathrm{x}}(\|\xi\|,k) + \gamma'_{\mathrm{w}}(\|\boldsymbol{w}_{[k-1]}\|) + \gamma'_{\mathrm{e}}(\|\boldsymbol{e}_{[k-2]}\|) + \gamma'_{\epsilon}(\|\boldsymbol{\epsilon}_{[-1,k-2]}\|) \tag{C.141}$$

with $\gamma'_{\mathrm{w}}(r) = \gamma_{\mathrm{x}}(3(1+L_U\|B\|)\,r)$, $\gamma'_{\mathrm{e}}(r) = \gamma_{\mathrm{x}}(3L_U\|A\|\|B\|\,r)$, and $\gamma'_{\epsilon}(r) = \gamma_{\mathrm{x}}(3\|B\|\,r)$. We made use of the fact that $\gamma_{\mathrm{x}}(r_1+\cdots+r_n) \le \gamma_{\mathrm{x}}(nr_1)+\cdots+\gamma_{\mathrm{x}}(nr_n)$, which holds for every $\gamma_{\mathrm{x}} \in \mathcal{K}$ for all $r_1,\ldots,r_n \in \mathbb{R}_+$ (Rawlings and Ji, 2012). Note also that (C.141) nicely illustrates that the optimization error effects the system state with a delay of two sampling instants. This is due to the fact that according to (C.136a) $\epsilon(k)$ affects $U(k+1)$ and then $U(k+1)$ affects $x(k+2)$. Of course, this heavily depends on the used notation, in particular on the way how the optimization error is defined. However, it should be noted that, as stated in the theorem, the initial optimization error $\epsilon(-1) := U(0) - \hat{U}^*(x(0))$ needs to be included into the sequence $\boldsymbol{\epsilon} = \{\epsilon(-1),\epsilon(0),\ldots\}$. Let us now consider the optimizer dynamics. To this end, note that (C.136b) and again $\hat{x}^+(k) = A\big(x(k)+e(k)\big) + B\Pi_0\,U(k) = x(k+1) + Ae(k) - w(k)$ directly imply that

$$U(k) = \hat{U}^*(x^+(k-1)) + \epsilon(k-1) = \hat{U}^*\big((x(k) + Ae(k-1) - w(k-1)\big) + \epsilon(k-1)\,. \tag{C.142}$$

Lipschitz continuity of the exact optimizer and $\hat{U}^*(0) = 0$ ensure that $\|\hat{U}^*(x)\| \leq L_U \|x\|$ for any $x \in \mathbb{R}^n$. Thus,

$$\begin{aligned} \|U(k)\| &= \|\hat{U}^*((x(k) + Ae(k-1) - w(k-1)) + \epsilon(k-1)\| && \text{(C.143a)} \\ &\leq L_U \|x(k) + Ae(k-1) - w(k-1)\| + \|\epsilon(k-1)\| && \text{(C.143b)} \\ &\leq L_U \|x(k)\| + L_U \|A\| \|e(k-1)\| + L_U \|w(k-1)\| + \|\epsilon(k-1)\|. && \text{(C.143c)} \end{aligned}$$

Inserting now the upper bound for $\|x(k)\|$ given in (C.141), we obtain that for $k \in \mathbb{N}_+$

$$\|U(k, \xi, \boldsymbol{w}, \boldsymbol{e}, \boldsymbol{\epsilon})\| \leq \beta_{\mathrm{u}}(\|\xi\|, k) + \gamma''_{\mathrm{w}}(\|\boldsymbol{w}_{[k-1]}\|) + \gamma''_{\mathrm{e}}(\|\boldsymbol{e}_{[k-1]}\|) + \gamma''_{\epsilon}(\|\boldsymbol{\epsilon}_{[-1,k-1]}\|), \quad \text{(C.144)}$$

where $\beta_{\mathrm{u}}(r,s) = L_U \beta_{\mathrm{x}}(r,s)$ and $\gamma''_{\mathrm{w}}(r) = L_U(\gamma'_{\mathrm{w}}(r) + r)$, $\gamma''_{\mathrm{e}}(r) = L_U(\gamma'_{\mathrm{e}}(r) + \|A\| r)$, $\gamma''_{\epsilon}(r) = L_U \gamma'_{\epsilon}(r) + r$. By combining (C.141) and (C.144) and noting that $\|\xi\| = \|x(0)\| \leq \|\zeta\| = \|(x(0), U(0))\|$, we then finally get the claimed result. In particular, we may conclude that for any $k \in \mathbb{N}_+$

$$\|(x(k), U(k))\| \leq \beta(\|\zeta\|, k) + \gamma_{\mathrm{w}}(\|\boldsymbol{w}_{[k-1]}\|) + \gamma_{\mathrm{e}}(\|\boldsymbol{e}_{[k-1]}\|) + \gamma_{\epsilon}(\|\boldsymbol{\epsilon}_{[-1,k-1]}\|) \quad \text{(C.145)}$$

with $\beta(r,s) = \beta_{\mathrm{x}}(r,s) + \beta_{\mathrm{u}}(r,s)$, $\gamma_{\mathrm{w}}(r) = \gamma'_{\mathrm{w}}(r) + \gamma''_{\mathrm{w}}(r)$, $\gamma_{\mathrm{v}}(r) = \gamma'_{\mathrm{v}}(r) + \gamma''_{\mathrm{v}}(r)$, and $\gamma_{\epsilon}(r) = \gamma'_{\epsilon}(r) + \gamma''_{\epsilon}(r)$. As outlined above, the optimization error sequence needs to include the initial error $\varepsilon(-1) = U(0) - \hat{U}^*(x(0))$. This completes the proof.

C.25 Proof of Lemma 4.1

It is a well-known fact that a feasibly initialized Newton method leads also in the equality constrained case to a descent direction with respect to the underlying cost function (Boyd and Vandenberghe, 2004, Section 10.2). In the following, we show this result for the relaxed barrier function based problem at hand and derive an explicit bound for the minimally required step size.

Let us for the Newton method search direction given in (4.51) define the associated squared Newton decrement as follows

$$\lambda(w,x)^2 := \Delta w^\top \nabla^2_w \hat{J}_N(w,x) \Delta w\,, \quad \text{(C.146)}$$

which allows us to rewrite the Armijo-type condition (4.57) as

$$\hat{J}_N(w + s\Delta w, x) \leq \hat{J}_N(w,x) - c_1 s\, \lambda(w,x)^2\,. \quad \text{(C.147)}$$

On the other hand, applying Taylor's Theorem to the left-hand side of (C.147), we get

$$\hat{J}_N(w + s\Delta w, x) = \hat{J}_N(w,x) + s \nabla_w \hat{J}_N(w,x)^\top \Delta w + \frac{1}{2} s^2 \Delta w^\top \nabla^2_w \hat{J}_N(w + \tau s \Delta w, x) \Delta w \quad \text{(C.148)}$$

for some $\tau \in (0,1)$, see for example Theorem 2.1 in (Nocedal and Wright, 1999). Exploiting now the upper and lower bounds from (4.48a), we obtain

$$\hat{J}_N(w + s\Delta w, x) \leq \hat{J}_N(w,x) + s \nabla_w \hat{J}_N(w,x)^\top \Delta w + \frac{s^2 L}{2\mu} \lambda(w,x)^2\,. \quad \text{(C.149)}$$

Furthermore, (4.50) and (4.51) imply $\nabla_w \hat{J}_N(w,x) = r_d - C^\top \nu = -\nabla_w^2 \hat{J}_N(w,x)\Delta w - C^\top \tilde{\nu}$ with $\tilde{\nu} := \nu + \Delta\nu$, as well as $C\Delta w = -r_p = -(Cw - b(x))$. Thus, together with (C.149),

$$\hat{J}_N(w + s\Delta w, x) \leq \hat{J}_N(w,x) - s\Big(1 - \frac{L}{2\mu}s\Big)\lambda(w,x)^2 + s\tilde{\nu}^\top\left(Cw - b(x)\right). \tag{C.150}$$

However, since we assumed that (w,x) satisfies the underlying equality constraints (4.47b), the last term on the right-hand side vanishes, which allows us to conclude that

$$\hat{J}_N(w + s\Delta w, x) \leq \hat{J}_N(w,x) - s\Big(1 - \frac{L}{2\mu}s\Big)\lambda(w,x)^2. \tag{C.151}$$

Being in fact equivalent to the result that we obtained for the condensed case in Appendix B.10, (C.151) directly implies that (C.147), and hence (4.57), will be satisfied for any step size satisfying $s \leq \bar{s} := 2\mu(1 - c_1)/L$. Note that a corresponding backtracking line search will therefore again always terminate with $s = \rho^j \geq \rho\bar{s}$, which shows that the maximal number of backtracking iteration is bounded as $j \leq \lceil 1 + \log_\rho(2\mu(1 - c_1)/L)\rceil$. Furthermore, (C.150) nicely reveals that primal feasibility of (w,x) is in fact crucial in order to ensure a guaranteed decay in the associated cost function.

C.26 Proof of Theorem 4.8

Due to the design of the underlying primordial relaxed logarithmic barrier function $\hat{B}(\cdot)$, the applied recentering, and $Q \in \mathbb{S}^n_{++}$, $R \in \mathbb{S}^n_{++}$, $P \in \mathbb{S}^n_{++}$, both the barrier function based stage cost $\hat{\ell} : \mathbb{R}^n \times \mathbb{R}^m \to \mathbb{R}_+$ and the quadratic terminal cost $\hat{F} : \mathbb{R}^n \to \mathbb{R}_+$ are guaranteed to be continuous, positive definite, and radially unbounded. As a direct consequence, $\hat{J}_N : \mathbb{R}^{n_w} \times \mathbb{R}^n \to \mathbb{R}_+$ given in (4.47) is a continuous, positive definite, and radially unbounded function. In view of (Khalil, 2002, Lemma 4.3), this implies that there exist $\underline{\alpha}, \bar{\alpha} \in \mathcal{K}_\infty$ such that

$$\underline{\alpha}(\|(w,x)\|) \leq \hat{J}_N(w,x) \leq \bar{\alpha}(\|(w,x)\|), \tag{C.152}$$

for all $(w,x) \in \mathbb{R}^{n_w} \times \mathbb{R}^n$. The remainder of the proof is structured as follows. In the first part, we show that for a feasible initialization (w_0, x_0) (meaning that $Cw_0 = b(x_0)$), the cost function will in addition to (C.152) always decrease along trajectories of the closed-loop system. In the second part, we then use this result for proving the claimed stability result for arbitrary initializations $(w_0, x_0) \in \mathbb{R}^{n_w} \times \mathbb{R}^n$. The main insight underlying the following arguments is that the closed-loop trajectories resulting from an infeasible initialization (w_0, x_0) will for $k \geq 1$ always be identical to those associated to a suitably chosen feasible initialization, i.e., the projection of (w_0, x_0) onto the set $\mathcal{H} := \{(w,x) \mid Cw = b(x)\}$.

Part 1. Consider an arbitrary but fixed initial condition $(w_0, x_0) \in \mathbb{R}^{n_w} \times \mathbb{R}^n$ and assume for the moment that $Cw_0 = b(x_0)$. Consider the successor state $x_0^+ = Ax_0 + B\Pi_0 w_0$ and let the corresponding warm-start solution $w_0^+ = \Psi_s(w_0, x_0)$ be computed via the shift operator $\Psi_s(\cdot,\cdot)$ given in (4.55). As w_0 satisfies the underlying equality constraints related to the linear system dynamics, w_0^+ is nothing else than the stacked vector of the input and state sequences associated to the shifted (suboptimal) candidate solution $\hat{u}^+(x_0)$ that we applied in the stability proofs of Theorems 3.3 and 4.3. As a consequence, we observe that

$$\begin{aligned} &\hat{J}_N(w_0^+, x_0^+) - \hat{J}_N(w_0, x_0) && \text{(C.153a)}\\ &\quad = -\hat{\ell}(x_0, \Pi_0 w_0) + \|A_K x_N\|_P^2 - \|x_N\|_P^2 + \|x_N\|_Q^2 + \|K x_N\|_R^2 + \varepsilon \hat{B}_K(x_N) && \text{(C.153b)}\\ &\quad \leq -\hat{\ell}(x_0, \Pi_0 w_0), && \text{(C.153c)} \end{aligned}$$

where we again made use of a telescoping sum and exploited the global quadratic bound on $\hat{B}_K(\cdot)$ from Lemma 3.4 as well as the choice of P according to (4.5). Similar to the condensed case, w_0^+ therefore represents a globally valid warm-start solution for the next system state that ensures $Cw_0^+ = b(x_0^+)$ and which will in addition always lead to a decrease in the cost function. Furthermore, satisfaction of the Armijo-type condition (4.57) ensures that the optimization algorithm update, i.e., applying the Newton-based operator $\Psi_\text{o}(\cdot,\cdot)$ given in (4.56), does not only preserve primal feasibility but will also lead to an even further cost function decrease. In particular, for any number $i \in \mathbb{N}_+$ of Newton iterations, it holds that

$$\begin{aligned} \hat{J}_N(\Phi^i(w_0, x_0), x_0^+) &\leq \hat{J}_N(w_0^+, x_0^+) - c_1 s \|\Delta w(w_0^+, x_0^+)\|^2_{\nabla_w^2 \hat{J}_N(w_0^+, x_0^+)} && \text{(C.154a)}\\ &\leq \hat{J}_N(w_0, x_0) - \hat{\ell}(x_0, \Pi_0 w_0) - c_1 s \mu \|\Delta w(w_0^+, x_0^+)\|^2, && \text{(C.154b)} \end{aligned}$$

where $\mu \in \mathbb{R}_{++}$ is given as in Corollary 4.3. As the above holds for any $(w_0, x_0) \in \mathbb{R}^{n_w} \times \mathbb{R}^n$ satisfying $Cw_0 = b(x_0)$, we conclude that the extended state of system (4.53) satisfies in this case for any $k \in \mathbb{N}$ the decrease condition

$$\hat{J}_N(w(k+1), x(k+1)) - \hat{J}_N(w(k), x(k)) \leq -\hat{\ell}(x(k), \Pi_0 w(k)) - \bar{\mu}\|\Delta w(w^+(k), x^+(k))\|^2, \tag{C.155}$$

where $w^+(k) := \Psi_\text{s}(w(k), x(k))$, $x^+(k) = Ax(k) + B\Pi_0 w(k)$, and $\bar{\mu} := c_1 s \mu \in \mathbb{R}_{++}$. As both terms on the right-hand side are nonpositive, this implies that $(w, x) = (0, 0)$ is stable in the sense that the combined plant and optimizer state satisfies $\|(w(k), x(k))\| \leq \underline{\alpha}^{-1}(\bar{\alpha}(\|(w(0), x(0))\|))$ for any $(w(0), x(0)) \in \mathbb{R}^{n_w} \times \mathbb{R}^n$ satisfying $Cw(0) = b(x(0))$. Moreover, (C.155) ensures that all trajectories of system (4.53) will under the above assumptions asymptotically converge to the largest invariant set in

$$\Omega = \left\{ (w, x) \in \mathbb{R}^{n_w} \times \mathbb{R}^n : \hat{\ell}(x, \Pi_0 w) = 0,\ \Delta w(\Psi_\text{s}(w, x), Ax + B\Pi_0 w) = 0 \right\}, \tag{C.156}$$

i.e., to the largest invariance set in which the right-hand side of (C.155) vanishes. The following arguments show that the only solution that can stay identically in the set Ω is given by the origin, i.e., $\Omega = (0, 0)$. First, note that positive definiteness of $\hat{\ell}(\cdot,\cdot)$ together with $\hat{\ell}(x, \Pi_0 w) = 0$ implies that $x = 0$, $\Pi_0 w = 0$. Thus, $x^+ = Ax + B\Pi_0 w = 0$. Consequently, the second condition reveals that $\Delta w(\Psi_\text{s}(w, 0), 0) = 0$ and hence, by virtue of (4.50) and (4.51),

$$\nabla_w \hat{J}_N(\Psi_\text{s}(w, 0), 0) + C^\top(\nu + \Delta\nu) = 0. \tag{C.157}$$

As this is nothing else than the KKT condition (4.50a) for $\tilde{\nu} = \nu + \Delta\nu$ and $x = 0$, we can conclude that $\Psi_s(w, x) = \hat{w}^*(0) = 0$. Together with the definition of the shift update operator given in (4.55), this immediately implies that $u_k = 0$ for $k = 1, \dots, N-1$. Thus, as in the condensed case, we have that $x = 0$ and $u_k = 0$, $k = 0, \dots, N-1$ (note that $u_0 = \Pi_0\, w = 0$). Moreover, as we know that $Cw = b(x) = 0$, the structure of the matrix C (see Appendix B.9) reveals that all the predicted system states are equal to zero as well, i.e., $x_k = 0$ for $k = 1, \dots, N$. Hence, we conclude that $\Omega = (0, 0)$, which reveals that all trajectories of system (4.53) will asymptotically converge to the origin. However, note that we can not yet conclude the desired general *global* asymptotic stability result as the above results only hold for primal feasible initializations satisfying $Cw(0) = b(x(0))$.

Part 2. Let us now turn to the case of an arbitrary initialization (w_0, x_0) that does not necessarily need to satisfy the underlying linear equality constraints. Given an arbitrary but fixed $(w_0, x_0) \in \mathbb{R}^{n_w} \times \mathbb{R}^n$, consider $(\tilde{w}_0, x_0)$ with

$$\tilde{w}_0 := \begin{bmatrix} u_0 \\ Ax_0 + Bu_0 \\ u_1 \\ A^2x_0 + ABu_0 + Bu_1 \\ \vdots \\ u_{N-1} \\ A^Nx_0 + A^{N-1}Bu_0 + \dots + Bu_{N-1} \end{bmatrix} = \tilde{\Gamma}_{\mathrm{w}} w_0 + \tilde{\Gamma}_{\mathrm{x}} x_0\,, \tag{C.158}$$

which can be seen as a projection of the initialization (w_0, x_0) onto the feasible subspace $\mathcal{H} := \{(w, x)\,|\, Cw = b(x)\}$. The underlying matrices $\tilde{\Gamma}_{\mathrm{w}} \in \mathbb{R}^{n_w \times n_w}$ and $\tilde{\Gamma}_{\mathrm{x}} \in \mathbb{R}^{n_w \times n}$ can easily be derived as

$$\tilde{\Gamma}_{\mathrm{w}} = \begin{bmatrix} I & 0 & 0 & 0 & \cdots & 0 & 0 \\ B & 0 & 0 & 0 & \cdots & 0 & 0 \\ 0 & 0 & I & 0 & \cdots & 0 & 0 \\ AB & 0 & B & 0 & \cdots & 0 & 0 \\ \vdots & \vdots & \vdots & \ddots & \ddots & \vdots & \vdots \\ 0 & 0 & 0 & 0 & \cdots & I & 0 \\ A^{N-1}B & 0 & A^{N-2}B & 0 & \cdots & B & 0 \end{bmatrix}, \qquad \tilde{\Gamma}_{\mathrm{x}} = \begin{bmatrix} 0 \\ A \\ 0 \\ A^2 \\ \vdots \\ 0 \\ A^N \end{bmatrix}. \tag{C.159}$$

Note that the above projection preserves both the initial system state x_0 and the values of the initial control input u_0. Consequently, $\tilde{x}_0^+ := Ax_0 + B\Pi_0\,\tilde{w}_0 = Ax_0 + B\Pi_0\, w_0 = x_0^+$ and $\Psi_s(\tilde{w}_0, x_0) = \Psi_s(w_0, x_0)$, which implies that the iterates of the optimization algorithm operator (4.54) will in fact be identical for both initializations. A direct consequence of this observation is that the resulting closed-loop trajectories of system (4.53) will for the potentially different initializations $(\tilde{w}_0, x_0)$ and (w_0, x_0) only differ in the initial extended state $(w(0), x(0))$, whereas they will actually be identical for all $k \geq 1$. This result is made more precise by the following Lemma, which will prove to be useful for the remaining theoretical results from this section.

Lemma C.2. *Let all assumptions of Theorem 4.8 hold and consider the closed-loop system (4.53) with the operator $\Phi^{i_T(k)}$ being defined according to (4.54)–(4.56). Given an arbitrary initialization $(w_0, x_0) \in \mathbb{R}^{n_w} \times \mathbb{R}^n$, let $(\tilde{w}_0, x_0)$ denote its feasible projection as given in (C.158). Then, the trajectories of system (4.53) with initialization (w_0, x_0) coincide for all $k \in \mathbb{N}_+$ with those for the initialization $(\tilde{w}_0, x_0)$. Furthermore, the closed-loop realizations of plant states $x(k)$ and control inputs $u(k) = \Pi_0 w(k)$ coincide for both initializations for all $k \in \mathbb{N}$.*

Proof. As discussed above, the fact that both the initial system state x_0 and the initial control input u_0 are preserved under the considered projection, ensures that the initializations (w_0, x_0) and $(\tilde{w}_0, x_0)$ will always lead to the same initial plant state $x(0)$ and the same initial control input $u(0) = \Pi_0 w(0)$ and, thus, also to the same predicted successor state $x_0^+ = \tilde{x}_0^+$. Furthermore, also the shifted warm start solutions which are used for initialization of the corresponding Newton iteration will be identical, that is,

$$w_0^+ = \Upsilon_s(w_0, x_0) = \Upsilon_s(\tilde{w}_0, x_0) = \tilde{w}_0^+ . \tag{C.160}$$

As a consequence, the underlying Newton iteration is in both cases initialized with the same warm-start solutions $(w_0^+, x_0^+) = (\tilde{w}_0^+, \tilde{x}_0^+)$, which reveals that all subsequent search directions and Newton-based optimizer updates will also be identical. This immediately implies that

$$w(1) = \Phi^{i_T(k)}(w_0, x_0) = \Phi^{i_T(k)}(\tilde{w}_0, x_0) = \tilde{w}(1) , \tag{C.161}$$

which shows that the optimizer states will be identical for $k = 1$. In addition, as both initializations possess the same initial system state x_0 and apply the same initial control input $u(0) = u_0$, we also have that

$$x(1) = Ax_0 + B\Pi_0 w_0 = Ax_0 + B\Pi_0 \tilde{w}_0 = \tilde{x}(1) . \tag{C.162}$$

Thus, the overall state of system (4.53) will for $k = 1$ coincide for both initializations, which automatically implies that the respective closed-loop trajectories will also coincide for all $k \geq 1$, that is, for all $k \in \mathbb{N}_+$.
A direct consequence of this result is that also the plant states $x(k)$ and the control input $u(k) = \Pi_0 w(k)$ will coincide for all $k \in \mathbb{N}_+$. However, as discussed above, this latter claim also holds at $k = 0$, which finally shows that the respective closed-loop state and input trajectories will be identical for all $k \in \mathbb{N}$. ■

An intuitive interpretation of the above result is that the initial values for the optimizer states x_i, $i = 1, \ldots, N$, which are obviously related to the predicted system states, will always be eliminated by the first application of the shift operator and will therefore not affect the closed-loop behavior for $k \geq 1$. Thus, although the cost function may in the general case not necessarily decrease during the transition from $(w(0), x(0))$ to $(w(1), x(1))$, we can nevertheless make use of the above arguments from the feasible initialization case in order to conclude global asymptotic stability of the origin. In particular, since the closed-loop trajectories will be identical for all $k \geq 1$, asymptotic convergence to the origin can immediately be guaranteed also for arbitrary (w_0, x_0). In addition, we know that the resulting extended closed-loop system state will for any $k \geq 1$ and any arbitrary initialization $(w_0, x_0) \in \mathbb{R}^{n_w} \times \mathbb{R}^n$ satisfy the stability gain relation

$$\|(w(k), x(k))\| \leq \underline{\alpha}^{-1}(\bar{\alpha}(\|(\tilde{w}_0, x_0)\|)), \tag{C.163}$$

where $(\tilde{w}_0, x_0)$ is the projected *virtual* initialization defined according to (C.158). However, as it obviously holds that $\|(w(0), x(0))\| = \|(w_0, x_0)\|$, this directly implies that for any $k \in \mathbb{N}$

$$\|(w(k), x(k))\| \leq \tilde{\alpha}(\|(w_0, x_0)\|), \tag{C.164}$$

with $\tilde{\alpha} \in \mathcal{K}_\infty$ being defined as $\tilde{\alpha}(r) = \max\left(\underline{\alpha}^{-1}(\bar{\alpha}(\|\tilde{\Gamma}\| r)), r\right)$ and $\tilde{\Gamma} = \begin{bmatrix}\tilde{\Gamma}_\mathrm{w} & \tilde{\Gamma}_\mathrm{x}\end{bmatrix}$. Thus, we can not only show that all trajectories of system (4.53) will asymptotically converge to the origin for any arbitrary initialization $(w_0, x_0) \in \mathbb{R}^{n_w} \times \mathbb{R}^n$ but also that the origin $(w, x) = (0, 0)$ is stable in the sense of (C.164). This shows that $(w, x) = (0, 0)$ is globally asymptotically stable as claimed, which completes the proof.
Note that a more intuitive interpretation of the above arguments is that the extended system state $(w(k), x(k))$ converges in one step to the primal feasible subspace $\mathcal{H} := \{(w, x) \,|\, Cw = b(x)\} \subseteq \mathbb{R}^{n_w} \times \mathbb{R}^n$, stays on $\mathcal{H}$ for all further iterations, and asymptotically converges to the origin by generating monotonically decreasing values of the relaxed barrier function based cost function. Furthermore, any arbitrary initialization can be projected onto $\mathcal{H}$ by a matrix multiplication with finite induced norm, cf. (C.158).

C.27 Proof of Theorem 4.9

The proof is very similar to that of Theorem 4.4 and exploits again the fact that the cost function is guaranteed to decay over time – except perhaps during the transition from $k = 0$ to $k = 1$ in case of an initialization that does not satisfy the underlying linear equality constraints. However, as revealed by Lemma C.2 in the proof of Theorem 4.8 above, the resulting state and input realizations $x(k)$ and $u(k)$ will for any arbitrary initialization $(w_0, x_0) \in \mathbb{R}^{n_w} \times \mathbb{R}^n$ be identical to those for the associated virtual initialization $(\tilde{w}_0, x_0)$. Thus, the constraint satisfaction properties of the proposed MPC iteration scheme can for any $(w_0, x_0) \in \mathbb{R}^{n_w} \times \mathbb{R}^n$ always be analyzed by studying them instead for the associated virtual feasible initialization $(\tilde{w}_0, x_0)$ according to (C.158).

The remainder of the proof consists of three parts. First, we show that when considering the projected feasible initialization $(w(0), x(0)) = (\tilde{w}_0, x_0)$, the relaxed barrier functions for the input and state constraints satisfy for any $k \in \mathbb{N}$

$$\varepsilon \hat{B}_\mathrm{x}(x(k)) \leq \hat{J}_N(w(k), x(k)) - x(k)^\top P_\mathrm{uc}^* x(k) \tag{C.165a}$$

$$\varepsilon \hat{B}_\mathrm{u}(u(k)) \leq \hat{J}_N(w(k), x(k)) - x(k)^\top P_\mathrm{uc}^* x(k), \tag{C.165b}$$

where $P_\mathrm{uc}^* \in \mathbb{S}_{++}^n$ is the solution to the discrete-time algebraic Riccati equation related to the infinite-horizon LQR problem for the setup (A, B, Q, R). In a second step, we then use (C.165) for deriving a time-varying upper bound on the maximally possible constraint violations for all $k \in \mathbb{N}$. Finally, we exploit the guaranteed decrease of the cost function in order to show the claimed monotonicity result.

Part 1. Based on the proof of Theorem 4.8 and the fact that $u(i) = \Pi_0\, w(i)$, we know that

$$\hat{J}_N(w(i+1), x(i+1)) - \hat{J}_N(w(i), x(i)) \leq -\hat{\ell}(x(i), u(i)), \tag{C.166}$$

for any $i \in \mathbb{N}$. Furthermore, we know that the closed-loop system is asymptotically stable and that $\lim_{i\to\infty} \hat{J}_N(w(i), x(i)) = 0$, see Theorem 4.8. Summing up over all future sampling instants beginning at $i = k$ and using a telescoping sum on the left-hand side, we get from (C.166) that $\hat{J}_N(w(k), x(k)) \geq \sum_{i=k}^{\infty} \hat{\ell}(x(i), u(i))$ for any $k \in \mathbb{N}$. Furthermore, it holds that $\sum_{i=k}^{\infty} \hat{\ell}(x(i), u(i)) \geq x(k)^\top P_{\mathrm{uc}}^* x(k) + \varepsilon \sum_{i=k}^{\infty} \hat{B}_{\mathrm{x}}(x(i)) + \hat{B}_{\mathrm{u}}(u(i))$ due to the definition of $\hat{\ell}(\cdot,\cdot)$ and the optimality of the unconstrained infinite-horizon LQR solution. In combination, this yields

$$\hat{J}_N(w(k), x(k)) \geq x(k)^\top P_{\mathrm{uc}}^* x(k) + \varepsilon \sum_{i=k}^{\infty} \hat{B}_{\mathrm{x}}(x(i)) + \hat{B}_{\mathrm{u}}(u(i)) \tag{C.167}$$

for any $k \in \mathbb{N}$. Since due to the recentering of the underlying barrier functions all terms in the sum on the right-hand side are guaranteed to be positive, we finally obtain (C.165).

Part 2. Based on these observations, let us for the ease of notation define the time-varying upper bound on the right-hand side of (C.165) as

$$\hat{\alpha}(k) := \hat{J}_N(w(k), x(k)) - x(k)^\top P_{\mathrm{uc}}^* x(k). \tag{C.168}$$

Note that $\hat{\alpha}(k) \in \mathbb{R}_+$ by definition, see (C.167). It then follows immediately from (C.165) and (C.168) that upper bounds for the maximal violations of state and input constraints are for any $k \in \mathbb{N}$ given by

$$\hat{z}_{\mathrm{x}}^i(k) = \max_{\xi} \left\{ C_{\mathrm{x}}^i \xi - d_{\mathrm{x}}^i \mid \varepsilon \hat{B}_{\mathrm{x}}(\xi) \leq \hat{\alpha}(k) \right\}, \quad i = 1, \dots, q_{\mathrm{x}} \tag{C.169a}$$

$$\hat{z}_{\mathrm{u}}^j(k) = \max_{v} \left\{ C_{\mathrm{u}}^j v - d_{\mathrm{u}}^j \mid \varepsilon \hat{B}_{\mathrm{u}}(v) \leq \hat{\alpha}(k) \right\}, \quad j = 1, \dots, q_{\mathrm{u}}. \tag{C.169b}$$

Thus, (4.58) holds with the elements of $\hat{z}_{\mathrm{x}}(k) \in \mathbb{R}^{q_{\mathrm{x}}}$ and $\hat{z}_{\mathrm{u}}(k) \in \mathbb{R}^{q_{\mathrm{u}}}$ being given by (C.169). Note that while the above result holds by virtue of Lemma C.2 for any arbitrary initialization (w_0, x_0), it is crucial to make at $k = 0$ use of the virtual initialization $(w(0), x(0)) = (\tilde{w}_0, x_0)$ for computing the upper bound given in (C.169). The optimization problems in (C.169) are convex as the involved barrier functions are convex and the objective functions are affine in the respective variables.

Part 3. Let us now show that $\hat{\alpha}(k)$ given in (C.168) is strictly monotonically decreasing and that $\lim_{k\to\infty} \hat{\alpha}(k) = 0$. First, note that the cost function decrease implies that

$$\hat{\alpha}(k+1) - \hat{\alpha}(k) \leq -\hat{\ell}(x(k), u(k)) - x(k+1)^\top P_{\mathrm{uc}}^* x(k+1) + x(k)^\top P_{\mathrm{uc}}^* x(k). \tag{C.170}$$

Now recall $\hat{\ell}(x(k),u(k)) = \|x(k)\|_Q^2 + \|u(k)\|_R^2 + \varepsilon\hat{B}_\mathrm{x}(x(k)) + \varepsilon\hat{B}_\mathrm{u}(u(k))$ and note that $\|x(k)\|_Q^2 + \|u(k)\|_R^2 \geq x(k)^\top P_\mathrm{uc}^* x(k) - x(k+1)^\top P_\mathrm{uc}^* x(k+1)$ due to the principle of optimality. Inserting these relations into (C.170) finally reveals that for any $k \in \mathbb{N}$

$$\hat{\alpha}(k+1) - \hat{\alpha}(k) \leq -\varepsilon\big(\hat{B}_\mathrm{x}(x(k)) + \hat{B}_\mathrm{u}(u(k))\big)\,. \tag{C.171}$$

Since $\hat{B}_\mathrm{x}(\cdot)$ and $\hat{B}_\mathrm{u}(\cdot)$ are positive definite, this ensures that $\alpha(k)$ decreases monotonically over time. Moreover, it follows directly from (C.171) and the convergence of the closed-loop system state to the origin that $\lim_{k\to\infty}\hat{\alpha}(k) = 0$. As $\hat{\alpha}(k)$ decreases monotonically, the same holds for the upper bounds $\hat{z}_\mathrm{x}^i(k)$ and $\hat{z}_\mathrm{u}^j(k)$ given in (C.169), which proves the claimed strict monotonicity result for any $k \in \mathbb{N}$. Finally, as both $x(k)$ and $u(k)$ asymptotically converge to the origin (see Theorem 4.8), there always exists a finite $k_0 \in \mathbb{N}$ such that $\hat{z}_\mathrm{x}(k) \leq 0$, $\hat{z}_\mathrm{u}(k) \leq 0$ for any $k \geq k_0$.

As discussed above, making use of the virtual initialization $(\tilde{w}_0, x_0)$ does not mean that the presented result only holds for feasible initializations satisfying $Cw_0 = b(x_0)$. Instead, considering $(\tilde{w}_0, x_0)$ is rather a trick that allows us simplify the analysis and to show in particular that there exists for any arbitrary initialization an upper bound on the constraint violation that is monotonically decaying for any $k \in \mathbb{N}$ (and not only for $k \geq 1$). Thus, the main difference to the condensed case is that in order to derive an a priori estimate of the maximally possible constraint violation, we need to solve (C.169) based on $\hat{\alpha}(0) = \hat{J}_N(\tilde{w}_0, x_0) - x_0^\top P_\mathrm{uc}^* x_0$, that is, by making use of the virtual initialization given in (C.158).

C.28 Proof of Theorem 4.10

Similar to the condensed case, we prove the result for the case of zero constraint violations, i.e. $\hat{z}_\mathrm{x,tol} = 0$, $\hat{z}_\mathrm{u,tol} = 0$, which then also comprises all cases with positive tolerances. As also the proof of Theorem 4.5, the proof consists of two parts. First, we show that there exists for any $\delta \in \mathbb{R}_{++}$ a compact and nonempty set $\hat{\mathcal{Z}}_N(\delta) \subseteq \mathbb{R}^{n_w} \times \mathbb{R}^n$ such that for any $(w_0, x_0) \in \hat{\mathcal{Z}}_N(\delta)$ with $Cw_0 = b(x_0)$, the state and input constraints will be satisfied also for all future steps, i.e., (4.60) will hold with $\hat{z}_\mathrm{x,tol} = 0$, $\hat{z}_\mathrm{u,tol} = 0$. In the second part, we then show that for any compact set $\mathcal{X}_0 \subseteq \mathcal{X}_N^\circ$ there exists $\bar{\delta}_0 \in \mathbb{R}_{++}$ and $\gamma \in \mathcal{K}_\infty$ such that if $x_0 \in \mathcal{X}_0$, $\delta \leq \bar{\delta}_0$, and $w_0 \in \mathcal{B}^{n_w}_{\gamma(\delta)}(\hat{w}^*(x_0))$, then $(w_0, x_0) \in \hat{\mathcal{Z}}_N(\delta)$.

Part 1. Following the key idea behind the proof of Theorem 4.5, let us for a given relaxation parameter $\delta \in \mathbb{R}_{++}$ define the scalars $\bar{\beta}_\mathrm{x}(\delta), \bar{\beta}_\mathrm{u}(\delta), \bar{\beta}(\delta) \in \mathbb{R}_{++}$ as

$$\bar{\beta}_\mathrm{x}(\delta) := \min_{i,\xi}\{\hat{B}_\mathrm{x}(\xi) \,|\, C_\mathrm{x}\xi \leq d_\mathrm{x},\ C_\mathrm{x}^i\xi = d_\mathrm{x}^i\},\ i = 1,\dots,q_\mathrm{x}\,, \tag{C.172a}$$

$$\bar{\beta}_\mathrm{u}(\delta) := \min_{j,v}\{\hat{B}_\mathrm{u}(v) \,|\, C_\mathrm{u}v \leq d_\mathrm{u},\ C_\mathrm{u}^j v = d_\mathrm{u}^j\},\ j = 1,\dots,q_\mathrm{u}\,, \tag{C.172b}$$

$$\bar{\beta}(\delta) := \min\{\bar{\beta}_\mathrm{x}(\delta), \bar{\beta}_\mathrm{u}(\delta)\}\,, \tag{C.172c}$$

which can be interpreted as lower bounds for the values that are attained by the relaxed barrier functions on the boundaries of the constraint sets $\mathcal{X}$ and $\mathcal{U}$, respectively. As shown by Lemma 3.1 in Section 3.5, the $\bar{\beta}(\delta)$-sublevel sets of $\hat{B}_{\mathrm{x}}(\cdot)$ and $\hat{B}_{\mathrm{u}}(\cdot)$ will always be contained within the sets $\mathcal{X}, \mathcal{U}$. Based on this observation and inspired by the proof of Theorem 4.9, we introduce

$$\hat{\alpha}(w,x) := \hat{J}_N(w,x) - x^\top P^*_{\mathrm{uc}} x \tag{C.173}$$

and define the set $\hat{\mathcal{Z}}_N(\delta) \subset \mathbb{R}^{n_w} \times \mathbb{R}^n$ as

$$\hat{\mathcal{Z}}_N(\delta) := \left\{(w,x) \in \mathbb{R}^{n_w} \times \mathbb{R}^n \,\middle|\, \hat{\alpha}(w,x) \leq \varepsilon\bar{\beta}(\delta)\right\}, \tag{C.174}$$

where $\bar{\beta}(\delta)$ is given by (C.172) and $P^*_{\mathrm{uc}} \in \mathbb{S}^n_{++}$ is again the solution to the DARE related to the infinite-horizon LQR problem for (A,B,Q,R). Suppose now that $(w_0,x_0) \in \hat{\mathcal{Z}}_N(\delta)$ and that, as stated in the theorem, $Cw_0 = b(x_0)$. Based on the proof of Theorem 4.9, we know that $\varepsilon\hat{B}_{\mathrm{x}}(x(k)) \leq \hat{\alpha}(w(k),x(k))$, $\varepsilon\hat{B}_{\mathrm{u}}(u(k)) \leq \hat{\alpha}(w(k),x(k))$, see (C.165). In addition, $\hat{\alpha}(w(k),x(k)) \leq \hat{\alpha}(w_0,x_0)$ for all $k \in \mathbb{N}$, cf. (C.171). So if $(w_0,x_0) \in \hat{\mathcal{Z}}_N(\delta)$, then $\hat{B}_{\mathrm{x}}(x(k)) \leq \bar{\beta}(\delta)$, $\hat{B}_{\mathrm{u}}(u(k)) \leq \bar{\beta}(\delta)$, which by the definition of $\bar{\beta}(\delta)$ and Lemma 3.1 implies that $x(k) \in \mathcal{X}$, $u(k) \in \mathcal{U}$ for all $k \in \mathbb{N}$. Thus, for given $\delta \in \mathbb{R}_{++}$, (4.60) holds with $\hat{z}_{\mathrm{x,tol}} = 0$, $\hat{z}_{\mathrm{u,tol}} = 0$ whenever (w_0,x_0) satisfies $Cw_0 = b(x_0)$ as well as $(w_0,x_0) \in \hat{\mathcal{Z}}_N(\delta)$. Note that an equivalent result can also be formulated for arbitrary initializations that do not necessarily satisfy the underlying linear equality constraints. In this case, the projected virtual initialization $(\tilde{w}_0,x_0)$ needs to satisfy $(\tilde{w}_0,x_0) \in \hat{\mathcal{Z}}_N(\delta)$. Note that this may not always be the case for $(w_0,x_0) \in \hat{\mathcal{Z}}_N(\delta)$ as the projection defined in (C.158) is in general not orthogonal. We will discuss this issue in more detail at the end of the proof.

Part 2. In the following, we are going to prove our second claim, namely the existence of $\bar{\delta}_0 \in \mathbb{R}_{++}$ and $\gamma \in \mathcal{K}_\infty$ with the properties specified above. As we need to vary the relaxation parameter, we make the influence of δ on the cost function more explicit. In particular, we will use $\hat{J}_N(w,x;\delta)$ to denote the value of the cost function for $(w,x) \in \mathbb{R}^{n_w} \times \mathbb{R}^n$ and a given relaxation parameter $\delta \in \mathbb{R}_{++}$. In accordance with (C.126), we further define $\hat{\alpha}(w,x;\delta) := \hat{J}_N(w,x;\delta) - x^\top P^*_{\mathrm{uc}} x$. Based on this notation, it obviously holds that

$$(w_0,x_0) \in \hat{\mathcal{Z}}_N(\delta) \Leftrightarrow \hat{\alpha}(w_0,x_0;\delta) \leq \varepsilon\bar{\beta}(\delta)\,. \tag{C.175}$$

We now proceed as in the proof of Theorem 4.5: First we show in *a)* that for any *fixed* $x_0 \in \mathcal{X}_0 \subseteq \mathcal{X}^\circ_N$ and a corresponding suitably chosen w_0 there exists a $\delta'_0(x_0)$ such that (C.175) holds for all $\delta \leq \delta'_0(x_0)$; in part *b)*, we then show the claimed result for $\mathcal{X}_0$ by *uniformly* choosing $\bar{\delta}_0 = \min_{x_0 \in \mathcal{X}_0} \delta'_0(x_0)$.
a) The key idea behind the following arguments is to show that for any $x_0 \in \mathcal{X}_0 \subseteq \mathcal{X}^\circ_N$ and suitably chosen w_0, the term $\hat{\alpha}(w_0,x_0;\delta)$ will stay bounded when δ is decreased below a certain threshold. As, on the other hand, $\bar{\beta}(\delta)$ grows without bound for $\delta \to 0$, the right-hand side of (C.128) can therefore always be satisfied by choosing δ sufficiently small.
Consider a fixed $x_0 \in \mathcal{X}_0 \subseteq \mathcal{X}^\circ_N$ and let $\tilde{w}^*(x_0)$ and $\tilde{J}^*_N(x_0) = \tilde{J}_N(\tilde{w}^*(x_0),x_0)$ denote the optimal solution and value function associated to the related *nonrelaxed* barrier function based problem formulation with additional zero terminal state constraint, cf. (C.46) in the proof of Theorem 3.7. As can be deduced from Lemma B.5 in Appendix B.7, the

associated optimal solution $\tilde{w}^*(x_0)$ is unique for any $x_0 \in \mathcal{X}_N^\circ$ and $\tilde{w}^* : \mathcal{X}_N^\circ \to \mathbb{R}^{n_w}$ is a continuous map (follows directly from continuity of $\tilde{U}^* : \mathcal{X}_N^\circ \to \mathbb{R}^{Nm}$ and linearity of the underlying equality constraints, which are necessarily satisfied at the optimal solution). As before, the additional constraint $x_N = 0$ is required in order to ensure that $\tilde{J}_N^*(x_0)$ does not depend on δ via the quadratic terminal cost that is used in the relaxed formulation. By definition, $\tilde{w}^*(x_0)$ exists for any $x_0 \in \mathcal{X}_0$ and results in strictly feasible state and input sequences $\{\tilde{x}_0, \tilde{x}_1, \ldots \tilde{x}_N\}$ and $\{\tilde{u}_0, \tilde{u}_1, \ldots \tilde{u}_{N-1}\}$, respectively. Now define

$$\delta_0(x_0) := \min_{i,j,k} \left\{ -C_{\mathrm{x}}^i \tilde{x}_k + d_{\mathrm{x}}^i,\ -C_{\mathrm{u}}^j \tilde{u}_k + d_{\mathrm{u}}^j \right\} \tag{C.176}$$

for $i = 1, \ldots, q_{\mathrm{x}}$, $j = 1, \ldots, q_{\mathrm{u}}$, $k = 0, \ldots, N-1$, which characterizes the minimal distance of the nonrelaxed open-loop trajectories to the boundaries of the respective constraint sets. Note that $\hat{J}_N(\tilde{w}^*(x_0), x_0; \delta) = \tilde{J}_N^*(x_0)$, whenever we choose $\delta \leq \delta_0(x_0)$, since in this case no constraint relaxation is active. In addition, it always holds due to optimality that $\hat{J}_N^*(x_0; \delta) \leq \hat{J}_N(\tilde{w}^*(x_0), x_0; \delta)$ and, thus, $\hat{J}_N^*(x_0; \delta) \leq \tilde{J}_N^*(x_0)$ for all $\delta \leq \delta_0(x_0)$. As $\tilde{w}^*(x_0)$ is a continuous map, $\delta_0(x_0)$ is also continuous.
Next, we derive an upper bound for the cost function $\hat{J}_N(w_0, x_0; \delta)$ for w_0 in the neighborhood of $\hat{w}^*(x_0)$. Assume $\delta \leq \delta_0(x_0)$ and consider $w_0 \in \mathcal{B}_{\gamma_0}^{n_w}(\hat{w}^*(x_0))$, that is, $w_0 = \hat{w}^*(x_0) + \gamma_0 v$, $\|v\| \leq 1$, for $\gamma_0 \in \mathbb{R}_{++}$. Due to Taylor's Theorem and the fact that the recentering ensures $\nabla_w \hat{J}_N(\hat{w}^*(x_0), x_0; \delta) = 0$, it then holds that

$$\hat{J}_N(w_0, x_0; \delta) = \hat{J}_N(\hat{w}^*(x_0), x_0; \delta) + \frac{1}{2}\gamma_0^2 v^\top \nabla_w^2 \hat{J}_N(\hat{w}^*(x_0) + sv, x_0; \delta) v \tag{C.177}$$

for some $s \in (0,1)$. However, together with $\hat{J}_N(\hat{w}^*(x_0), x_0; \delta) = \hat{J}_N^*(x_0; \delta) \leq \tilde{J}_N^*(x_0)$ from above, this implies

$$\hat{J}_N(w_0, x_0; \delta) \leq \tilde{J}_N^*(x_0) + \frac{\gamma_0^2}{2} \left\| H + \frac{\varepsilon}{\delta^2} G^\top \mathrm{diag}(w_r) G \right\|, \tag{C.178}$$

where we used $\|v\| \leq 1$ and the fact that $\nabla_w^2 \hat{J}_N(w, x; \delta) \preceq H + \frac{\varepsilon}{\delta^2} G^\top \mathrm{diag}(w_r) G$ for any $(w, x) \in \mathbb{R}^{n_w} \times \mathbb{R}^n$, see Appendix B.9 as well as Theorem 4.1 and the associated proof that is given in Section C.18 for the condensed formulation. Recall that $w_r \in \mathbb{R}_{++}^q$ denotes the underlying given recentering vector. Let us now choose $\gamma_0 = \gamma(\delta) = \delta / \|H\|^{1/2}$. Then, it follows from (C.178) that for any $x_0 \in \mathcal{X}_0$, any $\delta \leq \delta_0(x_0)$, and any $w_0 \in \mathcal{B}_{\gamma(\delta)}^{n_w}(\hat{w}^*(x_0))$

$$\hat{J}_N(w_0, x_0; \delta) \leq \tilde{J}_N^*(x_0) + \frac{1}{2}\left(\delta^2 + \frac{\varepsilon}{\|H\|} \left\| G^\top \mathrm{diag}(w_r) G \right\| \right). \tag{C.179}$$

Note that due to the fact that the underlying global quadratic upper bounds depend on the choice of the relaxation parameter, the matrix H, and hence also $\|H\|$, depend via the terminal cost matrix P given in (4.5) directly on δ. That is, $P = P(\delta)$ and $H = H(\delta)$. It is therefore not immediately clear how the right-hand side of (C.179) will behave for decreasing values of δ. However, we can state the following. As the (strong) solution of the

DARE (4.5) is continuous and monotonic in the δ-dependent matrices M_x and M_u (Lancaster et al., 1987), it follows that the solution $P \in \mathbb{S}^n_{++}$ is continuous and monotonic in δ and that, in particular, $P(\delta) \succeq P(\delta_0)$ for all $\delta \leq \delta_0$. Likewise, the matrix $H = H(\delta) \in \mathbb{S}^n_{++}$ as given in Appendix B.9 is continuous and monotonic in δ and $H(\delta) \succeq H(\delta_0)$ for all $\delta \leq \delta_0$. This observation has two important implications. First, the derived monotonicity relations imply together with (C.179) that we can state for any $w_0 \in \mathcal{B}^{n_w}_{\gamma_0}(\hat{w}^*(x_0))$ with $\delta \leq \delta_0(x_0)$ the following upper bound on the cost function

$$\hat{J}_N(w_0, x_0; \delta) \leq \tilde{J}^*_N(x_0) + \frac{1}{2}\left(\delta_0^2 + \frac{\varepsilon}{\|H(\delta_0)\|}\left\|G^\top \mathrm{diag}(w_r) G\right\|\right). \tag{C.180}$$

Here, we used that if $H(\delta) \succeq H(\delta_0)$ for all $\delta \leq \delta_0$, then $\|H(\delta)\| \geq \|H(\delta_0)\|$ for all $\delta \leq \delta_0$ due to monotonicity of the spectral matrix norm. Thus, we can conclude that if we choose the optimizer initialization w_0 sufficiently close to $\hat{w}^*(x_0)$, then the cost function value, and hence also $\hat{\alpha}(w_0, x_0; \delta)$ in (C.175), will stay bounded for any $\delta \leq \delta_0(x_0)$. Second, the fact that $H(\delta)$ is continuous and monotonically increasing for decreasing δ reveals that $\gamma : \delta \mapsto \delta / \|H(\delta)\|^{1/2}$ is continuous and strictly monotonically increasing. In addition, it is straightforward to show that $\lim_{\delta \to 0} \gamma(\delta) = 0$ and $\gamma(\delta) \to \infty$ for $\delta \to \infty$. Thus, $\gamma \in \mathcal{K}_\infty$. Let us now choose $\delta_0'(x_0) \leq \delta_0(x_0)$ in such a way that the right-hand side of (C.175) is satisfied for all $\delta \leq \delta_0'(x_0)$, which then implies that $(w_0, x_0) \in \hat{\mathcal{Z}}_N(\delta)$. In particular, let

$$\begin{aligned} \delta_0'(x_0) := \max \Big\{ \delta \in \mathbb{R}_{++} \,\Big|\, & \delta \leq \delta_0(x_0), \\ & \tilde{J}^*_N(x_0) + \frac{1}{2}\left(\delta_0(x_0)^2 + \frac{\varepsilon}{\|H(\delta_0(x_0))\|}\|G^\top \mathrm{diag}(w_r) G\|\right) - x_0^\top P^*_{\mathrm{uc}} x_0 \leq \varepsilon \bar{\beta}(\delta) \Big\}. \end{aligned} \tag{C.181}$$

In combination with the definition of $\hat{\alpha}(w, x; \delta)$, it follows from (C.180) and (C.181) that the right-hand side of (C.175) will be satisfied for any $\delta \leq \delta_0'(x_0)$ whenever $w_0 \in \mathcal{B}^{n_w}_{\gamma(\delta)}(\hat{w}^*(x_0))$. Hence, $(w_0, x_0) \in \hat{\mathcal{Z}}_N(\delta)$ for any $\delta \leq \delta_0'(x_0)$ whenever $w_0 \in \mathcal{B}^{n_w}_{\gamma(\delta)}(\hat{w}^*(x_0))$. Note that we may picture the construction of $\delta_0'(x_0)$ as decreasing δ from $\delta(x_0)$ until also the second condition in (C.181) is satisfied. As $\bar{\beta}(\delta)$ grows without bound for decreasing values of δ, this can always be achieved, which shows that $\delta'(x_0)$ exists for any $x_0 \in \mathcal{X}^\circ_N$. Moreover, since $\bar{\beta}(\delta)$ is continuous in δ and both $\delta_0(x_0)$ and $\tilde{J}^*_N(x_0)$ are continuous in x_0, the finally deduced relaxation parameter $\delta_0'(x_0)$ in (C.181) is also continuous. As already discussed in the proof of Theorem 3.7, continuity of $\bar{\beta}(\delta)$ follows from noting that the constraints in (C.172a) and (C.172b) are independent from δ and that the relaxed barrier functions $\hat{B}_x(x)$ and $\hat{B}_u(u)$ are continuous in both x (respectively u) and δ and possess, in addition, compact sublevel sets for any fixed value of δ, see (Fiacco, 1983, Theorem 2.2) as well as (Bonnans and Shapiro, 2000, Proposition 4.4).
b) Consider now the compact set $\mathcal{X}_0 \subseteq \mathcal{X}^\circ_N$ and define $\bar{\delta}_0 \in \mathbb{R}_{++}$ as $\bar{\delta}_0 = \min_{x_0 \in \mathcal{X}_0} \delta_0'(x_0)$. Due to the continuity of $\delta_0'(x_0)$ and the compactness of $\mathcal{X}_0$, this value always exists by virtue of the Weierstraß extreme value theorem. Then, by the above arguments, $(w_0, x_0) \in \hat{\mathcal{Z}}_N(\delta)$ for any $\delta \leq \bar{\delta}_0$ and any $x_0 \in \mathcal{X}_0$ if we choose w_0 as $w_0 \in \mathcal{B}^{n_w}_{\gamma(\delta)}(\hat{w}^*(x_0))$ with $\gamma(\delta) = \delta / \|H(\delta)\|^{1/2}$, $\gamma \in \mathcal{K}_\infty$, which proves our second claim.

Thus, combining finally the results from Part 1 and Part 2, we can conclude that there exist for any compact set $\mathcal{X}_0 \subseteq \mathcal{X}_N^\circ$ a sufficiently small relaxation parameter $\bar{\delta}_0 \in \mathbb{R}_{++}$ and a well-defined function $\gamma \in \mathcal{K}_\infty$ such that for any $0 < \delta \leq \bar{\delta}_0$ and any optimizer initialization $w_0 \in \mathcal{B}^{n_w}_{\gamma(\delta)}(\hat{w}^*(x_0))$ that satisfies in addition $Cw_0 = b(x_0)$, we can ensure that (4.60) will hold with $\hat{z}_{\text{x,tol}} = 0$, $\hat{z}_{\text{u,tol}} = 0$ for any $x_0 \in \mathcal{X}_0$. As mentioned above, the zero tolerance case comprises all other cases with nonzero tolerances $\hat{z}_{\text{x,tol}} \in \mathbb{R}^{q_x}_+$, $\hat{z}_{\text{u,tol}} \in \mathbb{R}^{q_u}_+$. This completes the proof.

Note that the additional assumption that the initialization is not only close to the exact optimal solution but also satisfies the linear equality constraints $Cw_0 = b(x_0)$ was mainly used in order to simplify the formulation of the theorem and the proof presented above. In fact, the whole result could also be formulated in a more general way by requiring instead that the *projection* of the optimizer initialization needs to be sufficiently close to the exact optimal solution, that is, $\tilde{w}_0 \in \mathcal{B}^{n_w}_{\gamma(\delta)}(\hat{w}^*(x_0))$ with $\tilde{w}_0$ given in (C.158). However, this would require to introduce the aforementioned projection already in the theorem, which we wanted to avoid for the sake of simplicity. Note that from a geometrical point of view, the additional assumption simply means that the discussed neighborhood around the exact optimal solution needs to be intersected with the hyperplane $\mathcal{H} := \{(w, x) \,|\, Cw = b(x)\}$. As it holds by definition that $(\hat{w}^*(x_0), x_0) \in \mathcal{H}$, this shows that there always exists a nonempty neighborhood of the exact optimal initialization for which we can still guarantee the constraint satisfaction properties stated in the theorem. Note furthermore that the assumption $Cw_0 = b(x_0)$ can in practical implementations be easily satisfied by choosing first some suitable values for the underlying control inputs $u_{i,0}$ and then simply computing the corresponding predicted system states $x_{i,0}$ contained in w_0 by a propagation of the linear system dynamics. In this case, all the required assumptions will hold whenever the respective control input vector is close enough to $\hat{U}^*(x_0)$, which is exactly the result that we obtained for the condensed formulation, cf. Theorem 4.5.

C.29 Proof of Theorem 4.11

The proof is basically identical to that of Theorem 4.6 for the condensed case. In particular, due to the assumption that the optimization algorithm operator $\Phi^{i_\text{T}(k)} : \mathbb{R}^{n_w} \times \mathbb{R}^n \to \mathbb{R}^{n_w}$ is continuous, the right hand side of (4.61) is continuous in all of its arguments for any given sequence $\boldsymbol{i}_\text{T} = \{i_\text{T}(0), i_\text{T}(1), \ldots\}$. Thus, the claimed IISS result follows directly from the fact that global asymptotic stability and integral input-to-state stability are in the discrete-time case equivalent, see the main result of Angeli (1999) as well as Theorem A.4 in Appendix A.2.

Note that the optimization algorithm operator and therefore also the right-hand side of (4.61b) may be discontinuous if $i_\text{T}(k)$ or the step size s are chosen based on $(w(k), x(k))$, e.g., if the optimization operator contains some form of gradient-based stopping criterion. However, as both s and $i_\text{T}(k)$ are guaranteed to be bounded, arguments like in Messina et al. (2005) may in this case be used to derive a similar IISS result, cf. Remark 4.5.

C.30 Proof of Theorem 4.12

The proof consists of two parts. First, we show that the optimizer $\hat{w}^* : \mathbb{R}^n \to \mathbb{R}^{n_w}$ is via the underlying linear dynamic constraints directly related to the optimizer $\hat{U}^* : \mathbb{R}^n \to \mathbb{R}^{Nm}$ of the condensed formulation, leading in particular to exactly the same nominal closed-loop behavior and satisfying an equivalent Lipschitz continuity condition. In the second part, we then exploit these insights by applying basically the same arguments as we did in the proof of Theorem 4.7 for the condensed case.

Part 1. As the optimal solution to problem (4.47) will by definition satisfy the underlying linear equality constraints, we can observe that the associated optimizer may for any $x \in \mathbb{R}^n$ be expressed as

$$\hat{w}^*(x) = \tilde{\Gamma}'_{\mathrm{u}} \hat{U}^*(x) + \tilde{\Gamma}'_{\mathrm{x}} x\,, \tag{C.182}$$

where $\hat{U}^*(x)$ is the optimizer of the corresponding equivalent condensed problem formulation (4.6) and the matrices on the right-hand side can easily be derived as

$$\tilde{\Gamma}'_{\mathrm{u}} = \begin{bmatrix} I & 0 & 0 & \cdots & 0 \\ B & 0 & 0 & \cdots & 0 \\ 0 & I & 0 & \cdots & 0 \\ AB & B & 0 & \cdots & 0 \\ \vdots & \vdots & \ddots & \ddots & \vdots \\ 0 & 0 & 0 & \cdots & I \\ A^{N-1}B & A^{N-2}B & A^{N-3}B & \cdots & B \end{bmatrix}, \qquad \tilde{\Gamma}'_{\mathrm{x}} = \begin{bmatrix} 0 \\ A \\ 0 \\ A^2 \\ \vdots \\ 0 \\ A^N \end{bmatrix}. \tag{C.183}$$

As a consequence, the optimal control input is at each sampling instant given by $u(k) = \Pi_0 \hat{w}^*(x(k)) = \hat{u}_0^*(x(k))$, which results in exactly the same closed-loop behavior as in the condensed case. Based on Theorem 3.8, this allows us to state the following corollary.

Corollary C.1. *Consider the relaxed barrier function based MPC formulation (4.3)–(4.5) and let $\hat{J}_N : \mathbb{R}^{n_w} \times \mathbb{R}^n \to \mathbb{R}_+$ given in (4.47a) refer to the cost function of the associated equality constrained formulation. Furthermore, let $u(k) = \Pi_0 \hat{w}^*(x(k))$ be the corresponding optimal feedback law. Then, the resulting closed-loop system*

$$x(k+1) = Ax(k) + B\Pi_0 \hat{w}^*(x(k)) + d(k) \tag{C.184}$$

is ISS with respect to the disturbance sequence $\mathbf{d} = \{d(0), d(1), \ldots\}$. Furthermore, the associated value function $\hat{J}_N^* : \mathbb{R}^n \to \mathbb{R}_+$ is a twice continuously differentiable ISS Lyapunov function.

Furthermore, (C.182) reveals in combination with Lipschitz continuity of the optimizer $\hat{U}^* : \mathbb{R}^n \to \mathbb{R}^{Nm}$ that also the optimizer of the now discussed sparse formulation is globally Lipschitz continuous. In particular, for any $x_1, x_2 \in \mathbb{R}^n$, it holds that

$$\|\hat{w}^*(x_1) - \hat{w}^*(x_2)\| \leq L_w \|x_1 - x_2\|\,, \qquad L_w := \|\tilde{\Gamma}'_{\mathrm{u}}\| L_U + \|\tilde{\Gamma}'_{\mathrm{x}}\|\,, \tag{C.185}$$

where $\tilde{\Gamma}'_{\mathrm{u}} \in \mathbb{R}^{n_w \times Nm}$ and $\tilde{\Gamma}'_{\mathrm{x}} \in \mathbb{R}^{n_w \times n}$ are from (C.182)–(C.183) and L_U is the Lipschitz constant provided by Lemma 3.8.

Part 2. Let us first reformulate the closed-loop (4.61) in such a way that the influence of the optimization error becomes more explicit. Let $\hat{x}(k) = x(k) + e(k)$ denote the inexact state estimate and let $\hat{x}^+(k) = A\hat{x}(k) + B\Pi_0 w(k)$ be the corresponding predicted next nominal system state. As can be seen from (4.61) and the definition of the optimization algorithm operator in (4.54) – (4.56), the iterative optimization will for each $k \in \mathbb{N}$ try to find the minimum of $\hat{J}_N(w, \hat{x}^+(k))$. Consequently, at each sampling step $k \in \mathbb{N}$, the optimization error may be written as

$$\epsilon(k) = \Phi^{i_T(k)}\left(w(k), \hat{x}(k)\right) - \hat{w}^*(\hat{x}^+(k)) = w(k+1) - \hat{w}^*(\hat{x}^+(k)), \qquad (C.186)$$

by which system (4.61) can be reformulated as

$$x(k+1) = A\,x(k) + B\Pi_0\, w(k) + d(k), \qquad (C.187a)$$

$$w(k+1) = \hat{w}^*\left(A\big(x(k) + e(k)\big) + B\Pi_0\, w(k)\right) + \epsilon(k). \qquad (C.187b)$$

The proof is then based on the following two auxiliary results from Part 1: *i*) when applying the exact optimal feedback $u(k) = \Pi_0\, \hat{w}^*(x(k))$, the associated closed-loop state dynamics (C.184) are ISS with with respect to arbitrary additive disturbances, and *ii*) the optimal solution $\hat{w}^*(x)$ related to the equality constrained problem formulation (4.47) is Lipschitz continuous in x, see (C.185). Note now that, based on (C.187b), the system state dynamics (C.187a) can be rewritten as

$$x(k+1) = Ax(k) + B\Pi_0\, \hat{w}^*(x(k)) + \tilde{d}(k) \qquad (C.188)$$

with $\tilde{d}(k)$ being defined as

$$\tilde{d}(k) = B\Pi_0\left(\hat{w}^*(\hat{x}^+(k-1)) - \hat{w}^*(x(k)) + \epsilon(k-1)\right) + d(k), \qquad (C.189)$$

where $\hat{x}^+(k-1) = A(x(k-1) + e(k-1)) + B\Pi_0 w(k-1)$. Due to the ISS result provided by Corollary C.1, there exist $\beta_x \in \mathcal{KL}$ and $\gamma_x \in \mathcal{K}$ such that for $\xi = x(0)$,

$$\|x(k, \xi, \tilde{\boldsymbol{d}})\| \leq \beta_x(\|\xi\|, k) + \gamma_x(\|\tilde{\boldsymbol{d}}_{[k-1]}\|). \qquad (C.190)$$

Furthermore, by noting that $\hat{x}^+(k-1)$ can via (C.187a) be expressed as $\hat{x}^+(k-1) = x(k) + Ae(k-1) - d(k-1)$ and by exploiting the Lipschitz continuity of $\hat{w}^*(\cdot)$, we get

$$\|\tilde{d}(k)\| \leq L_w\|A\|\|B\|\|e(k-1)\| + \|B\|\|\epsilon(k-1)\| + L_w\|B\|\|d(k-1)\| + \|d(k)\|, \quad (C.191)$$

which reveals that for any $k \in \mathbb{N}_+$

$$\|x(k, \xi, \boldsymbol{d}, \boldsymbol{e}, \boldsymbol{\epsilon})\| \leq \beta_x(\|\xi\|, k) + \gamma'_d(\|\boldsymbol{d}_{[k-1]}\|) + \gamma'_e(\|\boldsymbol{e}_{[k-2]}\|) + \gamma'_\epsilon(\|\boldsymbol{\epsilon}_{[-1,k-2]}\|) \qquad (C.192)$$

with $\gamma'_d(r) = \gamma_x(3(1 + L_w\|B\|)\,r)$, $\gamma'_e(r) = \gamma_x(3L_w\|A\|\|B\|\,r)$, and $\gamma'_\epsilon(r) = \gamma_x(3\|B\|\,r)$. We made use of the fact that $\gamma_x(r_1 + \cdots + r_n) \leq \gamma_x(nr_1) + \cdots + \gamma_x(nr_n)$, which holds for every $\gamma_x \in \mathcal{K}$ for all $r_1, \ldots, r_n \in \mathbb{R}_+$ (Rawlings and Ji, 2012). Note also that (C.192) again illustrates that the optimization error effects the system state with a delay of two sampling instants. This is due to the fact that according to (C.187a), $\epsilon(k)$ affects $w(k+1)$ and then $w(k+1)$ affects $x(k+2)$. Of course, this heavily depends on the used notation, in particular on the way how the optimization error is defined. However, it should be noted that, as stated in the theorem, the initial optimization error $\epsilon(-1) := w(0) - \hat{w}^*(x(0))$ needs to be included into the sequence $\boldsymbol{\epsilon} = \{\epsilon(-1), \epsilon(0), \ldots\}$. Let us now consider the optimizer dynamics. To this end, note that (C.187b) and again $\hat{x}^+(k) = A(x(k) + e(k)) + B\Pi_0 w(k) = x(k+1) + Ae(k) - d(k)$ directly imply that

$$w(k) = \hat{w}^*(x^+(k-1)) + \epsilon(k-1) = \hat{w}^*\left((x(k) + Ae(k-1) - d(k-1)\right) + \epsilon(k-1)\,. \quad \text{(C.193)}$$

Lipschitz continuity of the exact optimizer and $\hat{w}^*(0) = 0$ ensure that $\|\hat{w}^*(x)\| \leq L_w\|x\|$ for any $x \in \mathbb{R}^n$. Thus,

$$\begin{aligned} \|w(k)\| &= \|\hat{w}^*\left((x(k) + Ae(k-1) - d(k-1)\right) + \epsilon(k-1)\| && \text{(C.194a)} \\ &\leq L_w\|x(k) + Ae(k-1) - d(k-1)\| + \|\epsilon(k-1)\| && \text{(C.194b)} \\ &\leq L_w\|x(k)\| + L_w\|A\|\|e(k-1)\| + L_w\|d(k-1)\| + \|\epsilon(k-1)\|\,. && \text{(C.194c)} \end{aligned}$$

Inserting now the upper bound for $\|x(k)\|$ given in (C.192), we obtain that for $k \in \mathbb{N}_+$

$$\|w(k, \xi, \boldsymbol{d}, \boldsymbol{e}, \boldsymbol{\epsilon})\| \leq \beta_\text{w}(\|\xi\|, k) + \gamma''_\text{d}(\|\boldsymbol{d}_{[k-1]}\|) + \gamma''_\text{e}(\|\boldsymbol{e}_{[k-1]}\|) + \gamma''_\epsilon(\|\boldsymbol{\epsilon}_{[-1,k-1]}\|)\,, \quad \text{(C.195)}$$

where $\beta_\text{w}(r,s) = L_w\beta_\text{x}(r,s)$ and $\gamma''_\text{d}(r) = L_w(\gamma'_\text{d}(r) + r)$, $\gamma''_\text{e}(r) = L_w(\gamma'_\text{e}(r) + \|A\|r)$, $\gamma''_\epsilon(r) = L_w\gamma'_\epsilon(r) + r$. By combining (C.192) and (C.195) and noting that $\|\xi\| = \|x(0)\| \leq \|\zeta\| = \|(x(0), w(0))\|$, we then finally get the claimed result. In particular, we may conclude that for any $k \in \mathbb{N}_+$

$$\|(x(k), w(k))\| \leq \beta(\|\zeta\|, k) + \gamma_\text{d}(\|\boldsymbol{d}_{[k-1]}\|) + \gamma_\text{e}(\|\boldsymbol{e}_{[k-1]}\|) + \gamma_\epsilon(\|\boldsymbol{\epsilon}_{[-1,k-1]}\|) \quad \text{(C.196)}$$

with $\beta(r,s) = \beta_\text{x}(r,s) + \beta_\text{w}(r,s)$, $\gamma_\text{d}(r) = \gamma'_\text{d}(r) + \gamma''_\text{d}(r)$, $\gamma_\text{v}(r) = \gamma'_\text{v}(r) + \gamma''_\text{v}(r)$, and $\gamma_\epsilon(r) = \gamma'_\epsilon(r) + \gamma''_\epsilon(r)$. As outlined above, the optimization error sequence needs to include the initial error $\varepsilon(-1) = w(0) - \hat{w}^*(x(0))$. This completes the proof.

Notation

Sets

$\mathbb{N}$, $\mathbb{N}_+$	Sets of natural and strictly positive natural numbers, $0 \in \mathbb{N}$
$\mathbb{Z}$, $\mathbb{Z}_+$	Sets of integers and nonnegative integers, $\mathbb{Z}_+ = \mathbb{N}$
$\mathbb{R}$, $\mathbb{R}_+$, $\mathbb{R}_{++}$	Sets of real, nonnegative real, and strictly positive real numbers
$\mathbb{R}^n_+$	Set of n-dimensional vectors with nonnegative real entries
$\mathbb{R}^n_{++}$	Set of n-dimensional vectors with strictly positive real entries
$\mathbb{S}^n$	Set of $n \times n$-dimensional symmetric real matrices
$\mathbb{S}^n_+$	Set of $n \times n$-dimensional positive semi-definite matrices
$\mathbb{S}^n_{++}$	Set of $n \times n$-dimensional positive definite matrices
$[a, b)$	Interval $\mathcal{I} = \{x \in \mathbb{R} : a \leq x < b\}$ for $a, b \in \mathbb{R}$
Polytope	A compact set of the form $\mathcal{P} = \{x \in \mathbb{R}^n : Cx \leq d\}$
$\mathcal{S}^\circ$, $\bar{\mathcal{S}}$, $\partial\mathcal{S}$	Interior, closure, and boundary of a given set $\mathcal{S}$; $\partial\mathcal{S} = \bar{\mathcal{S}} \setminus \mathcal{S}^\circ$

Vectors, matrices, and norms

I_n, 0_n	Identity/zero matrix of dimension $n \in \mathbb{N}_+$[2]
$\mathbb{1}_n$, $0_{n\times m}$	$n \times 1$ vector of ones/$n \times m$ matrix of zeros for $n, m \in \mathbb{N}_+$
$\mathrm{diag}(x_1, \ldots, x_n)$	Diagonal matrix with diagonal entries $x_1, x_2, \ldots, x_n \in \mathbb{R}$
$(\cdot)^\top$	Transpose of a real vector or matrix
$(\cdot)^i$	i^{th} element or row of a real vector or matrix
$\lambda_{\min}(A)$	Minimal eigenvalue of a symmetric matrix $A \in \mathbb{S}^n$
$\lambda_{\max}(A)$	Maximal eigenvalue of a symmetric matrix $A \in \mathbb{S}^n$
$\|x\|$	Euclidean norm of vector $x \in \mathbb{R}^n$, i.e. $\|x\| := \sqrt{x^\top x}$
$\|x\|_M$	For $x \in \mathbb{R}^n$ and $M \in \mathbb{S}^n_+$, we define $\|x\|_M := \sqrt{x^\top M x}$
$\|A\|$	Induced matrix norm: $\|A\| = \|A\|_2 := \sqrt{\lambda_{\max}(A^\top A)}$, $A \in \mathbb{R}^{n\times m}$
$\|s\|$	Signal norm for $s := \{s_1, s_2 \ldots\}$ with $s_i \in \mathbb{R}^n$: $\|s\| = \max_i \|s_i\|$
$<, \leq, >, \geq$	Element-wise relation operators for real vectors, e.g., $a \leq b$ if $a^i \leq b^i$ for all $i = \ldots, n$ for given vectors $a, b \in \mathbb{R}^n$
$\prec, \preceq, \succ, \succeq$	Relation operator for real symmetric matrices, e.g., $A \succeq B$ if $A - B \in \mathbb{S}^n_+$ for given matrices $A, B \in \mathbb{S}^n$
$\otimes$	Kronecker matrix product

[2] Sometimes we also drop the index if the dimension is clear from the context.

Functions and derivatives

$f : \mathcal{D} \to C$	The function $f(\cdot)$ maps from its domain $\mathcal{D}$ into the set C
$\mathcal{C}^k$	The set of k-times continuously differentiable functions, $k \in \mathbb{N}$
$\mathcal{K}$-function	A function $\alpha : \mathbb{R}_+ \to \mathbb{R}_+$ is said to be a $\mathcal{K}$-function, or $\alpha \in \mathcal{K}$ for short, if it is continuous, strictly increasing, and $\alpha(0) = 0$.
$\mathcal{K}_\infty$-function	A function $\alpha : \mathbb{R}_+ \to \mathbb{R}_+$ is said to be a $\mathcal{K}_\infty$-function, or $\alpha \in \mathcal{K}_\infty$ for short, if it is a $\mathcal{K}$-function and in addition $\alpha(r) \to \infty$ for $r \to \infty$
$\mathcal{KL}$-function	A function $\beta : \mathbb{R}_+ \times \mathbb{R}_+ \to \mathbb{R}_+$ is said to be a $\mathcal{KL}$-function, or $\beta \in \mathcal{KL}$ for short, if $\beta(\cdot, s) \in \mathcal{K}$ for fixed $s \in \mathbb{R}_+$, while for fixed $r \in \mathbb{R}_+$, the function $\beta(r, \cdot)$ is decreasing and $\beta(r,s) \to 0$ as $s \to \infty$
$\nabla f(\bar{x})$	Gradient of a function $f(x), f : \mathcal{D} \to \mathbb{R}$, evaluated at $\bar{x} \in \mathcal{D}$
$\nabla^2 f(\bar{x})$	Hessian of a function $f(x), f : \mathcal{D} \to \mathbb{R}$, evaluated at $\bar{x} \in \mathcal{D}$
$\nabla_x f(\bar{x}, \bar{y})$	Gradient of a function $f(x,y), f : \mathcal{D}_1 \times \mathcal{D}_2 \to \mathbb{R}$, with respect to x, evaluated at $(\bar{x}, \bar{y}) \in \mathcal{D}_1 \times \mathcal{D}_2$ (similarly, define Hessian $\nabla_x^2 f(\bar{x}, \bar{y})$)
$\nabla_{xy} f(\bar{x}, \bar{y})$	Second derivative of a function $f(x,y), f : \mathcal{D}_1 \times \mathcal{D}_2 \to \mathbb{R}$, with respect to x and then y (in this order), evaluated at $(\bar{x}, \bar{y}) \in \mathcal{D}_1 \times \mathcal{D}_2$
$\lceil \cdot \rceil$	Ceiling function: $\lceil x \rceil := \min\{k \in \mathbb{Z} \mid k \geq x\}$ for $x \in \mathbb{R}$
$\lfloor \cdot \rfloor$	Floor function: $\lfloor x \rfloor := \max\{k \in \mathbb{Z} \mid k \leq x\}$ for $x \in \mathbb{R}$

Miscellaneous

$\boldsymbol{s}_{[l,k]}$	Sequence truncation for $l, k \in \mathbb{Z}$, $k \geq l$: $\boldsymbol{s}_{[l,k]} := \{s_l, s_{l+1}, \ldots, s_{k-1}, s_k\}$
$\boldsymbol{s}_{[k]}$	Sequence truncation starting at zero: $\boldsymbol{s}_{[k]} := \{s_0, s_1, \ldots, s_{k-1}, s_k\}$
$x_k(\boldsymbol{u}, x)$	Predicted system state at time instant $k = 0, \ldots, N$ for initial state $x_0 = x \in \mathbb{R}^n$ and given input sequence $\boldsymbol{u} = \{u_0, \ldots, u_{N-1}\}$
$x_k(U, x)$	Predicted system state for stacked input vector $U = [u_0^\top \; \cdots, u_{N-1}^\top]^\top$

Acronyms

DARE	Discrete algebraic Riccati equation
CLF	Control Lyapunov function
ISS/IISS	(Integral) Input-to-state stable
GAS/GES	Globally asymptotically/exponentially stable
KKT	Karush-Kuhn-Tucker
LP/QP/SDP	Linear/quadratic/semi-definite program
LQR	Linear quadratic regulator
LTI/LTV	Linear time-invariant/linear time-varying
MHE	Moving horizon estimation
(N)MPC	(Nonlinear) Model predictive control
ROA	Region of attraction

Bibliography

U. Ahrens, M. Diehl, and R. Schmehl. *Airborne wind energy*. Springer Science & Business Media, 2013.

M. Alamir. From certification of algorithms to certified MPC: The missing links. In *Proceedings of the 5th IFAC Conference on Nonlinear Model Predictive Control (NMPC'15)*, pages 65–72, 2015.

A. Alessandri, M. Baglietto, and G. Battistelli. Receding-horizon estimation for discrete-time linear systems. *IEEE Transactions on Automatic Control*, 48(3):473–478, 2003.

A. Alessio and A. Bemporad. A survey on explicit model predictive control. In *Nonlinear Model Predictive Control – Towards New Challenging Applications*, pages 345–369. 2009.

E. Anderson, Z. Bai, J. Dongarra, A. Greenbaum, A. McKenney, J. Du Croz, S. Hammerling, J. Demmel, C. Bischof, and D. Sorensen. LAPACK: A portable linear algebra library for high-performance computers. In *Proceedings of the 1990 ACM/IEEE Conference on Supercomputing*, pages 2–11, 1990.

E. Anderson, Z. Bai, C. Bischof, L. S. Blackford, J. Demmel, J. Dongarra, J. Du Croz, A. Greenbaum, S. Hammarling, A. McKenney, and D. Sorensen. *LAPACK Users' guide*. SIAM, 1999.

D. Angeli. Intrinsic robustness of global asymptotic stability. *Systems & Control Letters*, 38 (4):297–307, 1999.

D. Angeli, E. D. Sontag, and Y. Wang. A characterization of integral input-to-state stability. *IEEE Transactions on Automatic Control*, 45(6):1082–1097, 2000.

D. Axehill. Controlling the level of sparsity in MPC. *Systems & Control Letters*, 76:1–7, 2015.

R. A. Bartlett and L. T. Biegler. QPSchur: a dual, active-set, Schur-complement method for large-scale and structured convex quadratic programming. *Optimization and Engineering*, 7(1):5–32, 2006.

C. E. Beal and J. C. Gerdes. Model predictive control for vehicle stabilization at the limits of handling. *IEEE Transactions on Control Systems Technology*, 21(4):1258–1269, 2013.

A. Bemporad and A. Garulli. Output-feedback predictive control of constrained linear systems via set-membership state estimation. *International Journal of Control*, 73(8):655–665, 2000.

A. Bemporad, M. Morari, V. Dua, and E. N. Pistikopoulos. The explicit linear quadratic regulator for constrained systems. *Automatica*, 38(1):3–20, 2002.

A. Bemporad, L. Bellucci, and T. Gabbriellini. Dynamic option hedging via stochastic model predictive control based on scenario simulation. *Quantitative Finance*, 14(10): 1739–1751, 2014.

A. Bemporad, D. Bernardini, and P. Patrinos. A convex feasibility approach to anytime model predictive control. *arXiv:1502.07974*, 2015.

A. Ben Tal, M. Tsibulevskii, and I. Yusefovich. Modified barrier methods for constrained and minimax problems. Technical report, Optimization Laboratory, Technion, 1992.

F. Blanchini and S. Miani. *Set-Theoretic Methods in Control*. Birkhäuser, 2007.

J. F. Bonnans and A. Shapiro. *Perturbation analysis of optimization problems*. Springer, 2000.

S. Boyd and L. Vandenberghe. *Convex Optimization*. Cambridge University Press, 2004.

P. E. Caines and D. Q. Mayne. On the discrete time matrix Riccati equation of optimal control. *International Journal of Control*, 12(5):785–794, 1970.

A. Carvalho, Y. Gao, A. Gray, H. E. Tseng, and F. Borrelli. Predictive control of an autonomous ground vehicle using an iterative linearization approach. In *16th International IEEE Conference on Intelligent Transportation Systems*, pages 2335–2340, 2013.

A. Carvalho, S. Lefévre, G. Schildbach, J. Kong, and F. Borrelli. Automated driving: The role of forecasts and uncertainty – A control perspective. *European Journal of Control*, 24: 14–32, 2015.

L. Chisci and G. Zappa. Feasibility in predictive control of constrained linear systems: The output feedback case. *International Journal of Robust and Nonlinear Control*, 12(5): 465–487, 2002.

L. Chisci, J. A. Rossiter, and G. Zappa. Systems with persistent disturbances: predictive control with restricted constraints. *Automatica*, 37(7):1019–1028, 2001.

D. Chmielewski and V. Manousiouthakis. On constrained infinite-time linear quadratic optimal control. *Systems & Control Letters*, 29(3):121–129, 1996.

P. D. Christofides, R. Scattolini, D. Muñoz de la Peña, and J. Liu. Distributed model predictive control: A tutorial review and future research directions. *Computers & Chemical Engineering*, 51:21–41, 2013.

P. Cortés, M. P. Kazmierkowski, R. M. Kennel, D. E. Quevedo, and J. Rodríguez. Predictive control in power electronics and drives. *IEEE Transactions on Industrial Electronics*, 55(12): 4312–4324, 2008.

D. DeHaan and M. Guay. A real-time framework for model predictive control of continuous-time nonlinear systems. *IEEE Transactions on Automatic Control*, 52(11):2047–2057, 2007.

J. B. Dennis. *Mathematical programming and electrical networks*. PhD thesis, Massachusetts Institute of Technology, 1959.

S. Di Cairano, H. E. Tseng, D. Bernardini, and A. Bemporad. Vehicle yaw stability control by coordinated active front steering and differential braking in the tire sideslip angles domain. *IEEE Transactions on Control Systems Technology*, 21(4):1236–1248, 2013.

M. Diehl, R. Findeisen, F. Allgöwer, H. Bock, and J. Schlöder. Nominal stability of real-time iteration scheme for nonlinear model predictive control. *IEE proceedings – Control Theory and Applications*, 152(3):296–308, 2005.

M. Diehl, R. Findeisen, and F. Allgöwer. A stabilizing iteration scheme for nonlinear model predictive control. *In: Real-Time and Online PDE-Constrained Optimization*, pages 23–52, 2007.

M. Diehl, H. J. Ferreau, and N. Haverbeke. Efficient numerical methods for nonlinear MPC and moving horizon estimation. In *Nonlinear Model Predictive Control – Towards New Challenging Applications*, pages 391–417. Springer, 2009.

A. Domahidi and J. Jerez. FORCES Professional. embotech GmbH (`http://embotech.com/FORCES-Pro`), 2014.

A. Domahidi, A. U. Zgraggen, M. N. Zeilinger, M. Morari, and C. N. Jones. Efficient interior point methods for multistage problems arising in receding horizon control. In *Proceedings of the 51st IEEE Conference on Decision and Control*, pages 668–674, 2012.

P. Falcone. *Nonlinear model predictive control for autonomous vehicles*. PhD thesis, Università del Sannio, 2007.

P. Falcone, F. Borrelli, J. Asgari, H. E. Tseng, and D. Hrovat. Predictive active steering control for autonomous vehicle systems. *IEEE Transactions on Control Systems Technology*, 15(3):566–580, 2007.

P. Falcone, F. Borrelli, H. E. Tseng, J. Asgari, and D. Hrovat. Linear time-varying model predictive control and its application to active steering systems: Stability analysis and experimental validation. *International Journal of Robust and Nonlinear Control*, 18(8):862–875, 2008.

C. Feller and C. Ebenbauer. A barrier function based continuous-time algorithm for linear model predictive control. In *Proceedings of the 12th European Control Conference*, pages 19–26, 2013.

C. Feller and C. Ebenbauer. Continuous-time linear MPC algorithms based on relaxed logarithmic barrier functions. In *Proceedings of the 19th IFAC World Congress*, pages 2481–2488, 2014a.

C. Feller and C. Ebenbauer. Barrier function based linear model predictive control with polytopic terminal sets. In *Proceedings of the 53rd IEEE Conference on Decision and Control*, pages 6683–6688, 2014b.

C. Feller and C. Ebenbauer. Weight recentered barrier functions and smooth polytopic terminal set formulations for linear model predictive control. In *Proceedings of the 2015 American Control Conference*, pages 1647–1652, 2015a.

C. Feller and C. Ebenbauer. Input-to-state stability properties of relaxed barrier function based MPC. In *Proceedings of the 5th IFAC Conference on Nonlinear Model Predictive Control (NMPC'15)*, pages 302–307, 2015b.

C. Feller and C. Ebenbauer. Robust stability properties of MPC iteration schemes based on relaxed barrier functions. In *Proceedings of the 55th IEEE Conference on Decision and Control*, pages 1484–1489, 2016.

C. Feller and C. Ebenbauer. Relaxed logarithmic barrier function based model predictive control of linear systems. *IEEE Transactions on Automatic Control*, 62(3):1223–1238, 2017a.

C. Feller and C. Ebenbauer. A stabilizing iteration scheme for model predictive control based on relaxed barrier functions. *Automatica*, 80:328–339, 2017b.

C. Feller and T. A. Johansen. Explicit MPC of higher-order linear processes via combinatorial multi-parametric quadratic programming. In *Proceedings of the 12th European Control Conference*, pages 536–541, 2013.

C. Feller, T. A. Johansen, and S. Olaru. An improved algorithm for combinatorial multi-parametric quadratic programming. *Automatica*, 49(5):1370–1376, 2013.

C. Feller, M. Ouerghi, and C. Ebenbauer. Robust output feedback model predictive control based on relaxed barrier functions. In *Proceedings of the 55th IEEE Conference on Decision and Control*, pages 1477–1483, 2016.

A. Ferrante and L. Ntogramatzidis. Employing the algebraic Riccati equation for a parametrization of the solutions of the finite-horizon LQ problem: the discrete-time case. *Systems & Control Letters*, 54:693–703, 2005.

H. J. Ferreau, A. Potschka, and C. Kirches. qpOASES webpage. http://www.qpOASES.org/, 2007–2015.

H. J. Ferreau, H. G. Bock, and M. Diehl. An online active set strategy to overcome the limitations of explicit MPC. *International Journal of Robust and Nonlinear Control*, 18(8): 816–830, 2008.

H. J. Ferreau, C. Kirches, A. Potschka, H. G. Bock, and M. Diehl. qpOASES: A parametric active-set algorithm for quadratic programming. *Mathematical Programming Computation*, 6(4):327–363, 2014.

A. V. Fiacco. *Introduction to Sensitivity and Stability Analysis in Nonlinear Programming*. Academic Press, Inc., 1983.

D. Fontanelli, L. Greco, and A. Bicchi. Anytime control algorithms for embedded real-time systems. In *International Workshop on Hybrid Systems: Computation and Control*, pages 158–171, 2008.

M. Gharbi, C. Feller, and C. Ebenbauer. A first step toward moving horizon state estimation based on relaxed logarithmic barrier functions. In *Proceedings of the 56th IEEE Conference on Decision and Control*, 2017. Accepted for publication.

E. G. Gilbert and K. T. Tan. Linear systems with state and control constraints: The theory and application of maximal output admissible sets. *IEEE Transactions on Automatic Control*, 36(9):1008–1020, 1991.

P. Giselsson. Improved fast dual gradient methods for embedded model predictive control. In *Proceedings of the 19th IFAC World Congress*, pages 2303–2309, 2014.

J. Gondzio. Interior point methods 25 years later. *European Journal of Operational Research*, 218(3):587–601, 2012.

P. J. Goulart and E. C. Kerrigan. Output feedback receding horizon control of constrained systems. *International Journal of Control*, 80(1):8–20, 2007.

P. J. Goulart and E. C. Kerrigan. Input-to-state stability of robust receding horizon control with an expected value cost. *Automatica*, 44:1171–1174, 2008.

K. Graichen. A fixed-point iteration scheme for real-time model predictive control. *Automatica*, 48(7):1300–1305, 2012.

O. M. Grasselli, A. Isidori, and F. Nicolò. Dead-beat control of discrete-time bilinear systems. *International Journal of Control*, 32(1):31–39, 1980.

G. Grimm, M. J. Messina, S. E. Tuna, and A. R. Teel. Examples when nonlinear model predictive control is nonrobust. *Automatica*, 40(10):1729–1738, 2004.

G. Grimm, M. J. Messina, S. E. Tuna, and A. R. Teel. Model Predictive Control: For Want of a Local Control Lyapunov Function, All is Not Lost. *IEEE Transactions on Automatic Control*, 50(5):546–558, 2005.

S. Gros, M. Zanon, and M. Diehl. Control of airborne wind energy systems based on nonlinear model predictive control & moving horizon estimation. In *Proceedings of the 12th European Control Conference*, pages 1017–1022, 2013.

L. Grüne. NMPC without terminal constraints. In *Proceedings of the 4th IFAC Conference on Nonlinear Model Predictive Control (NMPC'12)*, pages 1–13, 2012.

L. Grüne and J. Pannek. *Nonlinear Model Predictive Control: Theory and Algorithms*. Springer Science & Business Media, 2011.

L. Grüne and A. Rantzer. On the infinite horizon performance of receding horizon controllers. *IEEE Transactions on Automatic Control*, 53(9):2100–2111, 2008.

L. Grüne, J. Pannek, M. Seehafer, and K. Worthmann. Analysis of unconstrained nonlinear MPC schemes with time varying control horizon. *SIAM Journal on Control and Optimization*, 48(8):4938–4962, 2010.

A. Guiggiani, I. Kolmanovsky, P. Patrinos, and A. Bemporad. Fixed-point constrained model predictive control of spacecraft attitude. In *Proceedings of the 2015 American Control Conference*, pages 2317–2322, 2015.

A. Gupta, S. Bhartiya, and P. S. V. Nataraj. A novel approach to multiparametric quadratic programming. *Automatica*, 47:2112–2117, 2011.

A. Halanay and V. Rasvan. *Stability and Stable Oscillations in Discrete Time Systems*. CRC Press, 2000.

A. Hansen and J. K. Hedrick. Receding horizon sliding control for linear and nonlinear systems. In *Proceedings of the 2015 American Control Conference*, pages 1629–1634, 2015.

J. Hauser and A. Saccon. A barrier function method for the optimization of trajectory functionals with constraints. In *Proceedings of the 45th IEEE Conference on Decision and Control*, pages 864–869, 2006.

R. Hovorka, V. Canonico, L. J. Chassin, U. Haueter, M. Massi-Benedetti, M. O. Federici, T. R. Pieber, H. C. Schaller, L. Schaupp, T. Vering, and M. E. Wilinska. Nonlinear model predictive control of glucose concentration in subjects with type 1 diabetes. *Physiological measurement*, 25(4):905–920, 2004.

R. Huang, S. C. Patwardhan, and L. T. Biegler. Robust stability of nonlinear model predictive control based on extended Kalman filter. *Journal of Process Control*, 22(1):82–89, 2012.

A. Jadbabaie and J. Hauser. Unconstrained receding horizon control of nonlinear systems. *IEEE Transaction on Automatic Control*, 46(5):776–783, 2001.

A. H. Jazwinski. *Stochastic Processes and Filtering*. Dover Publications, 1970.

Z.-P. Jiang and Y. Wang. Input-to-state stability for discrete-time nonlinear systems. *Automatica*, 37(6):857–869, 2001.

Z.-P. Jiang and Y. Wang. A converse Lyapunov theorem for discrete-time systems with disturbances. *Systems & Control letters*, 45(1):49–58, 2002.

T. A. Johansen, A. Cristofaro, and T. Perez. Ship collision avoidance using scenario-based model predictive control. In *Proceedings of the 10th IFAC Conference on Control Applications in Marine Systems*, pages 14–21, 2016.

R. E. Kalman. A new approach to linear filtering and prediction problems. *Transactions of the ASME–Journal of Basic Engineering*, 82:35–45, 1960.

S. S. Keerthi and E. G. Gilbert. Optimal infinite horizon feedback control laws for a general class of constrained discrete-time systems: stability and moving-horizon approximations. *Journal of Optimization Theory and Applications*, 57(2):265–293, 1988.

E. C. Kerrigan and J. M. Maciejowski. Soft constraints and exact penalty functions in model predictive control. In *UKACC International Conference (Control 2000)*, 2000.

E. C. Kerrigan, J. L. Jerez, S. Longo, and G. A. Constantinides. Number representation in predictive control. In *Proceedings of the 4th IFAC Conference on Nonlinear Model Predictive Control (NMPC'12)*, pages 60–67, 2012.

H. Khalil. *Nonlinear Systems*. Prentice Hall, 2002.

M. Kögel and R. Findeisen. Fast predictive control of linear systems combining Nesterov's gradient method and the method of multipliers. In *Proceedings of the 50th IEEE Conference on Decision and Control*, pages 501–506, 2011a.

M. Kögel and R. Findeisen. A fast gradient method for embedded linear predictive control. In *Proceedings of the 18th IFAC World Congress*, pages 1362–1367, 2011b.

J. Kong, M. Pfeiffer, G. Schildbach, and F. Borrelli. Kinematic and dynamic vehicle models for autonomous driving control design. In *2015 IEEE Intelligent Vehicles Symposium*, pages 1094–1099, 2015.

G. Kreisselmeier and R. Steinhauser. Application of vector performance optimization to a robust control loop design for a fighter aircraft. *International Journal of Control*, 37(2): 251–284, 1982.

M. Kvasnica, P. Grieder, and M. Baotić. Multi-Parametric Toolbox (MPT), 2004. URL `http://control.ee.ethz.ch/~mpt/`.

P. Lancaster, A. C. M. Ran, and L. Rodman. An existence and monotonicity theorem for the discrete algebraic matrix Riccati equation. *Linear and Multilinear Algebra*, 20(4): 353–361, 1987.

J. P. LaSalle. *The stability and control of discrete processes*, volume 62. Springer Science & Business Media, 1986.

M. Lazar, W. P. M. H. Heemels, and A. R. Teel. Lyapunov functions, stability and input-to-state stability subtleties for discrete-time discontinuous systems. *IEEE Transactions on Automatic Control*, 54(10):2421–2425, 2009.

Y. I. Lee and B. Kouvaritakis. Receding horizon output feedback control for linear systems with input saturation. *IEE proceedings – Control Theory and Applications*, 148(2):109–115, 2001.

D. Limón, T. Alamo, and E. F. Camacho. Input-to-state stable MPC for constrained discrete-time nonlinear systems with bounded additive uncertainties. In *Proceedings of the 41st IEEE Conference on Decision and Control*, pages 364–369, 2002.

D. Limón, T. Alamo, D. M. Raimondo, D. Muñoz de la Peña, J. M. Bravo, A. Ferramosca, and E. F. Camacho. Input-to-state stability: A unifying framework for robust model predictive control. In *Nonlinear Model Predictive Control – Towards New Challenging Applications*, pages 1–26. Springer, 2009.

J. Löfberg. YALMIP: A toolbox for modeling and optimization in MATLAB. In *IEEE International Symposium on Computer Aided Control Systems Design*, pages 284–289, 2004.

Y. Ma, A. Kelman, A. Daly, and F. Borrelli. Predictive control for energy efficient buildings with thermal storage: Modeling, simulation, and experiments. *IEEE Control Systems*, 32 (1):44–64, 2012.

J. M. Maciejowski. *Predictive Control with Constraints*. Pearson Education, 2002.

L. Magni, D. M. Raimondo, and R. Scattolini. Regional input-to-state stability for nonlinear model predictive control. *IEEE Transactions on Automatic Control*, 51(9):1548–1553, 2006.

V. Manikonda, P. Arambel, M. Gopinathan, R. Mehra, and F. Hadaegh. A model predictive control-based approach for spacecraft formation keeping and attitude control. In *Proceedings of the 1999 American Control Conference*, pages 4258–4262, 1999.

D. Q. Mayne. Model predictive control: Recent developments and future promise. *Automatica*, 50(12):2967–2986, 2014.

D. Q. Mayne and W. Langson. Robustifying model predictive control of constrained linear systems. *IET Electronics Letters*, 37(23):1422–1423, 2001.

D. Q. Mayne, J. B. Rawlings, C. V. Rao, and P. O. M. Scokaert. Constrained model predictive control: Stability and optimality. *Automatica*, 36(6):789–814, 2000.

D. Q. Mayne, M. M. Seron, and S. V. Raković. Robust model predictive control of constrained linear systems with bounded disturbances. *Automatica*, 41(5):219–224, 2005.

D. Q. Mayne, S. V. Raković, R. Findeisen, and F. Allgöwer. Robust output feedback model predictive control of constrained linear systems. *Automatica*, 42(7):1217–1222, 2006.

L. K. McGovern and E. Feron. Closed-loop stability of systems driven by real-time, dynamic optimization algorithms. In *Proceedings of the 38th IEEE Conference on Decision and Control*, volume 4, pages 3690–3696, 1999.

S. Mehrotra. On the implementation of a primal-dual interior point method. *SIAM Journal on Optimization*, 2(4):575–601, 1992.

M. J. Messina, S. E. Tuna, and A. R. Teel. Discrete-time certainty equivalence output feedback: Allowing discontinuous control laws including those from model predictive control. *Automatica*, 41(4):617–628, 2005.

M. Mönnigmann and M. Jost. Vertex based calculation of explicit MPC laws. In *Proceedings of the 2012 American Control Conference*, pages 423–428, 2012.

J. J. Moré and D. J. Thuente. Line search algorithms with guaranteed sufficient decrease. *ACM Transactions on Mathematical Software (TOMS)*, 20(3):286–307, 1994.

M. A. Müller and F. Allgöwer. Economic and distributed model predictive control: recent developments in optimization-based control. *SICE Journal of Control, Measurement, and System Integration*, 10(2):39–52, 2017.

S. G. Nash, R. Polyak, and A. Sofer. *Large Scale Optimization: State of the Art*, chapter A numerical comparison of barrier and modified barrier methods for large-scale bound-constrained optimization, pages 319–338. Kluwer Academic Publishers, 1994.

D. Nešić. *Dead-beat control for polynomial systems*. PhD thesis, Research School of Information Science and Engineering, Australian National University, 1996.

D. Nešić and I. M. Mareels. Dead beat controllability of polynomial systems: Symbolic computation approaches. *IEEE Transactions on Automatic Control*, 43(2):162–175, 1998.

Y. Nesterov. A method of solving a convex programming problem with convergence rate $\mathcal{O}(1/k^2)$. In *Soviet Mathematics Doklady*, volume 27, pages 372–376, 1983.

Y. Nesterov. *Introductory lectures on convex optimization: A basic course*. Kluwer Academic Publishers, Boston, 2004.

Y. Nesterov and A. Nemirovskii. *Interior-Point Polynomial Algorithms in Convex Programming*. SIAM, Philadelphia, 1994.

J. Nocedal and S. J. Wright. *Numerical Optimization*. Springer, 1999.

L. Ntogramatzidis. A simple solution to the finite-horizon LQ problem with zero terminal state. *Kybernetika*, 39(4):483–492, 2003.

T. Ohtsuka. A continuation/GMRES method for fast computation of nonlinear receding horizon control. *Automatica*, 40(4):563–574, 2004.

F. Oldewurtel, A. Parisio, C. N. Jones, D. Gyalistras, M. Gwerder, V. Stauch, B. Lehmann, and M. Morari. Use of model predictive control and weather forecasts for energy efficient building climate control. *Energy and Buildings*, 45:15–27, 2012.

J. M. Ortega and W. C. Rheinboldt. *Iterative solution of nonlinear equations in several variables.* SIAM, 1970.

P. R. Pagilla and Y. Zhu. Controller and observer design for Lipschitz nonlinear systems. In *Proceedings of the 2004 American Control Conference*, volume 3, pages 2379–2384, 2004.

G. Pannocchia. Distributed model predictive control. *Encyclopedia of Systems and Control*, pages 301–308, 2015.

G. Pannocchia, J. B. Rawlings, and S. J. Wright. Fast, large-scale model predictive control by partial enumeration. *Automatica*, 43(5):852–860, 2007.

G. Pannocchia, M. Laurino, and A. Landi. A model predictive control strategy toward optimal structured treatment interruptions in anti-HIV therapy. *IEEE Transactions on Biomedical Engineering*, 57(5):1040–1050, 2010.

G. Pannocchia, J. B. Rawlings, and S. J. Wright. Conditions under which suboptimal nonlinear MPC is inherently robust. *Systems & Control Letters*, 60(9):747–755, 2011.

P. Patrinos and A. Bemporad. An accelerated dual gradient-projection algorithm for embedded linear model predictive control. *IEEE Transactions on Automatic Control*, 59(1): 18–33, 2014.

B. T. Polyak. *Introduction to Optimization*. Optimization Software, Inc., Publications Division, 1987.

J. A. Primbs. Dynamic hedging of basket options under proportional transaction costs using receding horizon control. *International Journal of Control*, 82(10):1841–1855, 2009.

S. J. Qin and T. A. Badgwell. A survey of industrial model predictive control technology. *Control Engineering Practice*, 11(7):733–764, 2003.

R. Rajamani. *Vehicle dynamics and control*. Springer Science & Business Media, 2011.

C. V. Rao. *Moving horizon strategies for the constrained monitoring and control of nonlinear discrete-time systems.* PhD thesis, University of Wisconsin–Madison, 2000.

C. V. Rao, S. J. Wright, and J. B. Rawlings. Application of interior point methods to model predictive control. *Journal of Optimization Theory and Applications*, 99(3):723–757, 1998.

J. B. Rawlings and L. Ji. Optimization-based state estimation: Current status and some new results. *Journal of Process Control*, 22(8):1439–1444, 2012.

J. B. Rawlings and D. Q. Mayne. *Model predictive control: Theory and design*. Nob Hill Publishing, 2009.

J. B. Rawlings, D. Angeli, and C. N. Bates. Fundamentals of economic model predictive control. In *Proceedings of the 51st IEEE Conference on Decision and Control*, pages 3851–3861, 2012.

K. Reif and R. Unbehauen. The extended Kalman filter as an exponential observer for nonlinear systems. *IEEE Transactions on Signal Processing*, 47(8):2324–2328, 1999.

J. Renegar. *A Mathematical View of Interior-point Methods for Convex Optimization*. MPS/SIAM Series on Optimization, 2001.

S. Richter, C. N. Jones, and M. Morari. Real-time input-constrained MPC using fast gradient methods. In *Proceedings of the 48th IEEE Conference on Decision and Control*, pages 7387–7393, 2009.

S. Richter, C. N. Jones, and M. Morari. Computational complexity certification for real-time MPC with input constraints based on the fast gradient method. *IEEE Transactions on Automatic Control*, 57(6):1391–1403, 2012.

S. L. Richter and R. A. DeCarlo. Continuation methods – theory and applications. *IEEE Transactions on Automatic Control*, 28(6):660–665, 1983.

B. J. P. Roset, W. P. M. H. Heemels, M. Lazar, and H. Nijmeijer. On robustness of constrained discrete-time systems to state measurement errors. *Automatica*, 44(4):1161–1165, 2008.

M. Rubagotti, P. Patrinos, and A. Bemporad. Stabilizing linear model predictive control under inexact numerical optimization. *IEEE Transactions on Automatic Control*, 59(6): 1660–1666, 2014.

M. Saponara, V. Barrena, A. Bemporad, E. Hartley, J. M. Maciejowski, A. Richards, A. Tramutola, and P. Trodden. Model predictive control application to spacecraft rendezvous in Mars Sample & Return scenario. In *Proceedings of the 4th European Conference for Aerospace Sciences (EUCASS)*, 2011.

P. O. M. Scokaert and J. B. Rawlings. Constrained linear quadratic regulation. *IEEE Transactions on Automatic Control*, 43(8):1163–1169, 1998.

P. O. M. Scokaert and J. B. Rawlings. Feasibility issues in linear model predictive control. *AIChE Journal*, 45(8):1649–1659, 1999.

P. O. M. Scokaert, J. B. Rawling, and E. S. Meadows. Discrete-time stability with perturbations: Application to model predictive control. *Automatica*, 33(3):463–470, 1997.

P. O. M. Scokaert, D. Q. Mayne, and J. B. Rawlings. Suboptimal model predictive control (feasibility implies stability). *IEEE Transactions on Automatic Control*, 44(3):648–654, 1999.

M. M. Seron, G. C. Goodwin, and J. A. De Doná. Characterisation of receding horizon control for constrained linear systems. *Asian Journal of Control*, 5(2):271–286, 2003.

E. D. Sontag. Smooth stabilization implies coprime factorization. *IEEE Transactions on Automatic Control*, 34(4):435–443, 1989.

E. D. Sontag. Comments on integral variants of ISS. *Systems & Control Letters*, 34(1):93–100, 1998.

B. Srinivasan, L. T. Biegler, and D. Bonvin. Tracking the necessary conditions of optimality with changing set of active constraints using a barrier-penalty function. *Computers & Chemical Engineering*, 32(3):572–579, 2008.

D. Sui and T. A. Johansen. Linear constrained moving horizon estimator with pre-estimating observer. *Systems & Control Letters*, 67:40–45, 2014.

M. Sznaier and M. J. Damborg. Suboptimal control of linear systems with state and control inequality constraints. In *Proceedings of the 26th IEEE Conference on Decision and Control*, volume 26, pages 761–762, 1987.

R. Tapia, Y. Zhang, M. Saltzman, and A. Weiser. The Mehrotra predictor-corrector interior-point method as a perturbed composite Newton method. *SIAM Journal on Optimization*, 6(1):47–56, 1996.

P. Tøndel, T. A. Johansen, and A. Bemporad. An algorithm for multi-parametric quadratic programming and explicit MPC solutions. *Automatica*, 39:489–497, 2003.

S. E. Tuna, M. J. Messina, and A. R. Teel. Shorter horizons for model predictive control. In *Proceedings of the 2006 American Control Conference*, pages 863–868, 2006.

VEHICO GmbH. ISO Double Lane Change Test. accessed online (February 2017) `http://www.vehico.de/index.php/en/applications/iso-lane-change-test`, 2017.

S. Vichik and F. Borrelli. Solving linear and quadratic programs with an analog circuit. *Computers & Chemical Engineering*, 70:160–171, 2014.

Y. Wang and S. Boyd. Fast model predictive control using online optimization. *IEEE Transactions on Control Systems Technology*, 18(2):267–278, 2010.

P. Wenzelburger. Experimentelle Implementierung eines zeitkontinuierlichen Algorithmus für die modellprädiktive Regelung linearer Systeme. Student thesis (in German), Institute for Systems Theory and Automatic Control, University of Stuttgart, 2014.

A. G. Wills. *Barrier function based model predictive control*. PhD thesis, School of Electrical Engineering and Computer Science, University of Newcastle, Australia, 2003.

A. G. Wills and W. P. Heath. A recentred barrier for constrained receding horizon control. In *Proceedings of the 2002 American Control Conference*, volume 5, pages 4177–4182, 2002.

A. G. Wills and W. P. Heath. Barrier function based model predictive control. *Automatica*, 40:1415–1422, 2004.

A. G. Wills and W. P. Heath. Application of barrier function based model predictive control to an edible oil refining process. *Journal of Process Control*, 15(2):183–200, 2005.

S. J. Wright. Applying new optimization algorithms to model predictive control. In *AIChE Symposium Series*, volume 93, pages 147–155, 1997a.

S. J. Wright. *Primal-dual interior-point methods*. SIAM, 1997b.

V. M. Zavala and L. T. Biegler. The advanced-step NMPC controller: Optimality, stability and robustness. *Automatica*, 45(1):86–93, 2009.

M. N. Zeilinger and C. N. Jones. Real-time suboptimal model predictive control using a combination of explicit MPC and online optimization. *IEEE Transactions on Automatic Control*, 56(7):1524–1534, 2011.

M. N. Zeilinger, C. N. Jones, and M. Morari. Robust stability properties of soft constrained MPC. In *Proceedings of the 49th IEEE Conference on Decision and Control*, pages 5276–5282, 2010.

M. N. Zeilinger, M. Morari, and C. N. Jones. Soft constrained model predictive control with robust stability guarantees. *IEEE Transactions on Automatic Control*, 59(5):1190–1202, 2014a.

M. N. Zeilinger, D. M. Raimondo, A. Domahidi, M. Morari, and C. N. Jones. On real-time robust model predictive control. *Automatica*, 50(3):683–694, 2014b.

A. Zheng and M. Morari. Stability of model predictive control with mixed constraints. *IEEE Transactions on Automatic Control*, 40(10):1818–1823, 1995.